Chemical resistance of common types of gloves to various compounds

	Glove type		
Compound	Neoprene	Nitrile	Latex
Acetone	good	fair	good
Chloroform	good	poor	poor
Dichloromethane	fair	poor	poor
Diethyl ether	very good	good	poor
Ethanol	very good	excellent	excellent
Ethyl acetate	good	poor	fair
Hexane	excellent	excellent	poor
Methanol	very good	fair	fair
Nitric acid (conc.)	good	poor	poor
Sodium hydroxide	very good	excellent	excellent
Sulfuric acid (conc.)	good	poor	poor
Toluene	fair	fair	poor

Common organic solvents

Name	Boiling point (°C)	Density $(g \cdot ml^{-1})$	Flammable	Miscible with H_2O
Acetone (2-propanone)	56.5	0.792	yes	yes
Dichloromethane	39.8	1.326	no	no
Diethyl ether	34.6	0.713	yes	no
Ethanol (95% aq. azeotrope)	78.2	0.816	yes	yes
Ethanol (anhydrous)	78.5	0.789	yes	yes
Ethyl acetate	77	0.902	yes	slightly
Hexane	69	0.660	yes	no
Methanol	64.7	0.792	yes	yes
Pentane	36.1	0.626	yes	no
2-Propanol (Isopropyl alcohol)	82.5	0.785	yes	yes
Toluene	110.6	0.866	yes	no

Selected data on common acid and base solutions

Compound	Molarity	Density $(g \cdot ml^{-1})$	% by weight
Acetic acid (glacial)	17	1.05	100
Ammonia (concentrated)	15.3	0.90	28.4
Hydrobromic acid (concentrated)	8.9	1.49	48
Hydrochloric acid (concentrated)	12	1.18	37
Nitric acid (concentrated)	16	1.42	71
Phosphoric acid (concentrated)	14.7	1.70	85
Sodium hydroxide	6	1.22	20
Sulfuric acid (concentrated)	18	1.84	96.5

Techniques
in Organic Chemistry

Techniques in Organic Chemistry

Miniscale, Standard Taper Microscale, and Williamson Microscale

Second Edition

JERRY R. MOHRIG
Carleton College

CHRISTINA NORING HAMMOND
Vassar College

PAUL F. SCHATZ
University of Wisconsin, Madison

W. H. Freeman and Company
New York

Publisher: Craig Bleyer
Senior Acquisitions Editor: Clancy Marshall
Senior Marketing Manager: Krista Bettino
Media Editor: Victoria Anderson
Associate Editor: Amy Thorne
Assistant Editors: Jenness Crawford and Carrie Wright
Project Editor: Penelope Hull
Text Designer: Marcia Cohen
Cover Designers: Michael Jung and Cambraia Fernandes
Illustrator: Fine Line Illustrations
Illustration Coordinators: Shawn Churchman and Susan Timmins
Production Coordinator: Julia DeRosa
Compositor: Progressive Information Technologies
Printer and Binder: RR Donnelley

Library of Congress Control Number: 2005936592

ISBN: 07167-6935-2
EAN: 97807176769354

Printed in the United States of America
First printing

W. H. Freeman and Company
41 Madison Avenue, New York, NY 10010
Houndmills, Basingstoke RG21 6XS, England
www.whfreeman.com

Contents

PART 3   SPECTROSCOPIC METHODS / 227

Preface

Organic chemistry is an experimental science, and students learn its process in the laboratory. Teaching students how to interpret experimental results and draw reasonable conclusions from them is at the heart of the process of science. The primary goal in teaching organic chemistry laboratories should be to help students learn how our science is done, to take them beyond the "cookbook" approach. We should work to find opportunities that involve students in addressing questions whose answers come from their experimental data and to help them learn how to design and carry out organic syntheses in an environment where they can succeed. A well-written and comprehensive textbook on the techniques of organic chemistry is an important asset in reaching these goals.

Changes in the Second Edition

The Second Edition of *Techniques in Organic Chemistry* is a thorough revision of the First Edition. Entirely new sections discuss green chemistry, how to plan and carry out an organic reaction, and computational chemistry. In addition, the sections on chromatography and spectroscopy have been extensively rewritten to be more useful to students as they develop their experimental skills. The book is up-to-date, and it speaks directly to the practical needs of students as they encounter experimental organic chemistry. Subsections on sources of confusion walk students through the pitfalls that could easily discourage them if they did not have this practical support. For easy reference, commonly used data tables on solvents, acids, and bases appear inside the front cover. Data tables for IR and NMR spectroscopy appear inside the back cover and on the back foldout. In addition, quick references to frequently used techniques are located inside the back cover. You may want to point out these features to students because they will need to use this information frequently during their laboratory work.

Who Should Use This Book?

This book is intended to serve as a laboratory textbook of experimental techniques for all students of organic chemistry. It can be used in conjunction with any lab experiments to provide the background and skills necessary for mastering the organic chemistry laboratory. It can also serve as a useful reference for laboratory practitioners and instructors.

Flexibility

Techniques in Organic Chemistry offers a great deal of flexibility. It can be used in any organic laboratory with any glassware. The basic techniques for using standard taper miniscale glassware as well as 14/10 standard taper microscale and Williamson microscale glassware are all covered in this book. The miniscale glassware that is described is appropriate with virtually any 14/20 or 19/22 standard taper glassware kit.

Modern Instrumentation

Modern instrumental methods play a crucial role in supporting guided inquiry experiments, which provide the active learning opportunities many instructors seek for their students. We feature instrumental methods that provide quick, reliable, quantitative data. Modern spectroscopic techniques, as well as GC, are particularly important. The section on spectroscopic methods focuses on the kinds of information students need to use spectroscopy effectively in the organic laboratory, rather than only on textbook examples. The emphasis is on how to acquire good data and how to read spectra with real understanding. Chapters on NMR, IR, and mass spectrometry stress the practical interpretation of spectra and how they can be used to answer questions posed in an experimental context. They describe how to deal with real laboratory samples and include case studies of analyzed spectra.

Organization

The book is divided into three parts. Part 1, Basic Techniques, introduces the techniques for synthesis and purification of organic compounds, and it begins with a chapter on laboratory safety and protecting the environment, of paramount importance from the very beginning of any lab experience. Part 2, Chromatography, discusses the three chromatographic techniques widely used in the organic laboratory: thin-layer, gas-liquid, and liquid chromatography. Modern spectroscopy has revolutionized structure determination in organic chemistry, and Part 3, Spectroscopic Methods, discusses them in some detail. Three appendices cover computational chemistry, the sources needed for retrieval of chemical information, and integrated spectroscopy problems. Traditional organic qualitative analysis is available on our Web site: www.whfreeman.com/mohrig2e

Modern Projects and Experiments in Organic Chemistry

The accompanying laboratory manual, *Modern Projects and Experiments in Organic Chemistry,* comes in two complete versions:

*   *Modern Projects and Experiments in Organic Chemistry: Miniscale and Standard Taper Microscale* (ISBN 0-7167-9779-8)
*   *Modern Projects and Experiments in Organic Chemistry: Miniscale and Williamson Microscale* (ISBN 0-7167-3921-6)

Modern Projects and Experiments is a combination of inquiry-based and traditional experiments, plus multiweek inquiry-based projects. It is designed to provide quality content, student accessibility, and instructor flexibility. This laboratory manual introduces students to the way the contemporary organic lab actually functions and allows them to experience the process of science.

Custom Publishing

All experiments and projects are available through W. H. Freeman's custom publishing service. Instructors can use this service to create their own customized lab manual, selecting specific experiments and projects from *Modern Projects and Experiments* and incorporating their own material as well so that the manual is organized to suit their course. Visit http://custompub.whfreeman.com to learn more.

ACKNOWLEDGMENTS

We have benefited greatly from the insights and thoughtful critiques of the following reviewers:

Second Edition
Alfred Bacher, University of California Los Angeles
Debra L. Bautista, Eastern Kentucky University
Joyce Easter, Virginia Wesleyan College
Thomas G. Gardner, Muhlenberg College
Michael P. Haaf, Ithaca College
Margaret Kline, Santa Monica College
Melekeh Nasiri, Sacramento City College
Lynne O'Connell, Boston College
Jay Pike, Providence College
Harold R. Rogers, California State University–Fullerton
Charles E. Russell, Muhlenberg College
Janet Schrenk, Massachusetts Institute of Technology
Kelli M. Slunt, Mary Washington College
George H. Smith, Herkimer County Community College
Corey E. Stilts, Chatham College
Stephen Waratuke, Hamilton College

First Edition
Jamie L. Anderson, Santa Monica College
Luis Avila, Columbia University
Christine DiMeglio, Yale University
Mark Forman, St. Joseph's University
Alvan C. Hengge, Utah State University
Susan L. Knock, Texas A&M University, Galveston
Barbara Mayer, California State University, Fresno
Dan Steffek, North Carolina State University

   Alfred Bacher, Charles Russell, and Michael Haaf provided not only thoughtful critiques of the entire book but also many helpful suggestions for the improvement of Technique 19, Nuclear Magnetic Resonance Spectroscopy.

   We wish to thank Clancy Marshall, our editor at W. H. Freeman and Company, for her direction of this revision, and Penelope Hull, Project Editor, for her masterly orchestration of the production stages. We express heartfelt thanks for the patience and support of our spouses, Adrienne Mohrig, Bill Hammond, and Ellie Schatz during the writing of this book.

   We hope that teachers and students of organic chemistry find our approach to laboratory techniques effective, and we would be pleased to hear from those who use our book. Please write to us in care of the Chemistry Acquisitions Editor at W. H. Freeman and Company, 41 Madison Avenue, New York, NY 10010, or e-mail us at chemistry @whfreeman.com.

PART

1

Basic Techniques

Chemistry is an experimental science. The laboratory is where you learn its processes. Part of learning how to do organic chemistry in the laboratory includes learning how to do it safely. Technique 1 discusses laboratory safety and safe handling practices for the chemicals you use. We urge you to read it carefully before you begin laboratory work. Technique 1 also discusses green chemistry, doing chemistry while protecting the environment.

To learn experimental organic chemistry, you need to master an array of techniques for carrying out chemical reactions, separating products from their reaction mixtures, purifying them, and analyzing the results. Part 1 covers the basic techniques of the organic laboratory, including the following:

- Miniscale and microscale techniques
- Heating and cooling methods, and how to assemble glassware and equipment for doing organic chemistry
- Extraction, drying agents, recrystallization, distillation, and sublimation for separating and purifying organic compounds
- Melting points, polarimetry, and refractometry for analyzing products
- Planning organic reactions and references to literature and online resources of experimental organic chemistry

Part 2 discusses the three fundamental chromatographic methods used for separation and analysis in organic chemistry—thin-layer chromatography, gas-liquid chromatography, and liquid chromatography. The important modern spectroscopic methods for organic structure determination—infrared spectroscopy, nuclear magnetic resonance spectroscopy, and mass spectrometry—are discussed in Part 3. Integrated spectroscopy, computational chemistry, and the literature of organic chemistry are covered in appendices.

1

SAFETY IN THE LABORATORY AND PROTECTING THE ENVIRONMENT

As you begin your study of experimental organic chemistry, you need a basic understanding of safety principles for handling chemicals and equipment in the laboratory. However, what you do in the laboratory extends beyond the laboratory itself. Every person working in a laboratory must also be aware of the impact that he or she has on the environment outside the laboratory.

Consider this chapter to be required reading before you perform any experiments in the organic laboratory.

Safety in the Laboratory

The organic chemistry laboratory is a place where accidents can and do occur and where safety is everyone's business. While working in the laboratory, you are protected by the instructions in an experiment and by the laboratory itself, which is designed to safeguard you from most routine hazards. However, neither the experimental directions nor the laboratory facilities can protect you from the worst hazard—your own or your neighbors' carelessness.

In addition to knowledge of basic laboratory safety, you need to learn how to work safely with organic chemicals. Many organic compounds are flammable or toxic or can be absorbed through the skin; others are volatile and vaporize easily into the air in the laboratory. Despite the hazards, organic compounds can be handled with a minimum of risk if you are adequately informed about the hazards and necessary safe handling procedures and if you use common sense while you are in the laboratory.

At the first meeting of your lab class, local safety issues will be discussed—the chemistry department's policies on safety goggles and protective gloves, the location of safety showers and eye wash stations, and the procedures to be followed in emergency situations. The information in this chapter is intended to complement your instructor's safety rules and instructions.

1.1 Causes of Laboratory Accidents

Laboratory accidents are of three general types: accidents involving fires and explosions; accidents producing cuts or burns; and accidents occurring from inhalation, absorption through the skin, or ingestion of hazardous materials.

Fires and Explosions   Fire is the chemical union of a fuel with an oxidizing agent, usually oxygen, and is accompanied by the evolution of heat and a flame. Most fires involve ordinary combustible materials—hydrocarbons or their derivatives. These fires are *extinguished* by removing oxygen or the combustible material or by decreasing the heat of the fire. These fires are *prevented* by keeping flammable materials away from a flame source or from oxygen (obviously, the former is easier).

Four sources of ignition are present in the organic laboratory: *open flames, hot surfaces,* such as hot plates or hot heating mantles, *faulty electrical equipment,* and *chemicals.* The most obvious way to prevent a fire is to prevent ignition.

Open flames.   Open-flame ignition of organic vapors or liquids is easily prevented: **Never bring a lighted Bunsen burner or a match near a low-boiling-point flammable liquid.** Furthermore, because vapors from organic liquids can travel over long distances at bench or floor level (they are heavier than air), an open flame within 10 ft of diethyl ether, pentane, or other low-boiling-point (and therefore volatile) organic solvents is an unsafe practice. In fact, the use of a Bunsen burner or any other flame in an organic laboratory should be a rare occurrence and should be done only with the permission of your instructor.

Hot surfaces.   A hot surface, such as a hot plate or heating mantle, presents a trickier problem (Figure 1.1). An organic solvent spilled or heated recklessly on a hot-plate surface may burst into flames (Figure 1.2). The thermostat on most hot plates is not sealed, and it can spark when it cycles on and off. The spark can ignite flammable vapors from an open container such as a beaker. It is also possible for the vapors from a volatile organic solvent to be ignited by the hot surface of a hot plate or a heating mantle, even if you are not actually heating the solvent with the heating device. Remove any hot heating mantle or hot plate from the vicinity before pouring a volatile organic liquid.

FIGURE 1.1
Heating devices.

Ceramic heating mantle   Hot plate

FIGURE 1.2
Fire hazards with
a hot plate.

Flammable liquid spilled
on a hot plate

Flammable vapors emitted when
heating a volatile solvent
in an open container

Faulty electrical equipment. Electrical fires pose the same danger
in the laboratory that they do in the home. Do not use appliances
with frayed or damaged electrical cords.

Chemical fires. Chemical reactions sometimes produce enough
heat to cause a fire and explosion. Consider, for example, the reaction of metallic sodium with water. The hydrogen gas that forms
as a product of this reaction can explode and then ignite a volatile
solvent that happens to be nearby.

Cuts and Injuries

Cuts and mechanical injuries are hazards anywhere, including the
laboratory.

Breaking glass rods or tubing.   When you purposely break a glass rod
or a glass tube, do it correctly. Score (scratch) a small line on
one side of the tube with a file. Wet the scored line with a drop of
water. Then, holding the tube on both sides with a paper towel and
with the scored part away from you, quickly snap it by pulling the
ends toward you (Figure 1.3).

Inserting glass into stoppers.   Insert glass tubes or thermometers into
cork, rubber stoppers, and thermometer adapters carefully and
correctly. First, lubricate the end of the tube with a drop of water or
glycerol. Then, while holding it with a towel close to the lubricated
end, insert the tube slowly by firmly rotating it into the stopper.
Never hold the glass tube by the end far away from the stopper — it
may break and the shattered end may be driven into your hand.

FIGURE 1.3
Breaking a glass rod
properly.

Chipped glassware. Check the rim of beakers, flasks, and other glassware for chips. Discard any piece of glassware that is chipped, because you could be cut very easily by the sharp edge.

Inhalation, Ingestion, and Skin Absorption

Inhalation. The hoods in the laboratory protect you from inhalation of noxious fumes, hazardous vapors, or dust from finely powdered materials. A hood is an enclosed space with a continuous flow of air that sweeps over the bench top, removing vapors or fumes from the area.

Because a great many compounds are at least potentially dangerous, the best practice is to run every experiment in a hood, if possible. Your instructor will tell you when an experiment *must* be carried out in a hood. **Make sure that the hood is turned on before you use it.** Position the sash for the optimal airflow through the hood. If the optimum sash position is not indicated on the hoods in your laboratory, consult your instructor about how far to open the sash.

Ingestion. Ingestion of chemicals by mouth is easily prevented. **Never taste any substance or pipet any liquid by mouth.** Wash your hands with soap and water before you leave the laboratory. Food or drink of any sort should neither be brought into a laboratory nor eaten there.

Absorption through the skin. Many organic compounds are absorbed through the skin. Wear the appropriate gloves while handling these reagents and reaction mixtures. If you spill any substance on your skin, wash the affected area thoroughly with water for 10–15 min.

| 1.2 | **Safety Features in the Laboratory** |

Organic laboratories contain many safety features for the protection and comfort of the people who work in them. It is unlikely that you will have to use the safety features in your laboratory, but in the event that you do, you must know what and where they are and how they operate.

Fire Extinguishers

Colleges and universities all have standard policies regarding the handling of fires. Your instructor will inform you whether evacuation of the lab or the use of a fire extinguisher takes priority at your institution. **Learn where the exits from your laboratory are located.**

Fire extinguishers are strategically located in your laboratory. There may be several types, and your instructor may demonstrate their use.

Water as a fire extinguisher in an organic laboratory has a disadvantage. Because many flammable liquids are lighter than water, they rise to the surface and continue to burn. When water is splashed on this type of fire, it actually spreads it. Therefore, the lab is equipped with either class BC or class ABC dry chemical fire extinguishers suitable for solvent or electrical fires.

Fire Blankets

Fire blankets are used for one thing and one thing only—to smother a fire involving a person's clothing. Fire blankets are available in most labs. If a person's clothing catches fire, **drop the person to the floor** and **roll the person** to wrap the blanket tightly around the body. To wrap the blanket around a person who is standing may direct the flames toward the person's face.

Safety Showers

Safety showers are for acid burns and other spills of corrosive, irritating, or toxic chemicals on the skin or clothing. If a safety shower is nearby, it can also be used when a person's clothing or hair is ablaze. The typical safety shower dumps a huge volume of water in a short period of time and thus is effective for both fire and acid spills, when speed is of the essence. **Do not use the safety shower routinely, but do not hesitate to use it in an emergency.**

Eye Wash Stations

You should always wear safety goggles while working in a laboratory, but if you accidentally splash something in your eyes, immediately rinse them with copious quantities of tepid water for at least 15 min at an eye wash station. Learn the location of the eye wash stations in your laboratory and examine the instructions on them during the first (check-in) lab session. You need to keep your eyes open while using the eye wash. Because this is difficult, assistance may be required. Do not hesitate to call for help.

If you are wearing contact lenses, they must be removed for the use of an eye wash station to be effective, an operation that is extremely difficult if a chemical is causing severe discomfort to your eyes. Therefore, **it is prudent not to wear contact lenses in the laboratory.**

First Aid Kits

Your laboratory may contain a basic first aid kit consisting of such items as adhesive bandages, sterile pads, and adhesive tape for treating a small cut or burn. **All injuries, no matter how slight, should be reported to your instructor immediately.** Your instructor will send you to the school health service, accompanied by another person, after any accident except for the most trivial. However, some immediate first aid is almost always required. Your instructor will indicate the location of the first aid station and instruct you in its use.

1.3   Preventing Accidents

Accidents can largely be prevented by common sense and knowledge of simple safety rules.

Personal Safety

You need to constantly keep in mind how to protect yourself from the potential hazards posed by the chemicals and equipment you are using while you are working in the laboratory.

- Think about what you are doing as you work in the laboratory. Plan your experimental work ahead of the laboratory period and perform laboratory operations with careful forethought.

- It is a law in many states and common sense in the remainder to **wear safety glasses or goggles at all times in the laboratory.** Your institution may have a policy regarding wearing contact lenses in the laboratory; learn what it is.
- Wear clothing that covers and protects your body. Shorts, tank tops, and sandals (or bare feet) are not suitable attire for the lab. Avoid loose clothing and long hair, which are a fire hazard or could become entangled in an apparatus. Your instructor may require laboratory aprons or lab coats.
- **Never eat, chew gum, drink beverages, or apply cosmetics in the lab.**
- Be aware of what your neighbors are doing. Many accidents and injuries in the laboratory are caused by other people. Often the person hurt worst in an accident is the one standing next to the place where the accident occurred.
- **Never work alone in the laboratory.** Being alone in a situation in which you may be helpless is dangerous.
- Make yourself aware of the procedures that should be followed in case of any accident. [See Technique 1.4, What to Do If an Accident Occurs.]
- Women who are pregnant or who may become pregnant should discuss with the appropriate medical professionals the advisability of working in the organic laboratory.
- Always wash your hands with soap and water at the end of the laboratory period.

Precautions When Handling Reagents

The precautions listed here are general guidelines for all laboratory work. Sources of information for handling specific compounds are discussed in Technique 1.6.

- **Never taste, ingest, or sniff directly any chemical.** Always use the hood when working with volatile, toxic, or noxious materials. Handle all chemicals carefully, and remember that many chemicals can enter the body through the skin and eyes, as well as through the mouth and lungs.
- Disposable gloves are available in all laboratories. **Wear gloves** to prevent chemicals from coming into contact with your skin unnecessarily (see Table 1.1 on page 11). Wear a lab coat or apron when working with hazardous chemicals. Cotton is the preferred fabric because synthetic fabrics might melt in a fire or undergo a reaction that causes the fabric to adhere to the skin.
- If a chemical is spilled on the skin, wash it off immediately with soap and water. If you are handling a particularly hazardous compound, find out how to remove it from your skin before you begin the experiment. Consult your instructor if you are in doubt about the toxicity of the chemical.
- Flammable solvents with boiling points of less than 100°C, such as diethyl ether, methanol, pentane, hexane, ethanol, and acetone, should be distilled, heated, or evaporated on

a steam bath or heating mantle, **never with a Bunsen burner or a hot plate.**

- Use an Erlenmeyer flask fitted with a cork, **never an open beaker,** for temporarily storing flammable solvents at your work area.

Order in the Laboratory

Accidents and injuries may be prevented or minimized if the laboratory is kept in good order.

- Keep your laboratory space clean and clear of equipment not actually in use. In addition to your own bench area, the balance and chemical dispensing areas should be left clean and orderly. Reagents should be returned to their designated storage place after use.
- **If you spill anything while measuring out materials, notify your instructor and clean it up immediately.**
- After weighing chemicals, replace the caps on the containers and dispose of weighing papers in the appropriate receptacle.
- Keep gas and water valves closed whenever they are not in use.

Burns and Other Injuries

Equipment used in a laboratory can be a source of burns or other injuries.

- Familiarize yourself with the operation of heating devices and other electrical equipment before using them.
- **Remember that both glass and hot plates look the same when hot or when cold.** When heating glass, do not touch the hot spot. Do not lay hot glass on a bench where someone else might pick it up.
- **Steam and boiling water cause severe burns.** Turn off the steam source before removing containers from the top of a steam bath or steam cone. Handle containers of boiling water very carefully.
- **Never heat a closed system!** Also, never completely close off an apparatus in which a gas is being evolved; always provide a vent to prevent an explosion.
- Floors can become very slippery if water is spilled. Wipe up any spill immediately.

1.4   What to Do if an Accident Occurs

If an accident occurs, act quickly, but think first. The first few seconds after an accident may be crucial. Acquaint yourself with the following instructions so that you can be of immediate assistance.

Fire

In case of fire, get out of danger and then immediately notify your instructor. If possible, remove any containers of flammable solvents from the fire area.

Consult your laboratory instructor about the proper response to a fire. **It is important to know the policy of your institution** concerning when to evacuate the building and when to use a fire extinguisher.

Know the location of the fire extinguishers and how they operate. A fire extinguisher will always be available. Aim low and direct its nozzle first toward the edge of the fire and then toward the middle. Tap water is not always useful for extinguishing chemical fires and can actually make some fires worse, so always use the fire extinguisher.

If a person's clothes catch fire, smother the flames with a lab coat, a fire blanket, or a water shower. Be sure you know where the fire blanket and safety shower are located. If your clothing is on fire, do not run. Rapid movement fans flames.

Cuts and Burns

Learn the location of the first aid kit and the materials it contains for the treatment of simple cuts or burns. **Notify your instructor immediately if you are cut or burned. Seek immediate medical attention** for anything except the most trivial cut or burn.

Cuts.   Press on the cut to help slow the bleeding. Apply a bandage when the bleeding has ceased. If the cut is large or deep, seek immediate medical attention.

Heat burns.   Apply cold water for 15 min to any heat burn. Seek immediate medical attention for any extensive burn.

Chemical burns.   The first thing to do for any chemical wound, unless you have been specifically told otherwise, is to wash it well with water for 15 min. This treatment will rinse away the excess chemical reagent. For acids, bases, and toxic chemicals, thorough washing with water will save pain later. Skin contact with a strong base usually does not produce immediate pain or irritation, but serious tissue damage (especially to the eyes) can occur if the affected area is not immediately washed with copious amounts of water. Specific treatments for chemical burns are published in *The Merck Index* (Ref. 5).

Chemical splash in the eyes.   If a chemical gets into your eyes, use the eye wash station and wash your eyes immediately with a copious amount of water. Position your head so that the stream of water from the eye wash fountain is directed at your eyes. **Hold your eyes open** to allow the water to flush the eyeballs. Do not use very cold water, however, because it can also damage the eyeballs. Then seek medical treatment immediately.

General Policy Regarding Accidents

Always inform your instructor of any accident that has happened to you or your neighbors. **Let your instructor decide whether a physician's attention is needed.** If a physician's attention is necessary, the injured person should always be accompanied to the medical facility, even when he or she protests that he or she is fine. The injury may be more serious than it initially appears.

1.5   Chemical Toxicology

The toxicity of a compound generally refers to its ability to produce injury once it reaches a susceptible site in the body. The toxicity of a substance is related to its probability of causing injury and is a species-dependent term. What is toxic for people may not be toxic for other animals and vice versa.

A substance is *acutely toxic* if it causes a toxic effect in a short time; it is *chronically toxic* if it causes toxic effects with repeated small exposures over a long duration. A major concern in chemical toxicology is quantity or dosage. A *poison* is a substance that causes harm to living tissues when applied in small doses. *Poison* is a word that comes to us through many centuries of usage; it has meaning but is not a scientifically useful term. Remember, the dose makes the poison.

Most compounds in their pure form are toxic or poisonous in one way or another. For example, oxygen is strongly implicated in many forms of cell aging and cell destruction and is also important in cancer cell growth. Oxygen at a partial pressure of 2 atm is toxic; for this reason, scuba tanks are filled not with O_2 but with compressed air. Thus oxygen, although essential for life, is toxic to people if the concentration is sufficiently high. In the organic laboratory, most compounds you use are more toxic than oxygen, however, and you must be knowledgeable about how to handle them safely.

How Toxic Substances Enter the Body

Toxic substances enter the body through three main routes: the lungs, the mouth and gastrointestinal tract, and the skin. Through any of these routes, a toxic substance can reach the bloodstream, where it can be carried to other body tissues.

Many organic solvents are absorbed through the skin. Wearing gloves while working in the organic laboratory protects your hands. Table 1.1 lists commonly available types of gloves and their chemical resistance. The *CRC Handbook of Chemistry and Physics,* 85th edition, contains a much more extensive table (Ref. 6). The barrier properties of each glove may be affected by differences in glove thickness, a chemical's concentration, temperature, and length of exposure to the chemical.

Many organic solvents are also absorbed through the lungs, because most solvents have an appreciable vapor pressure at room temperature. Working in a properly ventilated hood can prevent exposure to chemical vapors.

Fortunately, not all poisons that accidentally enter the body reach a site where they can be deleterious. Even though a toxic substance is absorbed, it is often excreted rapidly. Our body protects us with various devices: the nose, scavenger cells, metabolism, cell replacement, and rapid exchange of good air for bad. Most toxic substances are detectably toxic only because they are present in our environment over long periods of time. Most foreign substances are detoxified and discharged from the body very quickly.

T A B L E 1 . 1	Chemical resistance of common types of gloves to various compounds		
		Glove type	
Compound	Neoprene	Nitrile	Latex
Acetone	good	fair	good
Chloroform	good	poor	poor
Dichloromethane	fair	poor	poor
Diethyl ether	very good	good	poor
Ethanol	very good	excellent	excellent
Ethyl acetate	good	poor	fair
Hexane	excellent	excellent	poor
Methanol	very good	fair	fair
Nitric acid (conc.)	good	poor	poor
Sodium hydroxide	very good	excellent	excellent
Sulfuric acid (conc.)	good	poor	poor
Toluene	fair	fair	poor

The information in this table is compiled from the Web site http://www.inform.umd.edu/CampusInfo/Departments/EnvirSafety/ls/gloves.html and from "Chemical Resistance and Barrier Guide for Nitrile and Natural Rubber Latex Gloves," Safeskin Corporation, San Diego, CA, 1996.

Action of Toxic Substances on the Body

The action of toxic substances varies from individual to individual. Although many substances are toxic to the entire system (arsenic, for example), many others act at specific sites; that is, they are site specific. Carbon monoxide, for example, forms a complex with blood hemoglobin, destroying the blood's ability to absorb and release oxygen. Acute benzene toxicity seems to affect only the platelet-producing bone marrow.

In certain instances (for example, methanol poisoning), the metabolites (in this case, formaldehyde and formic acid) are more deleterious than the original compound; formic acid affects the optic nerve, causing blindness. As far as the body is concerned, it does not matter whether the poison is the original substance or a metabolic product of it.

Toxicity Testing and Reporting

Consumers are protected by a series of laws that define toxicity, the legal limits and dosages of toxic materials, and the procedures for measuring toxicities.

Acute oral toxicity is measured in terms of LD_{50} (**LD** stands for "lethal dose"). The LD_{50} represents the dose, in milligrams per kilogram of body weight, that will be fatal to 50% of a certain population of experimental animals (for example, mice or rats). Other tests include dermal toxicity (skin sensitization) and irritation of the mucous membranes (eyes, nose).

The toxicity of almost every chemical compound commercially available has been reported, and every year the toxicities of many more compounds become known. A wall chart of toxicities for many common organic compounds may be hanging in your laboratory or near your stockroom. *The Merck Index* is a useful reference for toxicity and other hazard information

FIGURE 1.4

Page 111 of *The Merck Index of Chemicals, Drugs, and Biologicals,* 13th edition, Maryadele J. O'Neil, Ann Smith, Patricia E. Heckelman, John R. Obenchain Jr., Eds. (Merck & Co., Inc.: Whitehouse Station, NJ, USA, 2001).

(Reproduced with permission from *The Merck Index,* 13th edition. Copyright © 2001 by Merck & Co., Inc., Whitehouse Station, NJ, USA. All rights reserved.)

Aniline Mustard   662

Rhombic needles from petr ether, mp 86°, bp$_{0.02}$ 140°. [α]$_D^{25}$ −63.8° (methanol); −56.3° (chloroform). Very sol in alcohol, ether, chloroform, benzene, petr ether.

Hydrochloride. C$_{12}$H$_{15}$NO$_3$.HCl. Orthorhombic prisms, dec 255°; freely sol in hot water. Aq soln is neutral.

659. Anilazine. [101-05-3] 4,6-Dichloro-*N*-(2-chlorophenyl)-1,3,5-triazin-2-amine; 2,4-dichloro-6-(*o*-chloroanilino)-*s*-triazine; (*o*-chloroanilino)dichlorotriazine; Dyrene. C$_9$H$_5$Cl$_3$N$_4$; mol wt 275.53. C 39.23%, H 1.83%, Cl 38.60%, N 20.33%. Prepn: C. N. Wolf, **US 2720480** (1955 to Ethyl Corp.); E. G. Hill, E. Clinton, **US 2820032** (1958); K. H. Rattenburg *et al.,* **US 3074946** (1963 to Chemagro). Toxicity study: S. D. Cohen, S. D. Murphy, *J. Agr. Food Chem.* **21,** 140 (1973).

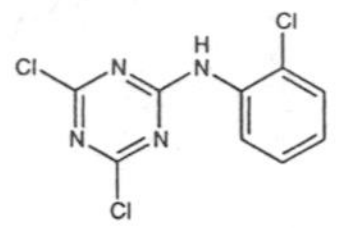

White to tan crystals, mp 159-160°. Insol in water. Soly at 30° (g/100 ml): toluene, 5; xylene, 4; acetone, 10. Subject to hydrolysis; not compatible with oils and alkaline materials. LD$_{50}$ orally in rats: >5000 mg/kg, Mobay Technical Information Sheet, Jan. 1979.

USE: Fungicide.

660. Anileridine. [144-14-9] 1-[2-(4-Aminophenyl)ethyl]-4-phenyl-4-piperidinecarboxylic acid ethyl ester; 1-(*p*-aminophenethyl)-4-phenylisonipecotic acid ethyl ester; ethyl 1-(4-aminophenethyl)-4-phenylisonipecotate; *N*-[β-(*p*-aminophenyl)ethyl]-4-phenyl-4-carbethoxypiperidine; *N*-β-(*p*-aminophenyl)ethylnormeperidine; Leritine. C$_{22}$H$_{28}$N$_2$O$_2$; mol wt 352.47. C 74.97%, H 8.01%, N 7.95%, O 9.08%. Synthesis: Weijlard *et al., J. Am. Chem. Soc.* **78,** 2342 (1956); **US 2966490** (1960 to Merck & Co.).

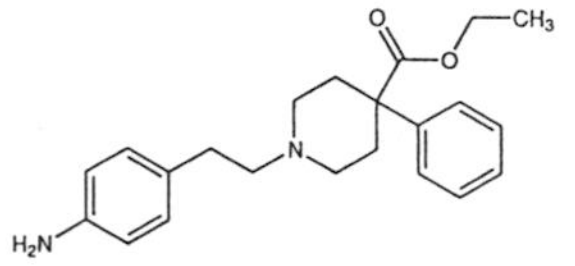

Mp 83°.

Dihydrochloride. [126-12-5] C$_{22}$H$_{28}$N$_2$O$_2$.2HCl. Crystals from methanol + ether, mp 280-287° (dec). Freely soluble in water, methanol. Solubility in ethanol: 8 mg/g. pH of aq solns 2.0 to 2.5. Solns are stable at pH 3.5 and below. At pH 4 and higher the insol free base is precipitated. uv max (pH 7 in 90% methanol contg phosphate buffer): 235, 289 nm (A$_{1cm}^{1\%}$ 293, 34.5). Distribution coefficient (water, pH 3.6/*n*-butanol): 0.9.

Note: This is a controlled substance (opiate): **21 CFR,** 1308.12.

THERAP CAT: Analgesic (narcotic).

661. Aniline. [62-53-3] Benzenamine; aniline oil; phenylamine; aminobenzene; aminophen; kyanol. C$_6$H$_7$N; mol wt 93.13. C 77.38%, H 7.58%, N 15.04%. First obtained in 1826 by Unverdorben from dry distillation of indigo. Runge found it in coal tar in 1834. Fritzsche, in 1841, prepared it from indigo and potash and gave it the name aniline. Manuf from nitrobenzene or chlorobenzene: *Faith, Keyes & Clark's Industrial Chemicals,* F. A. Lowenheim, M. K. Moran, Eds. (Wiley-Interscience, New York, 4th ed., 1975) pp 109-116. Procedures: A. I. Vogel, *Practical Organic Chemistry* (Longmans, London, 3rd ed., 1959) p 564; Gattermann-Wieland, *Praxis des organischen Chemikers* (de Gruyter, Berlin, 40th ed., 1961) p 148.

Brochure *"Aniline"* by Allied Chemical's National Aniline Division (New York, 1964) 109 pp, gives reactions and uses of aniline (877 references). Toxicity study: K. H. Jacobson, *Toxicol. Appl. Pharmacol.* **22,** 153 (1972).

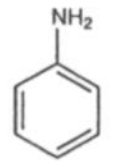

Oily liquid; colorless when freshly distilled, darkens on exposure to air and light. *Poisonous!* Characteristic odor and burning taste; combustible; volatile with steam. d$_{20}^{20}$ 1.022. bp 184-186°. Solidif −6°. Flash pt, closed cup: 169°F (76°C). n$_D^{20}$ 1.5863. pKb 9.30. pH of 0.2 molar aq soln 8.1. One gram dissolves in 28.6 ml water, 15.7 ml boil. water; misc with alcohol, benzene, chloroform, and most other organic solvents. Combines with acids to form salts. It dissolves alkali or alkaline earth metals with evolution of hydrogen and formation of anilides, e.g., C$_6$H$_5$NHNa. *Keep well closed and protected from light. Incompat.* Oxidizers, albumin, solns of Fe, Zn, Al, acids, and alkalies. LD$_{50}$ orally in rats: 0.44 g/kg (Jacobson).

Hydrobromide. C$_6$H$_7$N.HBr. White to slightly reddish, crystalline powder, mp 286°. Darkens in air and light. Sol in water, alc. *Protect from light.*

Hydrochloride. C$_6$H$_7$N.HCl. Crystals, mp 198°. d 1.222. Darkens in air and light. Sol in about 1 part water; freely sol in alc. *Protect from light.*

Hydrofluoride. C$_6$H$_7$N.HF. Crystalline powder. Turns gray on standing. Freely sol in water; slightly sol in cold, freely in hot alc.

Nitrate. C$_6$H$_7$N.HNO$_3$. Crystals, dec about 190°. d 1.36. Discolors in air and light. Sol in water, alc. *Protect from light.*

Hemisulfate. C$_3$H$_7$N.½H$_2$SO$_4$. Crystalline powder. d 1.38. Darkens on exposure to air and light. One gram dissolves in about 15 ml water; slightly sol in alc. Practically insol in ether. *Protect from light.*

Acetate. C$_6$H$_5$NH$_2$.HOOCCH$_3$. Prepd from aniline and acetic acid: Vignon, Evieux, *Bull. Soc. Chim. France* [4] **3,** 1012 (1908). Colorless liquid. d 1.070-1.072. Darkens with age; gradually converted to acetanilide on standing. Misc with water, alc.

Oxalate. C$_6$H$_5$NH$_2$.HOOCCOOH.H$_2$NC$_6$H$_5$. Prepd from aniline and oxalic acid in alc soln: Hofmann, *Ann.* **47,** 37 (1843). Triclinic rods from water, mp 174-175°. Readily sol in water; sparingly sol in abs alc. Practically insol in ether.

Caution: Poisoning may occur from inhalation, skin penetration or ingestion. Potential symptoms of overexposure are navy blue to black lips, tongue, mucous membranes; slate gray skin; headache, nausea, vomiting, confusion, ataxia, vertigo, tinnitus, weakness, disorientation, lethargy, drowsiness; dyspnea on effort; methemoglobinemia, cyanosis, coma; tachycardia, heart blocks, arryhthmia, shock; painful micturition, hematuria, hemoglobinuria, methemoglobinuria, oliguria, renal insufficiency; cirrhosis. Direct contact may cause eye irritation. Potential occupational carcinogen. *See NIOSH Pocket Guide to Chemical Hazards* (DHHS/NIOSH 97-140, 1997) p 18; *Clinical Toxicology of Commercial Products,* R. E. Gosselin *et al.,* Eds. (Williams & Wilkins, Baltimore, 5th ed., 1984) Section III, pp 31-36.

USE: Manuf dyes, medicinals, resins, varnishes, perfumes, shoe blacks; vulcanizing rubber; as solvent. Hydrochloride used in manuf of intermediates, aniline black and other dyes, in dyeing fabrics or wood black.

662. Aniline Mustard. [553-27-5] *N,N*-Bis(2-chloroethyl)benzenamine; *N,N*-bis(2-chloroethyl)aniline; phenylbis[2-chloroethylamine]; β,β'-dichlorodiethylaniline. C$_{10}$H$_{13}$Cl$_2$N; mol wt 218.13. C 55.06%, H 6.01%, Cl 32.51%, N 6.42%. Prepd by the action of phosphorus pentachloride on *N,N*-bis-[2-hydroxyethyl]aniline (phenyldiethanolamine): Robinson, Watt, *J. Chem. Soc.* **1934,** 1538; Korshak, Strepikheev, *J. Gen. Chem. USSR* **14,** 312 (1944).

Consult the Name Index before using this section.   Page 111

---

about organic compounds; a page from the 13th edition (2001) showing typical listings for organic compounds is reproduced in Figure 1.4.

1.6   Where to Find Chemical Safety Information

Material Safety Data Sheets

All laboratories must make available a Material Safety Data Sheet (MSDS) for every chemical used. Figure 1.5 shows the MSDS for dichloromethane (methylene chloride), detailing the hazards, safe handling practices, first aid information, storage, and disposal of dichloromethane.

```
Sigma Chemical Co.        Aldrich Chemical Co., Inc.    Fluka Chemical Corp.
P.O. Box 14508            1001 West St. Paul            980 South Second St.
St. Louis, MO 63178       Milwaukee, WI 53233           Ronkonkoma, NY 11779
Phone: 314-771-5765       Phone: 414-273-3850           Phone: 516-467-0980
                                               Emergency Phone: 516-467-3535

SECTION 1. - - - - - - - - - CHEMICAL IDENTIFICATION- - - - - - - -
    CATALOG #:             D65100
    NAME:                  DICHLOROMETHANE, 99.6%, A.C.S. REAGENT
SECTION 2. - - - - - COMPOSITION/INFORMATION ON INGREDIENTS - - - - - -
    CAS #:    75-09-2
    MF: CH2CL2
    EC NO: 200-838-9
  SYNONYMS
    AEROTHENE MM * CHLORURE DE METHYLENE (FRENCH) * DICHLOROMETHANE (DOT:
    OSHA) * METHANE DICHLORIDE * METHYLENE BICHLORIDE * METHYLENE
    CHLORIDE (ACGIH:DOT:OSHA) * METHYLENE DICHLORIDE * METYLENU CHLOREK
    (POLISH) * NARKOTIL * NCI-C50102 * R 30 * R30 (REFRIGERANT) * RCRA
    WASTE NUMBER U080 * SOLAESTHIN * SOLMETHINE * UN1593 (DOT) *
SECTION 3. - - - - - - - - - HAZARDS IDENTIFICATION - - - - - - - - -
    LABEL PRECAUTIONARY STATEMENTS
    TOXIC (USA)
    HARMFUL (EU)
    HARMFUL BY INHALATION, IN CONTACT WITH SKIN AND IF SWALLOWED.
    IRRITATING TO EYES, RESPIRATORY SYSTEM AND SKIN.
    POSSIBLE RISK OF IRREVERSIBLE EFFECTS.
    CALIF. PROP. 65 CARCINOGEN.
    POSSIBLE CARCINOGEN/MUTAGEN.
    NEUROLOGICAL HAZARD.
    READILY ABSORBED THROUGH SKIN.
    TARGET ORGAN(S):
    LIVER, PANCREAS
    IN CASE OF ACCIDENT OR IF YOU FEEL UNWELL, SEEK MEDICAL ADVICE
    IMMEDIATELY (SHOW THE LABEL WHERE POSSIBLE).
    IN CASE OF CONTACT WITH EYES, RINSE IMMEDIATELY WITH PLENTY OF
    WATER AND SEEK MEDICAL ADVICE.
    AFTER CONTACT WITH SKIN, WASH IMMEDIATELY WITH PLENTY OF WATER.
    WEAR SUITABLE PROTECTIVE CLOTHING, GLOVES AND EYE/FACE
    PROTECTION.
    HANDLE AND STORE UNDER NITROGEN.
SECTION 4. - - - - - - - - - FIRST-AID MEASURES- - - - - - - - - -
    IN CASE OF CONTACT, IMMEDIATELY WASH SKIN WITH SOAP AND COPIOUS
    AMOUNTS OF WATER.
    CONTAMINATION OF THE EYES SHOULD BE TREATED BY IMMEDIATE AND PROLONGED
    IRRIGATION WITH COPIOUS AMOUNTS OF WATER.
    ASSURE ADEQUATE FLUSHING OF THE EYES BY SEPARATING THE EYELIDS
    WITH FINGERS.
    IF INHALED, REMOVE TO FRESH AIR. IF NOT BREATHING GIVE ARTIFICIAL
    RESPIRATION. IF BREATHING IS DIFFICULT, GIVE OXYGEN.
    IF SWALLOWED, WASH OUT MOUTH WITH WATER PROVIDED PERSON IS CONSCIOUS.
    CALL A PHYSICIAN.
    WASH CONTAMINATED CLOTHING BEFORE REUSE.

                                                         Page 1
```

```
EMITS TOXIC FUMES UNDER FIRE CONDITIONS.
SECTION 6. - - - - - - - ACCIDENTAL RELEASE MEASURES- - - - - - - - -
    EVACUATE AREA.
    WEAR SELF-CONTAINED BREATHING APPARATUS, RUBBER BOOTS AND HEAVY
    RUBBER GLOVES.
    ABSORB ON SAND OR VERMICULITE AND PLACE IN CLOSED CONTAINERS FOR
    DISPOSAL.
    VENTILATE AREA AND WASH SPILL SITE AFTER MATERIAL PICKUP IS COMPLETE.
SECTION 7. - - - - - - - - - - HANDLING AND STORAGE- - - - - - - - - -
    REFER TO SECTION 8.
SECTION 8. - - - - - EXPOSURE CONTROLS/PERSONAL PROTECTION- - - - -
    WEAR APPROPRIATE NIOSH/MSHA-APPROVED RESPIRATOR, CHEMICAL-RESISTANT
    GLOVES, SAFETY GOGGLES, OTHER PROTECTIVE CLOTHING.
    USE ONLY IN A CHEMICAL FUME HOOD.
    SAFETY SHOWER AND EYE BATH.
    DO NOT BREATHE VAPOR.
    AVOID CONTACT WITH EYES, SKIN AND CLOTHING.
    AVOID PROLONGED OR REPEATED EXPOSURE.
    READILY ABSORBED THROUGH SKIN.
    WASH THOROUGHLY AFTER HANDLING.
    TOXIC.
    POSSIBLE CARCINOGEN.
    IRRITANT.
    NEUROLOGICAL HAZARD.
    POSSIBLE MUTAGEN.
    KEEP TIGHTLY CLOSED.
    KEEP AWAY FROM HEAT AND OPEN FLAME.
    STORE IN A COOL DRY PLACE.
SECTION 9. - - - - - - - PHYSICAL AND CHEMICAL PROPERTIES - - - - - - -
    APPEARANCE AND ODOR
    COLORLESS LIQUID
    PHYSICAL PROPERTIES
    BOILING POINT:           39.0 C TO 40 C
    MELTING POINT:           -97 C
    FLASHPOINT               NONE
    EXPLOSION LIMITS IN AIR:
      UPPER                               22%
      LOWER                               14%
    AUTOIGNITION TEMPERATURE:    1223 F        661C
    VAPOR PRESSURE:    6.83PSI 20 C    24.48PSI 55 C
    SPECIFIC GRAVITY:       1.325
    VAPOR DENSITY:     2.9
SECTION 10. - - - - - - - -STABILITY AND REACTIVITY - - - - - - - -
    INCOMPATIBILITIES
    ALKALI METALS
    ALUMINUM
                    Page 2
```

```
VAPOR OR MIST IS IRRITATING TO THE EYES, M
RESPIRATORY TRACT.
CAUSES SKIN IRRITATION.
DICHLOROMETHANE IS METABOLIZED IN THE BODY
WHICH INCREASES AND SUSTAINS CARBOXYHEMOGL
REDUCING THE OXYGEN-CARRYING CAPACITY OF T
EXPOSURE CAN CAUSE:
NAUSEA, DIZZINESS AND HEADACHE
MAY CAUSE NERVOUS SYSTEM DISTURBANCES.
CHRONIC EFFECTS
POSSIBLE CARCINOGEN.
LABORATORY EXPERIMENTS HAVE SHOWN MUTAGENI
TARGET ORGAN(S):
LIVER
PANCREAS
NERVES
CARDIOVASCULAR SYSTEM
TO THE BEST OF OUR KNOWLEDGE, THE CHEMICAL
TOXICOLOGICAL PROPERTIES HAVE NOT BEEN THO
RTECS #: PA8050000
METHANE, DICHLORO-
IRRITATION DATA
SKN-RBT 810 MG/24H SEV
SKN-RBT 100 MG/24H MOD
EYE-RBT 162 MG MOD
EYE-RBT 10 MG MLD
EYE-RBT 500 MG/24H MLD
TOXICITY DATA
ORL-HMN LDLO:357 MG/KG
ORL-RAT LD50:1600 MG/KG
IHL-RAT LC50:52 GM/M3
IPR-RAT LD50:916 MG/KG
UNR-RAT LD50:5350 MG/KG
IHL-MUS LC50:14400 PPM/7H
IPR-MUS LD50:437 MG/KG
SCU-MUS LD50:6460 MG/KG
UNR-MUS LD50:4770 MG/KG
UNR-RBT LD50:1225 MG/KG
TARGET ORGAN DATA
PERIPHERAL NERVE AND SENSATION (PARESTHESI
BEHAVIORAL (ALTERED SLEEP TIME)
BEHAVIORAL (EUPHORIA)
BEHAVIORAL (SOMNOLENCE)
BEHAVIORAL (CONVULSIONS OR EFFECT ON SEIZU
BEHAVIORAL (ATAXIA)
CARDIAC (CHANGE IN RATE)
                        Page 3
```

FIGURE 1.5

Material Safety Data Sheet (MSDS) for dichloromethane.

(Reprinted with permission from Aldrich Chemical Co., Inc., Milwaukee, WI.)

```
PRODUCT #: D65100      NAME: DICHLOROMETHANE, 99.6%, A.C.S. REAGENT
        MATERIAL SAFETY DATA SHEET, Valid 11/96 - 1/97
        Printed Thursday, November 21, 1996  4:38PM

  LUNGS, THORAX OR RESPIRATION (TUMORS)
  LIVER (LIVER FUNCTION TESTS IMPAIRED)
  SPECIFIC DEVELOPMENTAL ABNORMALITIES (MUSCULOSKELETAL SYSTEM)
  SPECIFIC DEVELOPMENTAL ABNORMALITIES (UROGENITAL SYSTEM)
  TUMORIGENIC (CARCINOGENIC BY RTECS CRITERIA)
  ONLY SELECTED REGISTRY OF TOXIC EFFECTS OF CHEMICAL SUBSTANCES
  (RTECS) DATA IS PRESENTED HERE. SEE ACTUAL ENTRY IN RTECS FOR
  COMPLETE INFORMATION.
SECTION 12. - - - - - - - - ECOLOGICAL INFORMATION - - - - - - - -
  DATA NOT YET AVAILABLE.
SECTION 13. - - - - - - - - DISPOSAL CONSIDERATIONS - - - - - - - -
  DISSOLVE OR MIX THE MATERIAL WITH A COMBUSTIBLE SOLVENT AND BURN IN A
  CHEMICAL INCINERATOR EQUIPPED WITH AN AFTERBURNER AND SCRUBBER.
  OBSERVE ALL FEDERAL, STATE AND LOCAL ENVIRONMENTAL REGULATIONS.
SECTION 14. - - - - - - - - TRANSPORT INFORMATION - - - - - - - -
  CONTACT ALDRICH CHEMICAL COMPANY FOR TRANSPORTATION INFORMATION.
SECTION 15. - - - - - - - - REGULATORY INFORMATION - - - - - - - -
  EUROPEAN INFORMATION
    EC INDEX NO:      602-004-00-3
    HARMFUL
    R 40
    POSSIBLE RISK OF IRREVERSIBLE EFFECTS.
    S 23
    DO NOT BREATHE VAPOR.
    S 24/25
    AVOID CONTACT WITH SKIN AND EYES.
    S 36/37
    WEAR SUITABLE PROTECTIVE CLOTHING AND GLOVES.
  REVIEWS, STANDARDS, AND REGULATIONS
    OEL=MAK
    ACGIH TLV-TWA 50 PPM                        85INA8 6,981,91
    ACGIH TLV-SUSPECTED HUMAN CARCINOGEN        85INA8 6,981,91
    IARC CANCER REVIEW:ANIMAL SUFFICIENT EVIDENCE IMEMDT 41,43,86
    IARC CANCER REVIEW:HUMAN INADEQUATE EVIDENCE  IMEMDT 41,43,86
    IARC CANCER REVIEW:GROUP 2B                 IMSUDL 7,194,87
    EPA FIFRA 1988 PESTICIDE SUBJECT TO REGISTRATION OR RE-REGISTRATION
      FEREAC 54,7740,89
    MSHA STANDARD-AIR:TWA 500 PPM (1750 MG/M3)
      DTLVS* 3,171,71
    OSHA PEL (GEN INDU):8H TWA 500 PPM;CL 1000 PPM;PK 2000 PPM/5M/2H
      CFRGBR 29,1910.1000,94
    OSHA PEL (CONSTRUC):SEE 56 FR 57036
      CFRGBR 29,1926.55,94
    OSHA PEL (SHIPYARD):8H TWA 500 PPM (1740 MG/M3)
      CFRGBR 29,1915.1000,93
    OSHA PEL (FED CONT):8H TWA 500 PPM (1740 MG/M3)
      CFRGBR 41,50-204.50,94
    OEL-AUSTRALIA:TWA 100 PPM (350 MG/M3);CARCINOGEN JAN93
    OEL-AUSTRIA:TWA 100 PPM (360 MG/M3) JAN93
    OEL-BELGIUM:TWA 50 PPM (174 MG/M3);CARCINOGEN JAN93
    OEL-DENMARK:TWA 50 PPM (175 MG/M3);SKIN;CARCINOGEN JAN93
    OEL-FINLAND:TWA 100 PPM (350 MG/M3);STEL 250 PPM (870 MG/M3) JAN93
    OEL-FRANCE:TWA 100 PPM (360 MG/M3);STEL 500 PPM (1800 MG/M3) JAN93
    OEL-GERMANY:TWA 100 PPM (360 MG/M3);CARCINOGEN JAN93

                            Page 4
```

```
PRODUCT #: D65100      NAME: DICHLOROMETHANE, 99.6%, A.C.S. REAGENT
        MATERIAL SAFETY DATA SHEET, Valid 11/96 - 1/97
        Printed Thursday, November 21, 1996  4:38PM

  OEL-HUNGARY:STEL 10 MG/M3;CARCINOGEN JAN93
  OEL-JAPAN:TWA 100 PPM (350 MG/M3) JAN93
  OEL-THE NETHERLANDS:TWA 100 PPM (350 MG/M3);STEL 500 PPM JAN93
  OEL-THE PHILIPINES:TWA 500 PPM (1740 MG/M3) JAN93
  OEL-POLAND:TWA 50 MG/M3 JAN93
  OEL-RUSSIA:TWA 100 PPM;STEL 50 MG/M3 JAN93
  OEL-SWEDEN:TWA 35 PPM (120 MG/M3);STEL 70 PPM (250 MG/M3);SKIN JAN93
  OEL-SWITZERLAND:TWA 100 PPM (360 MG/M3);STEL 500 PPM JAN93
  OEL-THAILAND:TWA 500 MG/M3;STEL 1000 MG/M3 JAN93
  OEL-TURKEY:TWA 500 PPM (1740 MG/M3) JAN93
  OEL-UNITED KINGDOM:TWA 100 PPM (350 MG/M3);STEL 250 PPM JAN93
  OEL IN BULGARIA, COLOMBIA, JORDAN, KOREA CHECK ACGIH TLV
  OEL IN NEW ZEALAND, SINGAPORE, VIETNAM CHECK ACGIH TLV
  NIOSH REL TO METHYLENE CHLORIDE-AIR:CA LOWEST FEASIBLE CONCENTRATION
    NIOSH* DHHS #92-100,92
  NOHS 1974: HZD 47270; NIS 374; TNF 89025; NOS 192; TNE 975696
  NOES 1983: HZD 47270; NIS 363; TNF 87086; NOS 212; TNE 1438196; TFE
    352536
  ATSDR TOXICOLOGY PROFILE (NTIS** PB/89/194468/AS)
  EPA GENETOX PROGRAM 1988, POSITIVE: CELL TRANSFORM.-RLV F344 RAT EMBRYO
  EPA GENETOX PROGRAM 1988, POSITIVE: HISTIDINE REVERSION-AMES TEST
  EPA GENETOX PROGRAM 1988, POSITIVE: S CEREVISIAE GENE CONVERSION; S
    CEREVISIAE-HOMOZYGOSIS
  EPA GENETOX PROGRAM 1988, POSITIVE: S CEREVISIAE-REVERSION
  EPA GENETOX PROGRAM 1988, NEGATIVE: D MELANOGASTER SEX-LINKED LETHAL
  EPA TSCA SECTION 8(B) CHEMICAL INVENTORY
  EPA TSCA 8(A) PRELIMINARY ASSESSMENT INFORMATION, FINAL RULE
    FEREAC 47,26992,82
  EPA TSCA SECTION 8(D) UNPUBLISHED HEALTH/SAFETY STUDIES
  ON EPA IRIS DATABASE
  EPA TSCA TEST SUBMISSION (TSCATS) DATA BASE, JULY 1996
  NIOSH CURRENT INTELLIGENCE BULLETIN 46, 1986
  NIOSH ANALYTICAL METHOD, 1994: METHYLENE CHLORIDE, 1005
  NTP CARCINOGENESIS STUDIES (INHALATION);CLEAR EVIDENCE:MOUSE,RAT
    NTPTR* NTP-TR-306,86
  NTP 7TH ANNUAL REPORT ON CARCINOGENS, 1992 : ANTICIPATED TO BE
    CARCINOGEN
  OSHA ANALYTICAL METHOD #ID-59
U.S. INFORMATION
    THIS PRODUCT IS SUBJECT TO SARA SECTION 313 REPORTING REQUIREMENTS.
SECTION 16. - - - - - - - - OTHER INFORMATION- - - - - - - - - - -
  THE ABOVE INFORMATION IS BELIEVED TO BE CORRECT BUT DOES NOT PURPORT TO
  BE ALL INCLUSIVE AND SHALL BE USED ONLY AS A GUIDE. SIGMA, ALDRICH,
  FLUKA SHALL NOT BE HELD LIABLE FOR ANY DAMAGE RESULTING FROM HANDLING
  OR FROM CONTACT WITH THE ABOVE PRODUCT. SEE REVERSE SIDE OF INVOICE OR
  PACKING SLIP FOR ADDITIONAL TERMS AND CONDITIONS OF SALE.
  COPYRIGHT 1996 SIGMA CHEMICAL CO., ALDRICH CHEMICAL CO., INC.,
  FLUKA CHEMIE AG
  LICENSE GRANTED TO MAKE UNLIMITED PAPER COPIES FOR INTERNAL USE ONLY

                            Page 5
```

FIGURE 1.5 *(continued)*

MSDS information for thousands of compounds can also be obtained on the Internet. The number of Web sites changes almost daily; the following list of sites provides access to MSDS information and is current at the time of publication.

http://www.ilpi.com/msds/
http://www.phys.ksu.edu/area/jrm/safety/msds/html
http://physchem.ox.ac.uk/MSDS/
http://toxnet.nlm.nih.gov/

An extensive collection of chemical safety information is found in the three-volume set *Sigma-Aldrich Library of Regulatory and Safety Data* (Aldrich Chemical Co., 1993).

Hazardous Materials Identification Systems

The labels on chemical containers carry warnings about the hazards involved in handling and shipping the compounds. The four-diamond symbol and a globally harmonized system of pictograms are the most commonly used hazardous materials identification systems.

Four-diamond symbol. Chemical suppliers put a color-coded, four-diamond symbol—developed by the National Fire Protection Association—on the container label of all compounds they sell (Figure 1.6). The four diamonds provide information concerning the hazards associated with handling the compounds:

fire hazard (top, red diamond)
reactivity hazard (right, yellow diamond)

FIGURE 1.6
Four-diamond label for chemical containers indicating health, fire, reactivity, and special hazards. The symbol in the special hazard diamond indicates that the compound is reactive with water and should not come into contact with it.

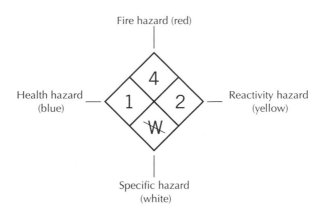

Fire hazard (red)

Health hazard (blue)

Reactivity hazard (yellow)

Specific hazard (white)

specific hazard (bottom, white diamond)
health hazard (left, blue diamond)

The numerical values in the diamonds range from 0 to 4 with 0 indicating no chemical hazard and 4 indicating extreme chemical hazard.

Globally Harmonized System (GHS) of pictograms. Many chemical suppliers also indicate hazards by printing the universally understandable pictograms approved at the UN-sponsored Rio Earth Summit in 1992 on the labels of their reagents (Figure 1.7).

Other warnings found on chemical labels. Chemical labels may also include warnings such as "Irritant," "Lachrymator," "Cancer suspect agent," "Mutagen," or "Teratogen." Definitions of these terms follow.

Irritant: Substance that causes irritation to skin, eyes, or mucous membranes.
Lachrymator: Substance that causes irritation and watering of the eyes (tears).
Cancer suspect agent: Substance that is carcinogenic in experimental animals at certain dose levels, by certain routes of administration, or by certain mechanisms considered relevant to human exposure. Available epidemiological data do not confirm an increased cancer risk in exposed humans.
Mutagen: Substance that induces genetic changes.
Teratogen: Substance that induces defects in a developing fetus.

FIGURE 1.7
Globally Harmonized System (GHS) pictograms indicating chemical hazards.

Explosive

Oxidizing

Highly flammable or extremely flammable

Toxic or very toxic

Harmful or irritant

Corrosive

Biohazard

Dangerous for the environment

Protecting the Environment

Protection in the laboratory does not stop with protection of the people who work there. Every student and industrial lab worker must be aware of the impact of what he or she does on others and on the environment outside the laboratory. Before disposing of anything in the lab, you should be conscious of how this disposal will affect the environment. Although zero waste is impossible, minimum waste is essential. Industries are now required to account for almost every gas, liquid, or solid waste they put into the environment. In the undergraduate laboratory, we should do likewise.

| 1.7 | Green Chemistry |

One way to protect the environment is to reduce or eliminate the waste and by-products from chemical reactions and manufacturing processes that use chemical reagents and solvents. The term *green chemistry* has been given to new chemical reactions and processes that replace existing methods with ones that have the following characteristics:

Use less and safer reagents and solvents.
Reduce energy requirements.
Utilize renewable resources whenever possible.
Minimize or prevent the formation of waste.

The goal of green chemistry is to be as environmentally friendly as possible in the synthesis and utilization of chemicals both in the laboratory and in industrial and manufacturing applications.

How can one go about changing an existing chemistry procedure to one that could be called green chemistry? The first step is to ascertain the safety information on the reagents and solvents that are currently being used, as well as information on any toxic by-products that would remain after completion of the reaction. The next steps are to consider what would be a safer, less toxic (or ideally nontoxic) alternative for the reagent or solvent and to ascertain whether another method would give the desired product using less hazardous materials.

For example, consider replacement solvents that pose fewer health and environmental hazards.

Water

In the quest for solvents that minimize health hazards and risks to the environment, water would appear to be ideal because it is readily available and nonhazardous. But the first requirement for a reaction solvent is that it dissolve the reagents used in the reaction, and a very large percentage of organic compounds are insoluble or only slightly soluble in water. However, reactions in aqueous reagent solutions can be promoted in several ways with water-insoluble organic compounds, such as using vigorous stirring or water-soluble catalysts.

Supercritical
Carbon Dioxide

Carbon dioxide is a gas under normal conditions. Solid CO_2 (dry ice) *sublimes,* or vaporizes from the solid to gaseous state without melting. When CO_2 is subjected to conditions of temperature and pressure that exceed its critical point, 31.1°C and 73 atm pressure, it becomes a single phase with properties intermediate between the properties of its gaseous and liquid states. A fluid above its critical-point temperature and pressure is called a *supercritical fluid.*

Supercritical CO_2 is a very good organic solvent with properties similar to many common organic solvents. The high-pressure equipment necessary to contain supercritical CO_2, however, makes its use in academic laboratories impractical. Applications using supercritical CO_2, which can replace traditional and hazardous solvents in industrial-scale chemical processes, include decaffeinating coffee, dry-cleaning clothing, cleaning electronic and industrial parts, and chemical reactions. At the end of these processes, the pressure is released and the escaping CO_2 gas can be easily recovered and recycled.

1.8   How Can a Laboratory Procedure Be Made Green?

The following examples illustrate how an organic lab procedure can be made "greener" by the use of alternative solvents and reagents.

Example 1.
Extraction of an
Organic Compound
from an Aqueous
Mixture

The organic chemist frequently needs to separate an organic compound from an aqueous mixture using the process of *extraction,* in which the higher solubility of the organic compound in an organic solvent selectively transfers it from the aqueous mixture. Consider a procedure that specifies dichloromethane as a solvent for extracting caffeine from tea leaves. Would ethyl acetate be a "greener" alternative, assuming that both solvents dissolve the caffeine adequately?

$$H_2C \overset{Cl}{\underset{Cl}{<}} \qquad H_3C-C \overset{O}{\underset{OCH_2CH_3}{<}}$$

Dichloromethane   Ethyl acetate

We need to ascertain and evaluate all the properties of ethyl acetate relative to those of dichloromethane to decide whether ethyl acetate would be a greener alternative.

Safety information. The safety information on the MSDS for dichloromethane indicates that it is a cancer-suspect agent, toxic, a neurological hazard, and an irritant to the skin, eyes, and mucous membranes. The MSDS for ethyl acetate states that it is an irritant to the skin, eyes, and mucous membranes. Ethyl acetate certainly looks safer.

Relative volatilities of dichloromethane and ethyl acetate. Dichloromethane has a high volatility (evaporation rate), related to its low boiling point (40°C). The boiling point of ethyl acetate is 77°C. The higher boiling point of ethyl acetate gives it a lower volatility than dichloromethane at room temperature, and it does not evaporate as readily during the handling and transfers that occur while the extraction is in progress. However, the higher boiling point of ethyl acetate means that it would require more heat (energy) to remove the solvent and recover the caffeine than would dichloromethane.

Solubility of water in the extraction solvent. For an extraction to be successful, the organic solvent and the aqueous phase must have a low solubility in one another. The solubility of water in ethyl acetate is five times greater than its solubility in dichloromethane. If we want to substitute ethyl acetate for dichloromethane as the extraction solvent, we need a way to decrease the solubility of water in ethyl acetate. The decrease can be accomplished by saturating the caffeine-containing aqueous mixture with sodium chloride, which reduces the amount of water that dissolves in ethyl acetate.

Relative costs of waste disposal. What happens to the solvent when the extraction of caffeine from tea is completed? It can be removed and recovered from the caffeine by distillation and possibly recycled for use in another application, but eventually the solvent becomes a waste that requires disposal either by burning in a process where the heat energy is recovered or by incineration where the heat is not recovered. Complete combustion of ethyl acetate produces carbon dioxide and water, whereas complete combustion of dichloromethane produces carbon dioxide, water, and hydrogen chloride (HCl). The HCl needs to be removed from the combustion gases before they are released to the atmosphere, a process that increases the disposal costs for all chlorinated compounds relative to nonhalogenated compounds.

Justification for the substitution of ethyl acetate for dichloromethane. Using ethyl acetate instead of dichloromethane is less hazardous to both the person doing the procedure and the environment. In addition, lower waste disposal costs make substitution of ethyl acetate a greener alternative than dichloromethane as the extraction solvent, despite the slightly higher energy costs incurred with ethyl acetate.

Example 2.
Oxidation of
Alcohols to Ketones

Chromium(VI) oxide (CrO_3) is a traditional reagent for oxidizing an alcohol to a ketone.

Alcohol   Carbonyl compound

The MSDS for CrO_3 indicates that it is highly toxic and a cancer suspect reagent. In addition, at the end of the reaction an equivalent amount of chromium(III) oxide is present as a by-product, requiring expensive disposal to prevent it from becoming an environmental contaminant. Household bleach, a 5.25% or 6.00% aqueous sodium hypochlorite (NaOCl) solution, is a green alternative for chromium(VI) oxide in oxidation reactions.

Oxidation of cyclohexanol.   The oxidation of cyclohexanol with aqueous sodium hypochlorite solution in the presence of acetic acid is an example of green chemistry oxidation (Ref. 14).

$$\text{Cyclohexanol} + \text{NaOCl} \xrightarrow[\text{H}_2\text{O}]{\text{acetic acid}} \text{Cyclohexanone} + \text{H}_2\text{O} + \text{NaCl}$$

Stirring to facilitate the reaction.   Cyclohexanol is a liquid at room temperature and is relatively insoluble in water. The water in the sodium hypochlorite solution provides the reaction medium. Even though cyclohexanol is largely insoluble in the aqueous sodium hypochlorite/acetic acid solution, vigorous stirring of the two phases increases the surface area of one liquid in contact with the other and greatly enhances the reaction rate.

Elimination of the extraction solvent.   Cyclohexanone has traditionally been recovered from the two-phase reaction mixture by extraction with an organic solvent, such as diethyl ether or dichloromethane. Steam distillation (codistillation of the organic compound with water) is a green alternative for separating the cyclohexanone from the inorganic salts in the aqueous reaction mixture. The tradeoffs for not using extractions to recover the product are a lower yield (50–60%) instead of the 70–80% that is possible using extractions, as well as higher energy costs, versus no organic solvent waste that would require disposal.

Nonhazardous by-products.   This synthesis also qualifies as green chemistry because the by-products of the reaction, water and sodium chloride, are nonhazardous wastes that can be washed down the sink. Any excess acetic acid remaining in the aqueous solution can be neutralized with sodium carbonate to form acetate ion, also a nonhazardous waste that can be washed down the sink.

Example 3.
Biochemical
Catalysis

Biochemical catalysis is a green alternative to traditional catalysts in organic synthesis. Using thiamine rather than potassium cyanide (KCN), the traditional catalyst in the condensation of two benzaldehyde molecules to form benzoin, is a green alternative.

Benzaldehyde   Benzoin

The MSDS for potassium cyanide indicates that it is highly toxic and readily absorbed through the skin. Its contact with acids produces highly toxic hydrogen cyanide gas. Vitamin B_1, in the form of thiamine, provides a far safer catalytic reagent for this reaction and eliminates the hazards and waste disposal costs of potassium cyanide (Ref. 15). Thiamine is a naturally occurring compound and a renewable resource. The MSDS for thiamine indicates that it may be harmful when ingested in high concentrations, and it may cause allergic reactions.

Overview of Greening a Chemical Process

These three examples are just a brief introduction to the ways chemical processes can be made greener. This endeavor is a continuing effort toward the goal of green chemistry—using chemistry in the synthesis and utilization of chemicals in as environmentally friendly a manner as possible. New manufacturing processes and chemical syntheses using green chemistry are being developed every day.

1.9   Atom Economy and Reaction Efficiency

In addition to finding greener alternatives for solvents and reagents, green chemistry is about finding ways to minimize or eliminate waste by generating fewer by-products in the course of a reaction. Chemists generally regard the percentage yield of a chemical reaction as the measure of its success. However, the percentage yield does not indicate how much atomic mass of the original reagents remains as by-products at the end of the reaction.

Atom Economy

The concept of atom economy was developed as a quantitative measure of how efficiently atoms of the starting materials and reagents are incorporated into the desired product (Ref. 16). Atom economy is defined as the percentage of atomic mass of all starting materials that appears in the final product, assuming 100% yield in the reaction. The balanced equation for the reaction is used in the calculation of atom economy.

Example 1. Consider the Williamson ether synthesis of 1-ethoxybutane, a substitution reaction in which the bromine atom of 1-bromobutane is replaced by an ethoxy group.

$$CH_3CH_2CH_2CH_2-Br + CH_3CH_2-O^-Na^+ \xrightarrow{CH_3CH_2OH}$$

1-Bromobutane   Sodium ethoxide
MW 137   MW 68.1

$$CH_3CH_2CH_2CH_2-O-CH_2CH_3 + NaBr$$

1-Ethoxybutane
MW 102

The atom economy for the reaction can be calculated as follows:

$$\text{atom economy} = \frac{MW_{1\text{-ethoxybutane}}}{MW_{1\text{-bromobutane}} + MW_{\text{sodium ethoxide}}} \times 100\% = \frac{102}{137 + 68} \times 100\% = 50\%$$

Thus, only 50% of the atomic mass of the starting materials is incorporated into the product. The other 50% of the atomic mass of the starting materials is the by-product, sodium bromide.

Example 2. Addition reactions are inherently high in atom economy because both reagents in the reaction are incorporated into the product. The Diels-Alder reaction is an example of an addition reaction.

2,3-Dimethyl-1,3-butadiene   Maleic anhydride   4,5-Dimethylcyclohex-4-ene-
MW 82.1   MW 98.1   *cis*-1,2-dicarboxylic anhydride
                                                             MW 180.2

The atom economy for this synthesis is 100% because the sum of atomic mass of the reagents (82.1 + 98.1) is equal to the atomic mass of the product (180.2).

Reaction Efficiency   The concept of reaction efficiency was developed as a measure of the mass of reactant atoms actually contained in the final product (Ref. 17). If the 1-ethoxybutane from the synthesis described in Example 1 were obtained in a 65% yield, the reaction efficiency would be

$$\text{reaction efficiency} = \% \text{ yield} \times \text{atom economy}$$
$$= 65\% \times 0.50 = 33\%$$

The reaction efficiency indicates that only 33% of the mass of reactants was recovered as product in the synthesis and the other 67% became waste, making the synthesis less than ideal from an environmental perspective.

If the yield for the Diels-Alder reaction in Example 2 were 80%, the reaction efficiency would also be 80%, indicating that only 20% of the total mass of reagents became waste in the synthesis, a much lower percentage than in the substitution reaction of Example 1.

One goal of green chemistry is to design synthetic pathways that improve both the atom economy of a reaction and the percentage yield to minimize the waste produced by chemical reactions.

References

Safety in the Laboratory

1. Lewis, Sr., R. J.; Sax, N. I. Sax's *Dangerous Properties of Industrial Materials;* 11th ed.; Wiley-Interscience: New York, 2004.
2. Furr, A. K. (Ed.) *CRC Handbook of Laboratory Safety;* 5th ed.; CRC Press: Boca Raton, FL, 2000.
3. The Manufacturing Chemists Association, Chemical Safety Data Sheets; Washington, DC.
4. U.S. Department of Labor, *Occupational Exposure to Hazardous Chemicals in Laboratories;* OSHA no. 95–33; U.S. Government Printing Office: Washington, DC, 1995.
5. O'Neill, M. J.; Smith, A.; Heckelman, P. A.; Oberchain, Jr., J. R. (Eds.) *The Merck Index of Chemicals, Drugs, and Biologicals;* 13th ed.; Merck & Co., Inc.: Whitehouse, NJ, 2001.
6. Lide, D. R. (Ed.) *CRC Handbook of Chemistry and Physics;* 85th ed.; CRC Press: Boca Raton, FL, 2004.
7. Lewis, Sr., R. J. *Rapid Guide to Hazardous Chemicals in the Workplace;* 4th ed.; Wiley-Interscience: New York, 2000.
8. Lenga, R. E.; Votoupal, K. L. (Eds.) *Sigma-Aldrich Library of Regulatory and Safety Data;* Aldrich Chemical Co.: Milwaukee, WI, 1993; 3 volumes.

9. National Research Council, *Prudent Practices in the Laboratory: Handling and Disposal of Chemicals;* National Academy Press: Washington, DC, 1995.
10. American Chemical Society, *Safety in Academic Chemistry Laboratories;* 7th ed.; American Chemical Society: Washington, DC, 2003.

Protecting the Environment

11. Anastas, P. T.; Warner, J. C. *Green Chemistry: Theory and Practice;* Oxford University Press: Oxford, 1998.
12. Doxee, K. M.; Hutchinson, J. E. *Green Organic Chemistry Strategies, Tools, and Laboratory Experiments;* Brooks/Cole: Belmont, CA, 2004.
13. American Chemical Society, *Less Is Better: Guide to Minimizing Waste in Laboratories;* American Chemical Society: Washington, DC, 2002.
14. Mohrig, J. R.; Neinhuis, D. M.; Linck, C. F.; Van Zoeren, C; Fox, B. G.; Mahaffy, P. G. *J. Chem. Educ.* **1985,** *62,* 519–521.
15. Mohrig, J. R.; Neckers, D. C. *Laboratory Experiments in Organic Chemistry;* 2nd ed.; Van Nostrand: New York, 1973, 184–187.
16. Trost, B. M. *Science* **1991,** *254,* 1471–1477.
17. Cann, M. C.; Dickneider, T. A. *J. Chem. Educ.* **2004,** *81,* 977–980.

TECHNIQUE

2

LABORATORY GLASSWARE

You will find an assortment of glassware and equipment in your laboratory desk, some items that will be familiar to you from your earlier lab experiences and other items that you may not have used previously. If your laboratory is equipped for miniscale experimentation, you will find specialized glassware that has carefully constructed ground glass joints designed to fit together tightly and interchangeably. This glassware is called **standard taper glassware,** and it comes in a variety of sizes. If you will be carrying out microscale experimentation, you will use scaled-down glassware designed for the milligram and milliliter quantities of reagents used in microscale work. There are two types of microscale glassware commonly used in the undergraduate organic laboratory—microscale standard taper glassware with threaded screw cap connectors and the Kontes/Williamson microscale glassware that fastens together with flexible elastomeric connectors.

SAFETY PRECAUTION

Before you use any glassware in an experiment, check it carefully for cracks or chips. Glassware with spherical surfaces, such as round-bottomed flasks, can develop small, star-shaped cracks. Replace damaged glassware. When cracked glassware is heated, it can break and ruin your experiment and possibly cause a serious spill or fire.

2.1 Desk Equipment

A typical student desk contains an assortment of beakers, Erlenmeyer flasks, filter flasks, thermometers, graduated cylinders, test tubes, funnels, and a variety of other items. Your desk or drawer will probably have most, if not all, of the equipment items shown in Figure 2.1. Make sure that all glassware is clean and has no chips or cracks. Replace damaged glassware.

2.2 Standard Taper Miniscale Glassware

Standard taper glassware is designated by the symbol ⊤. All the joints in standard taper glassware have been carefully ground so that they are exactly the same size, and all the pieces fit together interchangeably. We recommend the use of ⊤ 19/22 or ⊤ 14/20 glassware for miniscale experiments. The numbers represent, in millimeters, the diameter and the length of all ground glass surfaces (Figure 2.2). A typical set of ⊤ 19/22 glassware found in introductory organic laboratories is shown in Figure 2.3.

Greasing Ground Glass Joints

Because standard taper joints fit together tightly, they are not usually put together dry. The ground glass joints of ⊤ 19/22 or ⊤ 14/20 glassware are coated with a lubricating grease. The grease prevents interaction of the ground glass joints with the chemicals used in the experiment that can cause the joints to "freeze," or stick together. Taking apart stuck joints, although not impossible, is often not an easy task, and standard taper glassware (which is expensive) frequently is broken in the process. **Note:** Microscale glassware with ground glass joints is never greased unless the reaction involves strong bases such as sodium hydroxide or sodium methoxide.

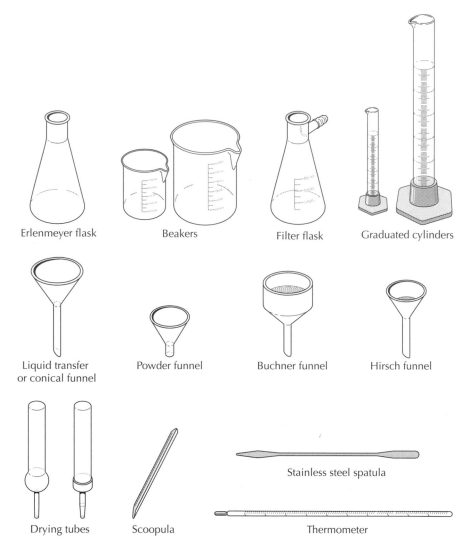

FIGURE 2.1
Typical equipment in a student desk.

Types of grease for ꙅ joints. Several greases are commercially available. For general purposes in an undergraduate laboratory, a hydrocarbon grease, such as Lubriseal, is preferred because it can be removed easily. Silicone greases have a very low vapor pressure

FIGURE 2.2
Standard taper
distillation head
and round-bottomed
flask showing the
dimensions of the
ground glass joints.

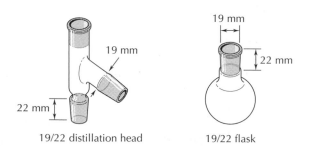

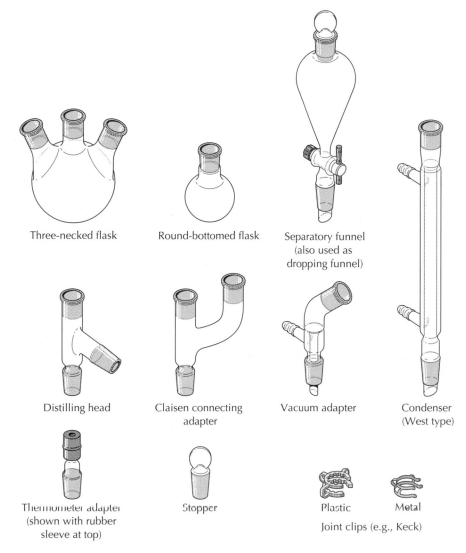

FIGURE 2.3
Standard taper glassware for miniscale experiments.

and are intended for sealing a system that will be under vacuum. Silicone greases are nearly impossible to remove completely because they do not dissolve in detergents or organic solvents.

Sealing a standard taper joint with grease. To seal a standard taper joint, apply two thin strips of grease the entire length of the inner joint about 180° apart, as shown in Figure 2.4. Gently insert the inner joint into the outer joint and rotate one of the pieces. The joint should rotate easily and the grease should become uniformly distributed so that the frosted surfaces appear clear.

Using excess grease is bad practice. Not only is it messy, but worse, it may contaminate the reaction or coat the inside of the

FIGURE 2.4
Apply two *thin* strips of grease the entire length of the inner joint about 180° apart.

flasks, making them difficult to clean. Just enough grease to coat the entire ground surface thinly is sufficient. If grease oozes above the top or below the bottom of the joint, you have used too much. Take the joint apart, wipe off the excess grease with a towel or tissue, and assemble the pieces again.

Removing grease from standard taper joints.   When you have finished the experiment, clean the grease from the joints by using a brush, detergent, and hot water. If this scrubbing does not remove all the grease, dry the joint and clean it with a towel (for example, a Kimwipe) moistened with toluene or hexane.

SAFETY PRECAUTION

Toluene and hexane are irritants and pose a fire hazard. Wear gloves and work in a hood.

2.3   Microscale Glassware

When the amounts of reagents used for experiments are in the 100–300-mg or 0.1–2.0-mL range, microscale glassware is used. Recovering any product from an operation at this scale would be difficult if you were using 19/22 or 14/20 standard taper glassware; most of the material would be lost on the glass surfaces. Two types of microscale glassware are commonly used in undergraduate organic laboratories—standard taper glassware with threaded screw cap connectors or Kontes/Williamson glassware that fastens together with flexible elastomeric connectors. Your instructor will tell you which type of microscale glassware is used in your laboratory.

Standard Taper Microscale Glassware

The pieces of microscale standard taper glassware needed for typical experiments in the introductory organic laboratory are shown in Figure 2.5. The pieces fit together with 14/10 standard taper joints.

Grease is NOT used with microscale glassware, except when the reaction mixture contains a strong base, because its presence could cause significant contamination of the reaction mixture. Instead, a threaded cap and O-ring ensure a tight seal and hold the pieces together, thus eliminating the use of clamps or clips. Place the threaded cap over the inner joint; then slip the O-ring over the tapered portion. Fit the inner joint inside the outer joint and screw the threaded cap tightly onto the outer joint (Figure 2.6). A securely screwed connection effectively prevents the escape of vapors and is also vacuum tight.

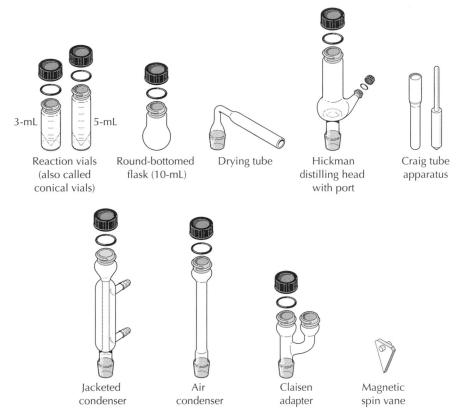

FIGURE 2.5
Standard taper microscale glassware.

Kontes/Williamson Microscale Glassware

The various pieces of Kontes/Williamson microscale glassware used in typical experiments in the organic laboratory are shown in Figure 2.7. This type of microscale glassware fits together with flexible elastomeric connectors that are heat and solvent resistant. **Grease is NOT used with this type of glassware connector.** A flexible connector with an aluminum support rod fastens two pieces of glassware together and provides a way of attaching the apparatus in a two-way clamp to a ring stand or vertical support rod. One piece of glassware is pushed into the connector, and then the second piece is pushed into the other end of the connector,

FIGURE 2.6
Assembling a standard taper joint on standard taper microscale glassware.

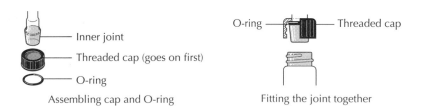

Short-necked and long-necked 5-mL flasks

Short-necked 10-mL flask*

Air condenser

Reaction tubes

Connecting adapter

Distillation head/ Claisen adapter

Flexible connector with support rod

Magnetic stirring bar

Flexible connector

Plastic funnel

Plastic Hirsch funnel with replaceable frit and 25-mL filter flask

10-mL Erlenmeyer flask

15-mL centrifuge tube with cap

Flexible thermometer adapter

8-mm sleeve stopper (fold-over rubber septum)

*Available from Kontes, Catalog no. 748008-9001.

FIGURE 2.7
Williamson microscale glassware and other microscale apparatus.
(Manufactured by Kontes Glass Co., Vineland, NJ.)

as shown in Figure 2.8. The flexible connector effectively seals the joint and prevents the escape of vapors.

2.4   Cleaning and Drying Laboratory Glassware

Part of careful laboratory technique includes cleaning the glassware before you leave the laboratory. Cleaning up immediately ultimately saves time and reduces everyone's exposure to chemicals. Clean glassware is essential for maximizing the yield in any organic reaction, and in many instances glassware also must be dry. Try not to have to wash something immediately before using it, because then you will waste precious time while it dries in the oven.

FIGURE 2.8
Assembling Williamson microscale glassware with a flexible connector.

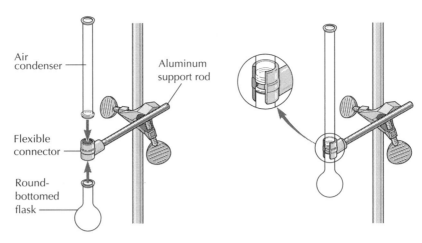

Fitting the glassware into the flexible connector one piece at a time

Cutaway showing the two pieces of glassware fastened in connector

Cleaning Glassware

Strong detergents and hot water are the ingredients needed to clean most glassware used for organic reactions. Scrubbing with a paste made from water and scouring powder, such as Ajax or Bon Ami, removes many organic residues from glassware. Organic solvents, such as acetone or hexane, help dissolve the polymeric tars that sometimes coat the inside of a flask after a distillation. You may want to wear gloves when cleaning glassware. A final rinse of clean glassware with distilled water prevents water spots.

SAFETY PRECAUTION

Solvents such as acetone and hexane are irritants and flammable. Wear gloves, use the solvents in a hood, and dispose of them in the flammable (nonhalogenated) waste container.

A solution of alcoholic sodium hydroxide* is our favorite cleanser for removing grease and organic residues from flasks and other glassware.

SAFETY PRECAUTION

Strong bases, such as sodium hydroxide, cause severe burns and eye damage. Skin contact with alkali solutions starts as a slippery feel to the skin followed by irritation. Wash the affected area with copious amounts of water. Wear gloves and eye protection while cleaning glassware with alcoholic NaOH solution.

Drying Glassware

Dry glassware is needed for most organic reactions because water acts as a contaminant or may inhibit a reaction from occurring at all. The easiest way to ensure dry glassware is to leave all glassware

*Made by dissolving 120 g of NaOH in 120 mL of water and diluting to 1 L with 95% ethanol.

washed and clean at the end of each lab session. It will be dry and ready to use by the next laboratory period.

Oven drying of glassware.   Wet glassware can be dried by heating it in an oven at 120°C for 20 min. Remove the dried glassware from the oven with tongs and allow it to cool to room temperature before using it for a reaction.

Drying wet glassware with acetone.   Glassware that is wet from washing can be dried quickly by rinsing it in a hood with a few milliliters of acetone. Acetone and water are completely miscible, so the water is removed from the glassware. The acetone is collected as flammable (nonhalogenated) waste; any residual acetone on the glassware is allowed to evaporate into the atmosphere. Thus there is an environmental cost, as well as the initial purchase and later waste disposal costs, in using acetone for drying glassware.

TECHNIQUE

3

THE LABORATORY NOTEBOOK

A few general comments are in order about the laboratory notebook that is the primary record of your experimental work. First, although many campus bookstores sell notebooks that are specifically designed as lab notebooks, it is often sufficient to use any notebook with tightly bound pages. Spiral and three-ring binders are inappropriate for lab notebooks because pages can be easily removed or torn out. **All entries about your work must be made directly in your laboratory notebook in ink.** Recording data on scraps of paper is an unacceptable practice because the papers may be lost; the practice is probably strictly forbidden in your laboratory.

3.1   Organization of the Laboratory Notebook

The notebook should begin with a table of contents; set aside the first two or three pages for this purpose. The rest of the pages should be numbered sequentially, and **no page should ever be torn out** of a laboratory notebook. Your notebook must be written with accuracy and completeness. It must be organized and legible, but it does not need to be a work of art.

Some flexibility in format and style may be allowed, but proper records of your experimental results must answer certain questions.

- *When* did you do the work?
- *What* are you trying to accomplish in the experiment?
- *How* did you do the experiment?
- *What* did you observe?
- *How* do you explain your observations?

A lab record needs to be written in three steps: prelab, during lab, and postlab. It should contain the following sections for each experiment you do.

To Be Done Before You Come to the Laboratory

The basic notebook setup discussed here is designed to help you prepare for an experiment in an effective and safe fashion. It includes the date and title of the experiment or project, the balanced chemical reaction you are studying, a statement of purpose, a table of reagents and solvents, the way you will calculate the percent yield, an outline of the procedure to be used, and answers to any prelab questions. Your instructor will undoubtedly provide specific guidelines for lab notebook procedures at your institution.

Title: Use a title that clearly identifies what you are doing in this experiment or project.

Date(s): Use the date on which an experiment is actually carried out. In some research labs, where patent issues are important, a witnessed signature of the date is required.

Balanced chemical reaction: Write balanced chemical equations that show the overall process. Any details of reaction mechanisms go into the summary.

Purpose statement: Write a brief statement of purpose for the synthesis or analysis, or state the question you are addressing, with a few words on major analytical or conceptual approaches.

Table: Include all reagents and solvents. The table normally lists molecular weights, the number of moles, and grams of reagents, as well as the densities of liquids you will be using, boiling points of compounds that are liquids at room temperature and melting points of all organic solids, and pertinent hazard warnings.

Method of yield calculation: Outline the computations to be used in a synthesis experiment, including calculation of the theoretical yield [see Technique 3.2].

Procedure and prelab questions: Write a procedural outline in sufficient detail so that the experiment could be done without reference to your lab textbook. This outline is especially important in experiments where you have designed the procedure. Answer any assigned prelab questions.

To Be Done During the Laboratory Session(s)

Recording observations during the experiment is a crucial part of your laboratory record. If your observations are incomplete, you cannot interpret the results of your experiments once you have left the laboratory. It is difficult, if not impossible, to reconstruct them at a later time.

Observations **must be recorded in your notebook in ink while you are doing an experiment.** You must record the **actual** quantities of all reagents as they are used, as well as the amounts of crude and purified products you obtain. Mention which measurements (temperature, time, melting point, and so on) you took and which spectra you recorded.

Because organic chemistry is primarily an experimental science, your observations are crucial to your success. Things that seem insignificant may be important in understanding and explaining your results later. Typical laboratory observations might be as follows:

- A white precipitate appeared, which dissolved when sulfuric acid was added.
- The solution turned cloudy when it was cooled to 10°C.
- An additional 10 mL of solvent were required to completely dissolve the yellow solid.
- The reaction was heated at 50°C for 25 min on a water bath.
- A small puff of white smoke appeared when sodium hydroxide was added to the reaction mixture.
- The NMR sample was prepared with [state amount of compound used], using 0.6 mL of $CDCl_3$.
- A capillary OV–101 GC column heated to 137°C was used.
- The infrared spectrum was run as a KBr pellet.

Your observations may be recorded in a variety of ways. They may be written on right-hand pages across from the corresponding section of the experimental outline on a left-hand page, or the page may be divided into columns with the left column used for procedure and the right column for observations. It is a good idea to cross-index your observations to specific steps in the procedure that you wrote out as part of your prelab preparation. Your instructor will probably provide specific advice on how you should record your observations during the laboratory.

To Be Done After the Experimental Work Has Been Completed

In this section of the notebook you evaluate and interpret your experimental results. Entries include a section on interpretation of physical and spectral data, a summary of your conclusions, calculation of the percent yield, and answers to any assigned postlab questions.

Conclusions and summary: In an inquiry-based project or experiment, return to the question being addressed and discuss the conclusions you can draw from analysis of your data. For both inquiry-based experiments and those where you learned about laboratory techniques and the design of organic syntheses, discuss how your experimental results support your conclusions. Include a thorough interpretation of NMR and IR spectra and other analytical results, such as TLC and GC analyses. Properly labeled spectra and chromatograms should be stapled into your notebook. Cite any reference sources that you used and include answers to any assigned postlab questions.

Percent yield: The single most important measure of success in a chemical synthesis is the quantity of desired product produced. To be sure, the purity of the product is also crucial, but if a synthetic method produces only small amounts of the needed product, it is not much good. Reactions on the pages of textbooks are often far more difficult to carry out in good yield than the books suggest. Calculation of the percent yield is discussed in Technique 3.2.

3.2　　Calculation of the Percent Yield

When you report the results of a synthesis reaction, the *percent yield* is always stated, a parameter defined as the ratio of product obtained to the *theoretical yield* (maximum amount expected) multiplied by 100:

$$\% \text{ yield } = \frac{\text{actual yield of product}}{\text{theoretical yield}} \times 100$$

You calculate the theoretical yield from the balanced chemical equation and the limiting reagent, assuming 100% conversion of the starting materials to product(s). For example, consider the Williamson ether synthesis of 1-ethoxybutane from 1-bromobutane and sodium ethoxide.

$$CH_3(CH_2)_3\!-\!Br + CH_3CH_2\!-\!O^-Na^+ \xrightarrow{\text{ethanol}} CH_3(CH_2)_3\!-\!O\!-\!CH_2CH_3 + NaBr$$

<table>
<tr><td>1-Bromobutane</td><td>Sodium ethoxide</td><td>1-Ethoxybutane</td></tr>
<tr><td>MW 137</td><td>MW 68.1</td><td>MW 102</td></tr>
<tr><td>density 1.27 g·mL^{-1}</td><td></td><td></td></tr>
</table>

The procedure specified 4.50 mL of 1-bromobutane, 3.70 g of sodium ethoxide, and 20 mL of anhydrous ethanol. To calculate the theoretical yield, it is necessary to ascertain whether 1-bromobutane or sodium ethoxide is the limiting reagent by calculating the moles of each reagent present in the reaction mixture:

$$\text{moles of 1-bromobutane} = \frac{4.50 \text{ mL} \times 1.27 \text{ g·mL}^{-1}}{137 \text{ g·M}^{-1}} = 0.0417 \text{ mol}$$

$$\text{moles of sodium ethoxide} = \frac{3.70 \text{ g}}{68.1 \text{ g·M}^{-1}} = 0.0543 \text{ mol}$$

Therefore, 1-bromobutane is the limiting reagent.

According to the balanced equation, equimolar amounts of the two reactants are required. Thus the theoretical yield, the maximum amount of product that is possible from the reaction assuming that it goes to completion and that no experimental losses occur, is 0.0417 mol or 4.25 g of ethoxybutane:

$$\text{theoretical yield} = 0.0417 \text{ mol} \times 102 \text{ g·M}^{-1} = 4.25 \text{ g of 1-ethoxybutane}$$

The percent yield for a synthesis that produced 2.7 g of 1-ethoxybutane is 64%:

$$\% \text{ yield} = \frac{2.7 \text{ g}}{4.2 \text{ g}} \times 100 = 64\%$$

4

USING HANDBOOKS AND ONLINE DATABASES

From time to time, you will need to find physical constants, such as melting and boiling points, densities, and other useful information about organic compounds. Compilations of physical constants are published in a number of handbooks. In addition, databases available on the Internet provide information about physical constants for many organic compounds.

4.1 Handbooks

Three handbooks are particularly useful for physical constants of organic compounds: the *Aldrich Catalog of Fine Chemicals*, the *CRC Handbook of Chemistry and Physics*, and *The Merck Index of Chemicals, Drugs, and Biologicals*.

Aldrich Catalog of Fine Chemicals

The *Aldrich Catalog of Fine Chemicals* is published annually by the Aldrich Chemical Company of Milwaukee, Wisconsin. It lists thousands of organic and inorganic compounds and includes the chemical structure for each one, a brief summary of its physical properties, references on IR, UV, and NMR spectra, plus safety and disposal information. There are also references to *Beilstein's Handbuch der Organischen Chemie* and to *Reagents for Organic Synthesis* by Fieser and Fieser (see Appendix C for more information about these reference works). Figure 4.1 shows a page from an *Aldrich Catalog*.

CRC Handbook of Chemistry and Physics

The *CRC Handbook of Chemistry and Physics* is a commonly used handbook and is also published annually. The *CRC Handbook* contains a wealth of information, including extensive tables of physical properties and solubilities, as well as structural formulas, for more than 12,000 organic and 2400 inorganic compounds. The arrangement of topics in recent editions of the *CRC Handbook* is different from that of earlier editions: the tables of organic compounds are now listed before the tables of inorganic compounds.

To locate an organic compound successfully, you must pay close attention to the nomenclature used in the tables. In general, IUPAC nomenclature is followed, but a compound usually known by its common name may be listed under both names or even only under the common name. For example, CH_3CO_2H, ethanoic acid, is listed under its common name of acetic acid as the primary name with ethanoic acid (its IUPAC name) given as the second name; no ethanoic acid entry occurs. Figure 4.2 shows a page from the 85th edition of the *CRC Handbook*. Note that the third entry (No. 1223) lists the IUPAC name, 1-bromohexane, as the primary name of the compound and hexyl bromide as the second name (synonym). In earlier

3,3-Dimethyl-1-butanol, 99%
[624-95-3] $(CH_3)_3CCH_2CH_2OH$ FW 102.17
Beil. **1**,IV,1729
bp . . . . 143 °C mp . . . . –60 °C
density . . . 0.844 g/mL 25 °C n_D^{20} 1.414
R: 10 S: 23-24/25; TSCA Fp: 47°C (117°F)

183105-1G	glass btl	1 g	15.10
183105-10G	glass btl	10 g	65.80
183105-50G	glass btl	50 g	232.00

3,3-Dimethyl-2-butanol, 98%
[464-07-3] $(CH_3)_3CCH(OH)CH_3$ FW 102.17
Beil. **1**,IV,1727
bp . . . . 119-121 °C mp . . . . .4.8 °C
density . . . 0.812 g/mL 25 °C n_D^{20} 1.415
R: 10 RTECS# EL2276000; TSCA Fp: 26°C (79°F)

| 136824-25G | glass btl | 25 g | 35.20 |
| 136824-100G | glass btl | 100 g | 114.00 |

3,3-Dimethyl-2-butanone, 98%
Pinacolone
[75-97-8] $CH_3COC(CH_3)_3$ FW 100.16
Merck **13**,7522; Beil. **1**,IV,3310
bp . . . . 106 °C n_D^{20} 1.396
density . . . 0.801 g/mL 25 °C
R: 11-22 S: 9-16-29-33 RTECS# EL7700000 Fp: 5°C (41°F)

P45605-5ML	glass btl	5 mL	15.90
P45605-100ML	glass btl	100 mL	24.10
P45605-500ML	glass btl	500 mL	99.60

2-Dimethyl-2-butene solution
[563-79-1] $(CH_3)_2C=C(CH_3)_2$ FW 84.16
1.0 M in tetrahydrofuran
Used for the preparation of thexylborane.
density . . . 0.857 g/mL 25 °C n_D^{20} . . . . 754 °F
R: 11-19-22-36/37/38 S: 16-26-36/37/39; TSCA Fp: -21°C (-5°F)

| 304034-100ML | Sure/Seal™ | 100 mL | 27.20 |
| 304034-800ML | Sure/Seal™ | 800 mL | 65.60 |

2,3-Dimethyl-1-butene, 97%
[563-78-0] $(CH_3)_2CHC(CH_3)=CH_2$ FW 84.16
Beil. **1**,IV,852
bp . . . . 56 °C vp . . . .412 mm Hg (37.7 °C)
density . . . . 0.68 g/mL 25 °C n_D^{20} 1.389
mp . . . . –158 °C ait . . . . 680 °F
R: 11-65 S: 16-33-62 Fp: -17°C (1°F)

| 190403-5G | glass btl | 5 g | 23.60 |
| 190403-25G | glass btl | 25 g | 79.40 |

2,3-Dimethyl-2-butene
[563-79-1] $(CH_3)_2C=C(CH_3)_2$ FW 84.16
Beil. **1**,IV,853
bp . . . . 73 °C vp . . . .215 mm Hg (37.7 °C)
density . . . 0.708 g/mL 25 °C n_D^{20} 1.412
mp . . . . –75 °C ait . . . . 754 °F
R: 11-65 S: 16-33-62; TSCA Fp: -8°C (18°F)

98%

129259-10ML	glass btl	10 mL	30.10
129259-25ML	glass btl	25 mL	60.20
129259-100ML	glass btl	100 mL	194.50
129259-1L	glass btl	1 L	894.00

≥99%

220159-10ML	ampules	10 mL	49.20
220159-25ML	ampules	25 mL	75.10
220159-100ML	ampules	100 mL	202.00

3,3-Dimethyl-1-butene, 95%
[558-37-2] $(CH_3)_3CCH=CH_2$ FW 84.16
Beil. **1**,IV,850
contains 50-150 ppm BHT as stabilizer
bp . . . . 41 °C vp . . . . . . 6.96 psi (20 °C)
density . . . 0.653 g/mL 25 °C vd . . . .>1 (vs air)
mp . . . . –115 °C n_D^{20} 1.376
R: 11-36/37/38-65 S: 16-26-33-36-62; TSCA Fp: -8°C (18°F)

| 119059-50ML | glass btl | 50 mL | 12.70 |
| 119059-250ML | glass btl | 250 mL | 51.10 |

⊠ Dimethyl(**tert**-butoxycarbonyl)methylphosphonate, see **tert**-Butyl
P,P-dimethylphosphonoacetate Page 528

1,3-Dimethylbutylamine, 98%
[108-09-8] $(CH_3)_2CHCH_2CH(CH_3)NH_2$ FW 101.19
Beil. **4**,191
bp . . . . 108-110 °C n_D^{20} 1.4085
density . . . 0.717 g/mL 25 °C
R: 11-22-34 S: 16-26-36/37/39-45 RTECS# EO4460000; TSCA
Fp: 13°C (55°F)

| 126411-5G | glass btl | 5 g | 28.70 |
| 126411-25G | glass btl | 25 g | 113.00 |

3,3-Dimethylbutylamine, 97%
[15673-00-4] $(CH_3)_3CCH_2CH_2NH_2$ FW 101.19
bp . . . . 114-116 °C n_D^{20} 1.4135
density . . . 0.752 g/mL 25 °C
R: 11-34 S: 16-26-27-36/37/39-45 Fp: 6°C (42°F)

| 183113-250MG | glass btl | 250 mg | 21.20 |
| 183113-1G | glass btl | 1 g | 63.80 |

N,N-Dimethylbutylamine, 99%
[927-62-8] $CH_3(CH_2)_3N(CH_3)_2$ FW 101.19
Beil. **4**,IV,546
bp . . . . 93.3 °C/750 mm Hg n_D^{20} 1.398
density . . . 0.721 g/mL 25 °C
R: 11-20/22-34 S: 16-26-36/37/39-45 RTECS# EJ4039250;
TSCA Fp: -5°C (23°F)

| 369527-250ML | glass btl | 250 mL | 24.90 |
| 369527-1L | glass btl | 1 L | 69.10 |

3,3-Dimethyl-1-butyne, 98%
tert-Butylacetylene
[917-92-0] $(CH_3)_3CC≡CH$ FW 82.14
Beil. **1**,IV,1022
bp . . . . 37-38 °C vp . . . . . . 7.88 psi (20 °C)
density . . . 0.667 g/mL 25 °C n_D^{20} 1.374
mp . . . . –78 °C
R: 12-36/37/38 S: 3-16-26-33-7/9 Fp: -25°C (-13°F)

| 244392-5G | ampules | 5 g | 47.70 |
| 244392-25G | ampules | 25 g | 165.50 |

3,3-Dimethylbutyraldehyde, 95%
tert-Butylacetaldehyde
[2987-16-8] $(CH_3)_3CCH_2CHO$ FW 100.16
Beil. **1**,III,2843
bp . . . . 104-106 °C n_D^{20} 1.397
density . . . 0.798 g/mL 25 °C
R: 11-36/37/38 S: 16-23 Fp: 13°C (55°F)

359904-1ML	ampules	1 mL	42.90
359904-5ML	ampules	5 mL	142.00
359904-25ML	ampules	25 mL	561.00

2,2-Dimethylbutyric acid, 96%
[595-37-9] $C_2H_5C(CH_3)_2CO_2H$ FW 116.16
Beil. **2**,335
bp . . . . . 94-96 °C/5 mm Hg n_D^{20} 1.4154
density . . . 0.928 g/mL 25 °C
R: 36/37/38 S: 26-36/37 Fp: 79°C (175°F)

| D152609-5ML | glass btl | 5 mL | 21.80 |
| D152609-25ML | glass btl | 25 mL | 32.70 |

FIGURE 4.1
Page 987 from the 2005–2006 *Aldrich Catalog*. Listings provide a summary of the physical properties for each compound. (Reprinted with permission from Aldrich Chemical Co., Inc., Milwaukee, WI.)

PHYSICAL CONSTANTS OF ORGANIC COMPOUNDS (continued)

No.	Name	Synonym	Mol. Form.	Mol. Wt.	CAS RN	Physical Form	mp/°C	bp/°C	den/g cm⁻³	n_D	Solubility
1221	1-Bromohexadecane		$C_{16}H_{33}Br$	305.337	112-82-3		18	336	0.9991^{20}	1.4618^{25}	i H_2O; s eth
1222	2-Bromohexadecanoic acid		$C_{16}H_{31}BrO_2$	335.320	18263-25-7		52.8				
1223	1-Bromohexane	Hexyl bromide	$C_6H_{13}Br$	165.071	111-25-1	liq	-83.7	155.3	1.1744^{20}	1.4478^{20}	i H_2O; msc EtOH, eth; s ace, vs chl
1224	2-Bromohexane		$C_6H_{13}Br$	165.071	3377-86-4			143; 78^{90}	1.1658^{20}	1.4832^{25}	i H_2O; vs EtOH; s eth, ace, sl chl
1225	3-Bromohexane		$C_6H_{13}Br$	165.071	3377-87-5			142	1.1799^{20}	1.4472^{20}	i H_2O; vs ace, eth, EtOH, chl
1226	2-Bromohexanoic acid, (±)		$C_6H_{11}BrO_2$	195.054	2681-83-6		2.0	240, 140^{23}	1.2810^{33}		s EtOH, chl
1227	6-Bromohexanoic acid		$C_6H_{11}BrO_2$	195.054	4224-70-8	cry (peth)	35	167^{20}			vs peth
1228	6-Bromohexanoyl chloride		$C_6H_{10}BrClO$	213.499	22809-37-6			101^6			
1229	1-Bromo-4-(hexyloxy)benzene		$C_{12}H_{17}BrO$	257.166	30752-19-3			156^{13}	1.2306^{20}	1.5262^{20}	
1230	5-Bromo-2-hydroxybenzaldehyde		$C_7H_5BrO_2$	201.018	1761-61-1	nd (al), lf (eth)	105.5				i H_2O; s EtOH, eth; sl chl
1231	4-Bromo-α-hydroxybenzeneacetic acid, (±)	p-Bromomandelic acid	$C_8H_7BrO_3$	231.044	7021-04-7		119				vs H_2O, EtOH, eth, bz, chl
1232	5-Bromo-2-hydroxybenzenemethanol	Bromosaligenin	$C_7H_7BrO_2$	231.034	2316-64-5	lf (bz)	113				vs bz, eth, EtOH, chl
1233	5-Bromo-2-hydroxybenzoic acid		$C_7H_5BrO_3$	217.017	89-55-4	nd (w, dil al)	169.8				sl H_2O, ace; vs EtOH, eth
1234	3-Bromo-4-hydroxy-5-methoxybenzaldehyde		$C_8H_7BrO_3$	231.044	2973-76-4	pl (HOAc), nd, pl (al)	167.0				i H_2O; s EtOH, DMSO; sl eth, bz
1235	1-Bromo-2-iodobenzene		C_6H_4BrI	282.904	583-55-1	nd	9.5	257, 120^{15}	2.2570^{25}	1.6618^{25}	i H_2O; sl EtOH, HOAc; s ace
1236	1-Bromo-3-iodobenzene		C_6H_4BrI	282.904	591-18-4	liq	-9.3	252, 120^{18}			i H_2O; sl EtOH, HOAc
1237	1-Bromo-4-iodobenzene		C_6H_4BrI	282.904	589-87-7	pr or pl (eth-al)	92	252			i H_2O; sl EtOH, chl, s eth
1238	Bromoiodomethane		CH_2BrI	220.835	557-68-6			139.5	2.926^{17}	1.6410^{20}	vs chl
1239	1-Bromo-4-isocyanatobenzene	p-Bromophenyl isocyanate	C_7H_4BrNO	198.017	2493-02-9	nd		226			s eth
1240	1-Bromo-4-isopropylbenzene		$C_9H_{11}Br$	199.087	586-61-8	liq	-22.5	218.7	1.3145^{20}	1.5569^{20}	i H_2O; s eth, bz, chl; sl ctc
1241	4-Bromoisoquinoline		C_9H_6BrN	208.055	1532-97-4	cry (peth)	41.5	282.5			vs eth
1242	Bromomethane	Methyl bromide	CH_3Br	94.939	74-83-9	col gas	-93.68	3.5	1.6755^{20}	1.4218^{20}	sl H_2O; msc EtOH, eth, chl, CS_2
1243	1-Bromo-2-methoxyethane		C_3H_7BrO	138.991	6482-24-2			110	1.4623^{20}	1.4475^{30}	
1244	Bromomethoxymethane		C_2H_5BrO	124.964	13057-17-5			87	1.5976^{20}	1.4562^{20}	
1245	2-Bromo-4-methylaniline		C_7H_8BrN	186.050	583-68-6	lf	26	240	1.510^{20}	1.5999^{20}	i H_2O; s EtOH, eth
1246	4-Bromo-2-methylaniline		C_7H_8BrN	186.050	583-75-5	cry (al)	59.5	240			sl H_2O, chl; s EtOH; vs eth, HOAc
1247	(Bromomethyl)benzene	Benzyl bromide	C_7H_7Br	171.035	100-39-0	liq	-1.5	201	1.4380^{25}	1.5752^{20}	i H_2O; msc EtOH, eth; s ctc
1248	4-(Bromomethyl)benzoic acid		$C_8H_7BrO_2$	215.045	6232-88-8		226.3				
1249	3-(Bromomethyl)benzonitrile		C_8H_6BrN	196.045	28188-41-2		96.5	130^4			
1250	4-(Bromomethyl)benzonitrile		C_8H_6BrN	196.045	17201-43-3	nd (lig)	114				
1251	1-Bromo-2-methylbutane, DL		$C_5H_{11}Br$	151.045	5973-11-5			119	1.2205^{20}	1.4452^{20}	i H_2O; s EtOH, eth; vs chl
1252	1-Bromo-3-methylbutane	Isopentyl bromide	$C_5H_{11}Br$	151.045	107-82-4	liq	-112	120.4	1.2071^{20}	1.4420^{20}	i H_2O; s EtOH, eth; sl ctc; vs chl
1253	2-Bromo-2-methylbutane	tert-Pentyl bromide	$C_5H_{11}Br$	151.045	507-36-8			108	1.197^{18}	1.4421	
1254	3-Bromo-3-methylbutanoic acid	β-Bromoisovaleric acid	$C_5H_9BrO_2$	181.028	5798-88-9	nd (lig)	74				vs bz, eth, EtOH
1255	1-Bromo-3-methyl-2-butene		C_5H_9Br	149.029	870-63-3			dec 131; 50^{40}	1.2930^{15}	1.4930^{15}	vs ace, bz, eth, EtOH
1256	1-(Bromomethyl)-2-chlorobenzene		C_7H_6BrCl	205.480	611-17-6			109^{10}			
1257	(Bromomethyl)chlorodimethylsilane		$C_3H_8BrClSi$	187.539	16532-02-8			131	1.375^{25}	1.4630^{25}	
1258	1-Bromo-3-methylcyclohexane	3-Methylcyclohexyl bromide	$C_7H_{13}Br$	177.082	13905-48-1			181; 60^{11}	1.2676^{15}	1.4979^{20}	i H_2O; vs eth; s bz
1259	(Bromomethyl)cyclohexane		$C_7H_{13}Br$	177.082	2550-36-9			76^{26}	1.283^{20}	1.4907^{30}	vs bz, eth, chl
1260	1-(Bromomethyl)-3-fluorobenzene		C_7H_6BrF	189.025	456-41-7			88^{20}		1.5474^{20}	
1261	3-(Bromomethyl)heptane		$C_8H_{17}Br$	193.125	18908-66-2			67^{10}			
1262	1-(Bromomethyl)-2-methylbenzene		C_8H_9Br	185.061	89-92-9	pr	21	217, 108^{16}	1.3811^{23}	1.5730^{20}	i H_2O; s EtOH, eth, ace, bz
1263	1-(Bromomethyl)-3-methylbenzene		C_8H_9Br	185.061	620-13-3			212.5	1.3711^{23}	1.5660^{20}	i H_2O; vs EtOH, eth

3-70

FIGURE 4.2

Page 3-70 from the 85th edition of the *CRC Handbook of Chemistry and Physics*. (Reprinted with permission from CRC Press, Inc., Boca Raton, FL.)

Rhombic needles from petr ether, mp 86°, bp$_{0.02}$ 140°. [α]$_D^{25}$ −63.8° (methanol); −56.3° (chloroform). Very sol in alcohol, ether, chloroform, benzene, petr ether.

Hydrochloride. $C_{12}H_{15}NO_3.HCl$. Orthorhombic prisms, dec 255°; freely sol in hot water. Aq soln is neutral.

659. Anilazine. [101-05-3] 4,6-Dichloro-*N*-(2-chlorophenyl)-1,3,5-triazin-2-amine; 2,4-dichloro-6-(*o*-chloroanilino)-*s*-triazine; (*o*-chloroanilino)dichlorotriazine; Dyrene. $C_9H_5Cl_3N_4$; mol wt 275.53. C 39.23%, H 1.83%, Cl 38.60%, N 20.33%. Prepn: C. N. Wolf, US 2720480 (1955 to Ethyl Corp.); E. G. Hill, E. Clinton, US 2820032 (1958); K. H. Rattenburg *et al.*, US 3074946 (1963 to Chemagro). Toxicity study: S. D. Cohen, S. D. Murphy, *J. Agr. Food Chem.* **21,** 140 (1973).

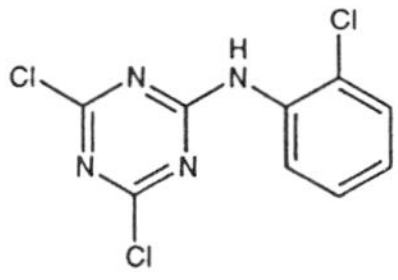

White to tan crystals, mp 159-160°. Insol in water. Soly at 30° (g/100 ml): toluene, 5; xylene, 4; acetone, 10. Subject to hydrolysis; not compatible with oils and alkaline materials. LD$_{50}$ orally in rats: >5000 mg/kg, Mobay Technical Information Sheet, Jan. 1979.

USE: Fungicide.

660. Anileridine. [144-14-9] 1-[2-(4-Aminophenyl)ethyl]-4-phenyl-4-piperidinecarboxylic acid ethyl ester; 1-(*p*-aminophenethyl)-4-phenylisonipecotic acid ethyl ester; ethyl 1-(4-aminophenethyl)-4-phenylisonipecotate; *N*-[β-(*p*-aminophenyl)ethyl]-4-phenyl-4-carbethoxypiperidine; *N*-β-(*p*-aminophenyl)ethylnormeperidine; Leritine. $C_{22}H_{28}N_2O_2$; mol wt 352.47. C 74.97%, H 8.01%, N 7.95%, O 9.08%. Synthesis: Weijlard *et al., J. Am. Chem. Soc.* **78,** 2342 (1956); US 2966490 (1960 to Merck & Co.).

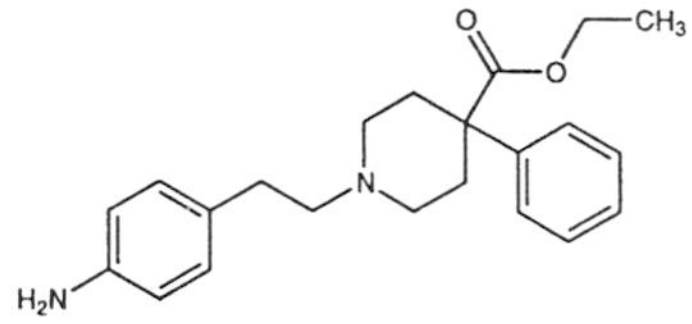

Mp 83°.

Dihydrochloride. [126-12-5] $C_{22}H_{28}N_2O_2.2HCl$. Crystals from methanol + ether, mp 280-287° (dec). Freely soluble in water, methanol. Solubility in ethanol: 8 mg/g. pH of aq solns 2.0 to 2.5. Solns are stable at pH 3.5 and below. At pH 4 and higher the insol free base is precipitated. uv max (pH 7 in 90% methanol contg phosphate buffer): 235, 289 nm (A$_{1cm}^{1\%}$ 293, 34.5). Distribution coefficient (water, pH 3.6/*n*-butanol): 0.9.

Note: This is a controlled substance (opiate): **21 CFR,** 1308.12.

THERAP CAT: Analgesic (narcotic).

661. Aniline. [62-53-3] Benzenamine; aniline oil; phenylamine; aminobenzene; aminophen; kyanol. C_6H_7N; mol wt 93.13. C 77.38%, H 7.58%, N 15.04%. First obtained in 1826 by Unverdorben from dry distillation of indigo. Runge found it in coal tar in 1834. Fritzsche, in 1841, prepared it from indigo and potash and gave it the name aniline. Manuf from nitrobenzene or chlorobenzene: *Faith, Keyes & Clark's Industrial Chemicals,* F. A. Lowenheim, M. K. Moran, Eds. (Wiley-Interscience, New York, 4th ed., 1975) pp 109-116. Procedures: A. I. Vogel, *Practical Organic Chemistry* (Longmans, London, 3rd ed., 1959) p 564; Gattermann-Wieland, *Praxis des organischen Chemikers* (de Gruyter, Berlin, 40th ed., 1961) p 148.

Brochure *"Aniline"* by Allied Chemical's National Aniline Division (New York, 1964) 109 pp, gives reactions and uses of aniline (877 references). Toxicity study: K. H. Jacobson, *Toxicol. Appl. Pharmacol.* **22,** 153 (1972).

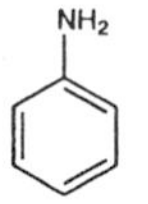

Oily liquid; colorless when freshly distilled, darkens on exposure to air and light. *Poisonous!* Characteristic odor and burning taste; combustible; volatile with steam. d$_{20}^{20}$ 1.022. bp 184-186°. Solidif −6°. Flash pt, closed cup: 169°F (76°C). n$_D^{20}$ 1.5863. pKb 9.30. pH of 0.2 molar aq soln 8.1. One gram dissolves in 28.6 ml water, 15.7 ml boil. water; misc with alcohol, benzene, chloroform, and most other organic solvents. Combines with acids to form salts. It dissolves alkali or alkaline earth metals with evolution of hydrogen and formation of anilides, e.g., C_6H_5NHNa. *Keep well closed and protected from light.* Incompat. Oxidizers, albumin, solns of Fe, Zn, Al, acids, and alkalies. LD$_{50}$ orally in rats: 0.44 g/kg (Jacobson).

Hydrobromide. $C_6H_7N.HBr$. White to slightly reddish, crystalline powder, mp 286°. Darkens in air and light. Sol in water, alc. *Protect from light.*

Hydrochloride. $C_6H_7N.HCl$. Crystals, mp 198°. d 1.222. Darkens in air and light. Sol in about 1 part water; freely sol in alc. *Protect from light.*

Hydrofluoride. $C_6H_7N.HF$. Crystalline powder. Turns gray on standing. Freely sol in water; slightly sol in cold, freely in hot alc.

Nitrate. $C_6H_7N.HNO_3$. Crystals, dec about 190°. d 1.36. Discolors in air and light. Sol in water, alc. *Protect from light.*

Hemisulfate. $C_3H_7N.\frac{1}{2}H_2SO_4$. Crystalline powder. d 1.38. Darkens on exposure to air and light. One gram dissolves in about 15 ml water; slightly sol in alc. Practically insol in ether. *Protect from light.*

Acetate. $C_6H_5NH_2.HOOCCH_3$. Prepd from aniline and acetic acid: Vignon, Evieux, *Bull. Soc. Chim. France* [4] **3,** 1012 (1908). Colorless liquid. d 1.070-1.072. Darkens with age; gradually converted to acetanilide on standing. Misc with water, alc.

Oxalate. $C_6H_5NH_2.HOOCCOOH.H_2NC_6H_5$. Prepd from aniline and oxalic acid in alc soln: Hofmann, *Ann.* **47,** 37 (1843). Triclinic rods from water, mp 174-175°. Readily sol in water; sparingly sol in abs alc. Practically insol in ether.

Caution: Poisoning may occur from inhalation, skin penetration or ingestion. Potential symptoms of overexposure are navy blue to black lips, tongue, mucous membranes; slate gray skin; headache, nausea, vomiting, confusion, ataxia, vertigo, tinnitus, weakness, disorientation, lethargy, drowsiness; dyspnea on effort; methemoglobinemia, cyanosis, coma; tachycardia, heart blocks, arryhthmia, shock; painful micturition, hematuria, hemoglobinuria, methemoglobinuria, oliguria, renal insufficiency; cirrhosis. Direct contact may cause eye irritation. Potential occupational carcinogen. *See NIOSH Pocket Guide to Chemical Hazards* (DHHS/NIOSH 97-140, 1997) p 18; *Clinical Toxicology of Commercial Products,* R. E. Gosselin *et al.,* Eds. (Williams & Wilkins, Baltimore, 5th ed., 1984) Section III, pp 31-36.

USE: Manuf dyes, medicinals, resins, varnishes, perfumes, shoe blacks; vulcanizing rubber; as solvent. Hydrochloride used in manuf of intermediates, aniline black and other dyes, in dyeing fabrics and wood black.

662. Aniline Mustard. [553-27-5] *N,N*-Bis(2-chloroethyl)benzenamine; *N,N*-bis(2-chloroethyl)aniline; phenylbis[2-chloroethylamine]; β,β'-dichlorodiethylaniline. $C_{10}H_{13}Cl_2N$; mol wt 218.13. C 55.06%, H 6.01%, Cl 32.51%, N 6.42%. Prepd by the action of phosphorus pentachloride on *N,N*-bis-[2-hydroxyethyl]aniline (phenyldiethanolamine): Robinson, Watt, *J. Chem. Soc.* **1934,** 1538; Korshak, Strepikheev, *J. Gen. Chem. USSR* **14,** 312 (1944).

Consult the Name Index before using this section. Page 111

FIGURE 4.3

Page 111 of *The Merck Index of Chemicals, Drugs, and Biologicals,* 13th edition, Maryadele J. O'Neil, Ann Smith, Patricia E. Heckelman, John R. Obenchain Jr., Eds. (Merck & Co., Inc.: Whitehouse Station, NJ, USA, 2001). (Reproduced with permission from *The Merck Index,* Thirteenth Edition. Copyright © 2001 by Merck & Co., Inc., Whitehouse Station, NJ, USA. All rights reserved.)

editions of the *CRC Handbook* substituted derivatives of compounds were listed under the heading of the parent compound rather than simply in alphabetical order by the first letter of the compound's name. For example, 1-bromohexane was listed under the parent alkane as "Hexane, 1-bromo-". A brief explanation of the nomenclature system, plus definitions of abbreviations and symbols, precedes the tables of organic compounds in all editions of the *CRC Handbook*.

The Merck Index of Chemicals, Drugs, and Biologicals

The Merck Index, currently in its 13th edition, has over 10,000 organic compound entries that give physical properties and solubilities as well as many references to the compounds' syntheses, safety information, and uses. *The Merck Index* is particularly comprehensive for organic compounds of medical and pharmaceutical importance. Figure 4.3 repeats the page from the 13th edition of *The Merck Index* shown in Technique 1.

4.2   Online Resources

The World Wide Web provides access to many sites that have information about organic compounds; the number of Web sites changes almost daily. At the time of publication, the following sites provided information on physical constants of organic compounds.

> http://ChemFinder.com (Database)
> http://www.liv.ac.uk/chemistry/links/REFdatabases.html
> http://www.sigma-aldrich.com (Online catalog identical to printed version)

A Web search using keywords such as "chemical compound databases" or "chemistry databases" will help you locate other current Web sites with information about organic compounds.

TECHNIQUE

5   MEASURING MASS AND VOLUME

Whether you are carrying out miniscale or microscale experiments, you need to accurately measure the reagents used in the reaction. The techniques for weighing solids and liquids, methods for measuring liquid volumes, and ways of transferring solids and liquids without loss are discussed in Technique 5.

5.1   Weighing

Which Type of Balance to Use?

Your laboratory is probably equipped with several types of electronic top-loading balances for weighing reagents. How do you decide which one to use to measure the mass of a reagent or product? As a general rule, a balance that weighs at least to the nearest centigram (0.01 g) is satisfactory for miniscale reactions using more than 2–3 g of a substance. However, for the small quantities of reagents used in microscale reactions (30–300 mg) and in miniscale reactions where quantities of less than 2 g are used, reagent quantities should be determined on a balance that weighs to the nearest milligram (0.001 g). A milligram balance has a draft shield to prevent air currents from disturbing the weighing pan while a sample is being weighed (Figure 5.1).

Care of the Balance

Electronic top-loading balances are expensive precision instruments that are easily rendered inaccurate by corrosion from spilled reagents. If anything spills on the balance or the weighing pan, clean it up immediately. Notify your instructor if the spill is extensive or the substance is corrosive.

Weighing Solids

No solid reagent should ever be weighed directly on the balance pan. Weigh the solid in a glass container (vial or beaker), in an aluminum or plastic weighing boat, or on glazed weighing paper. Then transfer it to the reaction vessel.

Tare mass. The mass of the container or weighing paper used to hold the sample being weighed is called the ***tare mass*** or just the ***tare.*** When weighing a specific quantity of reagent, the tare mass of the container or weighing paper is simply subtracted by pressing the tare or zero button *before* the sample is added. Then the solid is added until the desired mass appears on the readout screen.

If the mass of the container was not ***tared*** (subtracted) using the zero button before the sample was added, the container mass should be determined and recorded *after* the sample is transferred from it. A vial or flask that will be used to hold a product should be weighed with its cap or cork and labeled **before the product is placed in it.** Be sure to record the tare mass of the container in your notebook.

FIGURE 5.1
Milligram top-loading balance with draft shield.

Draft shield

How to weigh a sample. To weigh a specific quantity of a solid reagent, place the weighing container or paper on the balance pan and press the zero or tare button. Use a spatula to add small portions of the reagent until the desired mass (within 1–2%) is shown on the digital display. For example, the mass of a sample does not need to be exactly the 0.300 g specified, but normally it should be within ±0.005 g of that amount. **Record the actual amount you use** in your notebook. If the reagent you are weighing is the limiting factor, calculate the theoretical yield based on the actual amount used, not on the amount specified in the procedure.

Transferring Solids to the Reaction Vessel

Once the mass of a reagent has been determined, the reagent must be transferred to the reaction vessel without mishap.

Using a powder funnel. For reactions being run in miniscale round-bottomed flasks, a powder funnel aids in transferring the solid reagent from the weighing paper into the small neck of the flask (Figure 5.2a). The stem of a powder funnel has a larger diameter than that of a funnel used for liquid transfers; thus solids will not clog it. The powder funnel serves to keep the solid from spilling and prevents any solid from sticking to the inside of the joint at the top of the flask. Use of a powder funnel is essential with Williamson microscale glassware because of the very small opening at the top of the round-bottomed flasks and reaction tubes (Figure 5.2b).

Transferring solids to a standard taper microscale vial. To transfer a solid to a standard taper microscale vial, roll the weighing paper containing the solid into a cone by overlapping two corners on the same side of the paper. Place the rolled end well down into

FIGURE 5.2
Transferring solids with a powder funnel.

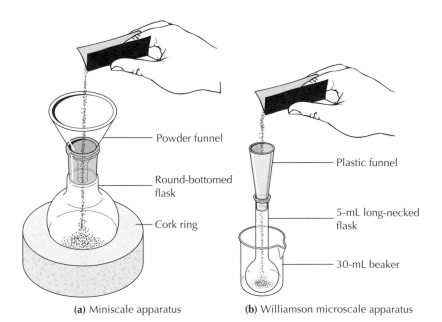

(a) Miniscale apparatus   (b) Williamson microscale apparatus

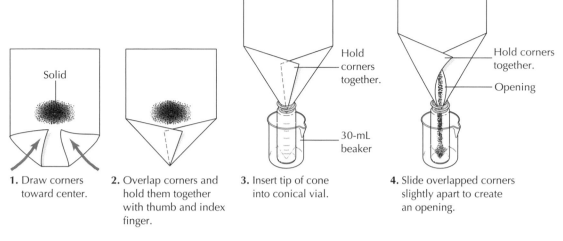

1. Draw corners toward center.

2. Overlap corners and hold them together with thumb and index finger.

3. Insert tip of cone into conical vial.

4. Slide overlapped corners slightly apart to create an opening.

Hold corners together.

30-mL beaker

Hold corners together.

Opening

FIGURE 5.3
Transferring solids with a weighing paper.

the neck of the microscale vial or flask and gently slide one corner of the paper away from the other to create an opening just large enough for the solid to fall into the reaction vessel, as shown in Figure 5.3. A microspatula can be used to gently push the solid through the opening in the weighed paper.

Weighing Liquids

To weigh a liquid, the mass of the container (tare) must be ascertained and recorded, or else subtracted by using the zero button on the balance, *before* the liquid is placed in it. If the liquid is volatile, a cap or cork for the container should be included in the tare so that the sample will not evaporate during the weighing process.

Be very careful that liquid does not spill on the balance while you are weighing a liquid sample. Should a spill occur, clean it up immediately.

To weigh a specific amount of a liquid compound, determine the volume of the required sample from its density and transfer that volume to a tared container. Ascertain the mass of the tared container plus the liquid to determine the mass of the liquid sample. If the mass of liquid needed is less than 1 g, an alternative to measuring the volume is to add the liquid dropwise to a tared container until the desired mass is obtained.

5.2   Measuring Volume

Several volume measuring devices are available in the laboratory, including graduated cylinders, assorted types of pipets, burets, and dispensing pumps, and even beakers and flasks with volume markings on them. The equipment used for measuring a specific volume of liquid depends on the accuracy with which the volume needs to be known. For example, the volume of a reagent that is the limiting factor in a reaction needs to be measured with a graduated pipet or a dispensing pump and then weighed to know the exact amount obtained. If the liquid is a solvent or present in excess of the limiting reagent, volume measurement should be done with a graduated pipet for microscale work and with a graduated pipet or cylinder for miniscale work. The volume markings on beakers and flasks can be used only to estimate an approximate volume and should never be used for measuring a reagent that will go into a reaction.

Graduated
Cylinders

Graduated cylinders do not provide high accuracy in volume measurement and should be used only to measure quantities of reagents other than the limiting reagent. The volume contained in a graduated cylinder is correctly read from the bottom of the meniscus, as shown in Figure 5.4.

Graduated cylinders are not used to measure reagents for microscale reactions. However, a 5- or 10-mL graduated cylinder can be used for measuring volumes of washing solvents greater than 1 mL in a microscale extraction.

Dispensing Pumps

Dispensing pumps fitted to glass bottles come in a variety of sizes designed to deliver a preset volume of liquid (>0.1 mL). Pumps in the 1-, 2-, and 5-mL range may sometimes be used in microscale work for dispensing solvents, but not reagents.

The operation of a dispensing pump consists of slowly pulling the plunger up until it reaches the preset volume stop (Figure 5.5). Hold the receiving container or reaction vessel under the spout and then gently push the plunger down as far as it will go to discharge the preset volume.

Be sure to check that the spout is filled with liquid and contains no air bubbles before you begin to draw up the plunger—air bubbles

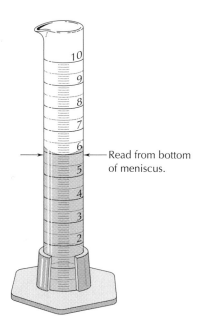

FIGURE 5.4
Graduated cylinder with liquid
showing a meniscus.

Read from bottom
of meniscus.

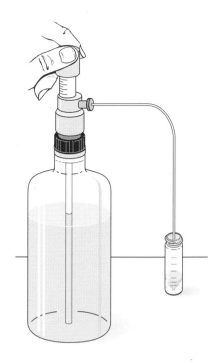

FIGURE 5.5
Dispensing pump.

will cause a volume less than the preset one to be delivered. If air bubbles are present in the spout, pull up the plunger and discharge one or two samples into another container until the spout is completely filled with liquid. (Place the discarded samples in the appropriate waste container.) Dispense a sample into a preweighed container and weigh the container and sample.

Graduated Pipets

The small volumes used in microscale and many miniscale reactions are conveniently and accurately measured with graduated pipets of 1.00-, 2.00-, and 5.00-mL size. A syringe attached to the pipet with a short piece of latex tubing or a pipet pump serves to fill the pipet and expel the requisite volume. The most accurate volumes are obtained by difference measurement, that is, filling the pipet to the zero mark, then discharging the liquid until the required volume has been dispensed. The excess remaining in the pipet should be placed in the appropriate waste container.

Two types of graduated pipets are available: one delivers its total capacity when the last drop is expelled (Figure 5.6a), and the other delivers its total capacity by stopping the delivery when the meniscus reaches the bottom graduation mark (Figure 5.6b). However, both kinds of graduated pipets are more frequently used to deliver a specific volume by stopping the delivery when the meniscus reaches the desired volume.

Automatic Delivery Pipet

Small volumes of 10–1000 μL (0.010–1.000 mL) can be measured very accurately and reproducibly with automatic delivery pipets or pipettors. Automatic pipets have disposable tips that hold the

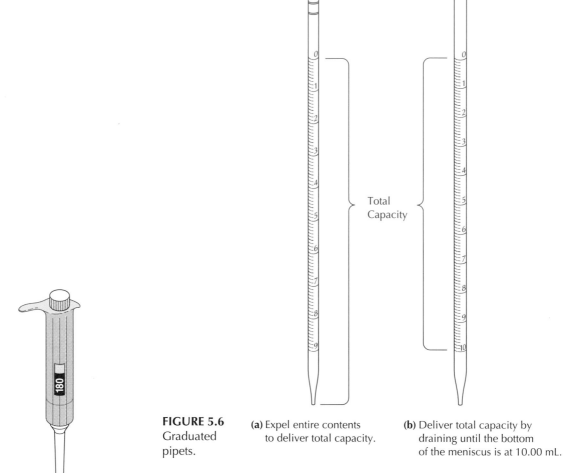

FIGURE 5.6 Graduated pipets.

(a) Expel entire contents to deliver total capacity.

(b) Deliver total capacity by draining until the bottom of the meniscus is at 10.00 mL.

FIGURE 5.7
Automatic delivery pipettor.

Disposable tip

preset volume of liquid; no liquid actually enters the pipet itself, and the pipet should never be used without a disposable tip in place (Figure 5.7). Automatic pipets are very expensive, and your instructor will demonstrate the specific operating technique for the type in your laboratory. Automatic pipets need to be properly calibrated before use; consult your instructor.

Pasteur Pipets and Plastic Transfer Pipets

Pasteur pipets are suitable for measuring only approximate volumes because they do not have volume markings. An approximate calibration is shown in Figure 5.8. Pasteur pipets are particularly useful for the transfer of liquids in microscale reactions or extractions and for transferring small volumes of liquid from one container to another. Attaching a syringe with a piece of latex tubing to a Pasteur pipet allows an approximate volume to be estimated

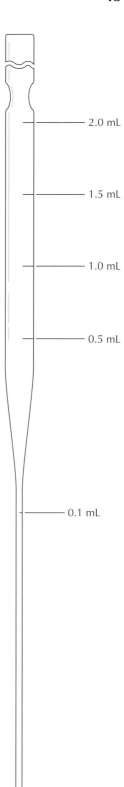

2.0 mL

1.5 mL

1.0 mL

0.5 mL

0.1 mL

FIGURE 5.8
Approximate calibration of a Pasteur pipet.

from the position of the plunger in the syringe as the liquid is drawn into the pipet (Figure 5.9).

Plastic transfer pipets are also useful for the transfer of liquids (Figure 5.10). Graduated plastic pipets, available in 1- and 2-mL sizes, are suitable for measuring the volume of aqueous washing solutions used for microscale extractions or for estimating the volume of solvent added in a recrystallization. Most plastic transfer pipets are made of polyethylene and are chemically impervious to aqueous acidic or basic solutions, alcohols such as methanol or ethanol, and diethyl ether. They are not suitable for halogenated hydrocarbons because the plasticizer leaches out of the polyethylene into the liquid being transferred.

Beakers, Erlenmeyer Flasks, Conical Vials, and Reaction Tubes

The volume markings found on beakers and Erlenmeyer flasks are only approximations and are not suitable for measuring any reagent that will be used in a reaction. The volume markings on conical vials or reaction tubes are also approximations and should be used only to estimate the volume of the contents, such as the final volume of a recrystallization solution, not for measuring the volume of a reagent used in a reaction.

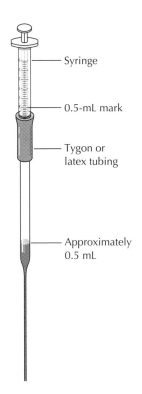

FIGURE 5.9
Using a syringe to estimate the volume of liquid drawn into a Pasteur pipet.

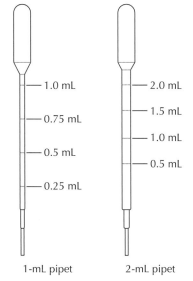

FIGURE 5.10
Graduated plastic transfer pipets.

6

HEATING AND COOLING METHODS

Many organic reactions do not occur spontaneously when the reactants are mixed together but require a period of heating to reach completion. Contrarily, exothermic organic reactions require removal of the heat generated during the reaction period by placing the reaction vessel in a cooling bath. Cooling baths are also used to ensure the maximum recovery of crystallized product from a solution or to cool the contents of a reaction flask.

6.1   Boiling Stones

Before describing various heating methods, let's consider the role of boiling stones (also called boiling chips) or boiling sticks. Liquids heated in laboratory glassware tend to boil by forming large bubbles of superheated vapor, a process called "bumping." The inside surface of the glass is so smooth that no tiny crevices exist where air bubbles can be trapped, unlike the surfaces of metal pans used for cooking. The addition of boiling stones supplies a porous surface. The liquid enters the vapor phase at the air-vapor interface of a pore in the boiling stone. As the volume of vapor nucleating at the pore increases, a bubble forms, is released, and continues to grow as it rises through the liquid. The sharp edges on boiling stones also catalyze bubble formation in complex ways not fully understood.

The boiling stones commonly used in the laboratory consist of small pieces of carborundum, a chemically inert compound of carbon and silicon. Their black color makes them easy to identify and remove from your product if you have not removed them earlier by filtration. Boiling sticks are short pieces of wooden applicator sticks. Boiling sticks should not be used in reaction mixtures or with any solvent that might react with wood or in a solution containing an acid.

One or two boiling stones suffice for smooth boiling of most liquids. **Boiling stones should always be added before heating the liquid.** Adding boiling stones to a hot liquid may cause the liquid to boil violently and erupt from the flask because superheated vapor trapped in the liquid is released all at once. If you forget to add boiling stones before heating, **the liquid must be cooled well below the boiling point before putting boiling stones into it.**

If a liquid you have boiled requires cooling and reheating, additional boiling stones should be added before reheating commences. Once boiling stones cool, their pores fill with liquid. The

You should always add boiling stones or a boiling stick to any unstirred liquid before boiling it—unless instructed otherwise.

liquid does not escape from the pores as readily as air does when the boiling stone is reheated, rendering the boiling stone less effective in promoting smooth boiling.

Mechanical or magnetic stirring agitates a boiling liquid enough to make boiling stones unnecessary, and stirring is frequently used instead of boiling stones.

6.2   Heating Devices

SAFETY PRECAUTION

These safety precautions pertain to all electrical heating devices.

1. The hot surface of a hot plate, the inside of a hot heating mantle, or the hot nozzle of a heat gun are fire hazards in the presence of volatile, flammable solvents. An organic solvent spilled on the hot surface can ignite if its flash point is exceeded. Remove any hot heating device from your work area before pouring a flammable liquid.

2. Never heat a flammable solvent in an open container on a hot plate; a buildup of flammable vapors around the hot plate could result. The thermostat on most laboratory hot plates is not sealed and it arcs each time it cycles on and off, providing an ignition source for flammable vapors. Steam baths, oil baths, or heating mantles are safer choices (Ref. 1).

Flash point or *autoignition temperature* is the minimum temperature at which a substance mixed with air ignites in the absence of a flame or spark.

Heating Mantles

Many reactions and other operations are carried out in round-bottomed flasks heated with electric heating mantles shaped to fit the bottom of the flask. Several types of heating mantles may be available in your laboratory. One type consists of woven fiberglass, with the heating element embedded between the layers of fabric. Fiberglass heating mantles come in a variety of sizes to fit specific sizes of round-bottomed flasks; one sized for a 100-mL flask will not work well with a flask of another size. A different type of heating mantle, called a Thermowell, has a metal housing and a ceramic well covering the heating element. Thermowell heating mantles can be used with flasks smaller than the designated size of the mantle because of radiant heating from the surface of the well.

Both types of heating mantles have no controls and **must** be plugged into a variable transformer (or rheostat) or other variable controller to adjust the rate of heating (Figure 6.1). The variable transformer is then plugged into a wall outlet.

Heating mantles are supported underneath a round-bottomed flask by an iron ring or lab jack [see Technique 6.4]. Fiberglass heating mantles should not be used on wooden surfaces because the bottom of the heating mantle can become hot enough to char the wood.

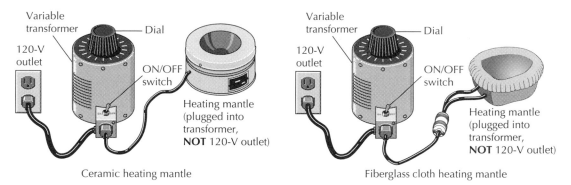

FIGURE 6.1
Heating mantle and variable transformer. (Note that the transformer dial is calibrated in percentage of line voltage, NOT in degrees.)

Hot Plates

Hot plates work well for heating flat-bottomed containers such as beakers, Erlenmeyer flasks, and crystallizing dishes used as water baths or sand baths.

Heating microscale glassware with aluminum blocks. Hot plates also serve to heat the aluminum blocks used with microscale glassware. Figure 6.2a shows a microscale setup for heating a standard taper conical vial fitted with an air condenser; Figure 6.2b shows a microscale setup for heating a Williamson reaction tube and

FIGURE. 6.2
Heating aluminum blocks used with microscale glassware.

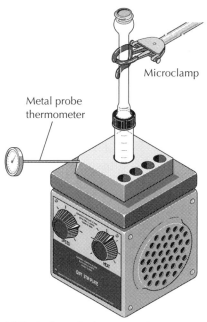
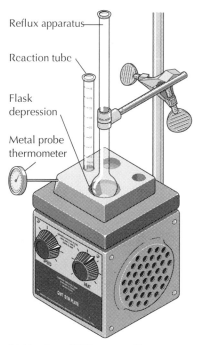

(a) Typical standard taper reaction apparatus with a conical vial and an air condenser

(b) Heating a Williamson reflux apparatus and a reaction tube

a round-bottomed flask fitted with an air condenser. Several types of aluminum heating blocks are available commercially. The blocks have holes sized to fit microscale reaction tubes or vials and a depression or hole for a 5- or 10-mL microscale round-bottomed flask. The blocks also have a hole designed to hold a metal probe thermometer so that the temperature of the block can be monitored.

SAFETY PRECAUTION

A mercury thermometer should not be used with an aluminum heating block because if it breaks, the heat will vaporize the mercury.

Auxiliary aluminum blocks designed in two sections can be placed on top of the aluminum block around a vial or round-bottomed flask to provide extra radiant heat, as shown in Figure 6.3.

Sand Baths

A sand bath provides another method for heating microscale reactions. Sand is a poor conductor of heat, so a temperature gradient

FIGURE 6.3
Using auxiliary aluminum blocks to provide extra radiant heat with microscale glassware.

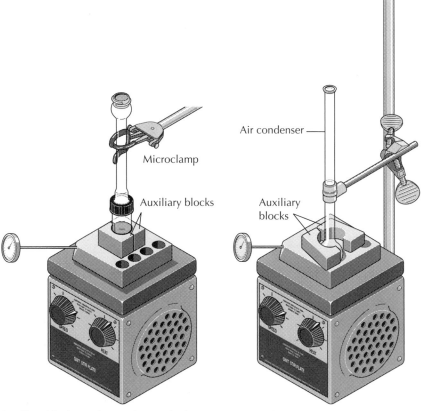

Auxiliary blocks used around a standard taper conical vial fitted with an air condenser

Auxiliary blocks used around a Williamson reflux apparatus for extra radiant heat

exists along the various depths of the sand, with the highest temperature occurring at the bottom of the sand and the lowest temperature near the top surface.

One method of preparing a sand bath uses a ceramic heating mantle, such as a Thermowell, about two-thirds full of washed sand (Figure 6.4a). A second method employs a crystallizing dish, heated on a hot plate, containing 1–1.5 cm of washed sand (Figure 6.4b); the sand in the dish should be level, not a mound. A thermometer is inserted in the sand so that the bulb is completely submerged at the same depth as the contents of the reaction vessel. The heating of a reaction vessel can be closely controlled by raising or lowering the vessel to a different depth in the sand, as well as by changing the heat supplied by the heating mantle or hot plate.

SAFETY PRECAUTION

Sand in a crystallizing dish should not be heated above 200°C, nor should the hot plate be turned to high heat settings. Either situation could cause the crystallizing dish to break.

Steam Baths

Steam baths or steam cones provide a safe and efficient way of heating low-boiling-point flammable organic liquids (Figure 6.5). Steam baths are extensively used in the organic laboratory for heating liquids below 100°C and in situations where precise temperature control is not required. The concentric rings on the top of the steam bath can be removed to accommodate containers of various sizes. A round-bottomed flask should be positioned so that the rings cover the flask to the level of the liquid it contains.

FIGURE 6.4
Sand baths.

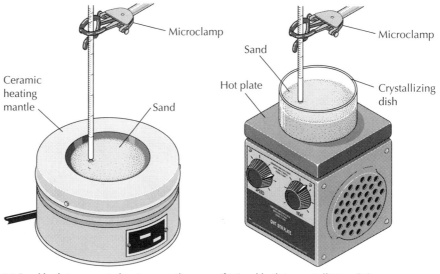

(a) Sand bath in ceramic heating mantle

(b) Sand bath in crystallizing dish on hot plate

Steam in

Steam out

Steam bath

Steam in

Drain

Steam cone

FIGURE 6.5
Steam baths.

For an Erlenmeyer flask, remove only enough rings to create an opening that is slightly larger than one-half of the bottom diameter of the flask.

SAFETY PRECAUTION

Steam is nearly invisible and causes severe burns. Turn off the steam before placing a flask on a steam bath or removing it. Grasp the neck of a hot flask with flask tongs. Do not use a test tube holder or a towel.

Steam baths operate at only one temperature, approximately 100°C. Increasing the rate of steam flow does not raise the temperature, but it does produce clouds of moisture within the laboratory or hood and in your sample. Adjust the steam valve for a **slow to moderate rate of steam flow** when using a steam bath. (**Caution:** The metal screw on the steam valve may be hot enough to cause burns.)

A steam bath has two disadvantages. First, it cannot be used to boil any liquid with a boiling point above 100°C. Second, water vapor from the steam may contaminate the sample being heated on the steam bath unless special precautions are taken to exclude moisture.

Water Baths

When a temperature of less than 100°C is needed, a water bath allows for closer temperature control than can be achieved with the heating methods discussed previously. The water bath can be contained in a beaker or crystallizing dish. Once the desired temperature of the water bath is reached, the water temperature can be maintained by using a low heat setting on a hot plate. Magnetic stirring of the water bath prevents temperature gradients and maintains a uniform water temperature.

The thermometer used to monitor the temperature of a water bath should **always** be clamped, as shown in Figure 6.6, so that it is not touching the wall or bottom of the vessel holding the water. It is very easy to bump a thermometer that is merely set in the beaker and propped against the lip, perhaps breaking it or upsetting the water bath. The reaction vessel should be submerged in the water bath no farther than the depth of the reaction mixture it contains.

If magnetic stirring of a reaction mixture is needed, the reaction vessel should be clamped as close to the stirring motor as possible and centered on the hot plate/stirrer surface. A crystallizing dish may be a better choice than a beaker for the water bath, particularly if the reaction vessel is a round-bottomed flask. The wide, shallow crystallizing dish allows a round-bottomed flask to be clamped closer to the magnetic stirrer than does a beaker.

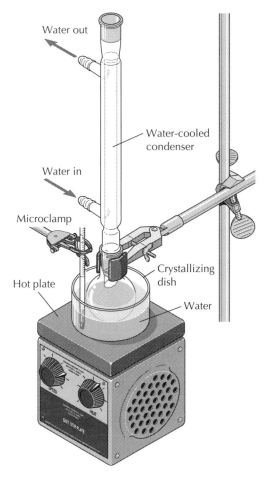
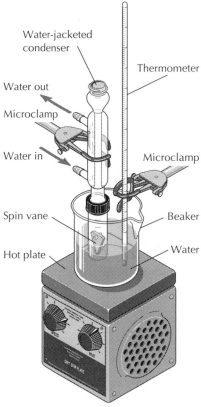

Heating miniscale reflux apparatus in a crystallizing dish

Heating microscale reflux apparatus in a water bath

FIGURE 6.6
Water baths.

Heat Guns

A heat gun allows hot air to be directed over a fairly narrow area (Figure 6.7a). A heat gun is particularly useful as a heat source for distillations because of the high temperature near the nozzle. Heat guns usually have two heat settings as well as a cool air setting. After use, the gun should be suspended in a ring clamp with the heat setting on cool for a few minutes to allow the nozzle to cool before the gun is set on the bench (Figure 6.7b).

Other uses of heat guns include (1) the rapid removal of moisture from glassware where dry but not strictly anhydrous conditions are needed and (2) the heating of thin-layer chromatographic plates after they have been dipped in a visualizing reagent that requires heat to develop the color.

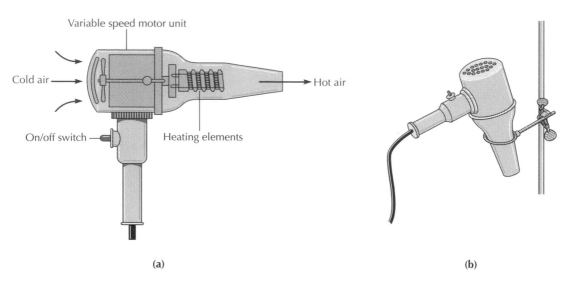

Variable speed motor unit

Cold air

Hot air

On/off switch

Heating elements

(a)

(b)

FIGURE 6.7
(a) Heat gun. (b) Heat gun cooling in a ring clamp after use.

Bunsen Burners

The use of Bunsen burners in the organic laboratory poses an extreme fire hazard because volatile vapors of organic compounds can ignite explosively when mixed with air. Use of a Bunsen burner or other source of open flame should be a very rare event in an organic laboratory and should never be undertaken without your instructor's supervision.

| 6.3 | **Cooling Methods** |

Cooling baths are frequently needed in the organic laboratory to control exothermic reactions, to cool reaction mixtures before the next step in a procedure, and to promote recovery of the maximum amount of crystalline solid from a recrystallization. Most commonly, cold tap water or an ice-water mixture serves as the coolant. Effective cooling with ice requires the addition of just enough water to provide complete contact between the flask or vial being cooled and the ice. Even crushed ice does not pack well enough against a flask for efficient cooling because the air in the spaces between the ice particles is a poor conductor of heat.

Temperatures from 0° to −10°C can be achieved by mixing solid sodium chloride into an ice-water mixture. The amount of water mixed with the ice should be only enough to make good contact with the vessel being cooled.

A cooling bath of 2-propanol and chunks of solid carbon dioxide (dry ice) can be used for temperatures from −10° to −78°C. (**Caution:** Foaming occurs as solid carbon dioxide chunks are added

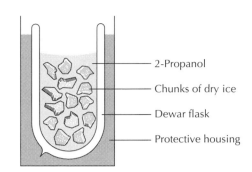

FIGURE 6.8
Dewar flask with a
mixture of 2-propanol
and dry ice.

to 2-propanol.) The 2-propanol/dry ice mixture is contained in a
Dewar flask, a double-walled vacuum chamber that insulates the
contents from ambient temperature (Figure 6.8).

6.4 Laboratory Jacks

Laboratory jacks are adjustable platforms that are useful for holding heating mantles, magnetic stirrers, and cooling baths under
reaction flasks (Figure 6.9). The reaction apparatus is assembled
with enough clearance between the bottom of the reaction or distillation flask and the bench top to position the heating or cooling
device under the flask by raising the platform of the lab jack. At the
end of the operation, the heating or cooling device can be removed
easily by lowering the lab jack.

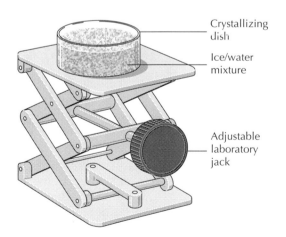

FIGURE 6.9
Laboratory jack with
ice bath.

References

1. James A. Kaufman & Associates. *The Laboratory Safety Newsletter* **1993**, *1*(4), 15.

2. Lodwig, S. N. *J. Chem. Educ.* **1989**, *66*, 77–84.

Questions

1. A student continually increases the heat applied to a flask set in a heating mantle until it appears that the liquid in the flask is very hot, yet no boiling is observed. The student's neighbor mentions that either a boiling stick or boiling stones are needed. What is the *first* thing the student should do to correct the situation? Why?

2. Why must a heating mantle be plugged into a variable transformer instead of directly into a 120-V outlet?

3. Steam causes more severe burns than boiling water, even though both have a temperature of 100°C. Explain.

4. Explain why it takes longer to carry out a reaction on a steam bath in Denver, Colorado, than in New York City.

7

ASSEMBLING A REACTION APPARATUS AND PLANNING A CHEMICAL REACTION

A reaction requires apparatus suitable for carrying it out. This technique describes first how the apparatus for a variety of reaction conditions is assembled and then the preparation that is necessary to plan and carry out a chemical reaction when you are not given explicit experimental directions.

Assembling a Reaction Apparatus

When carrying out organic reactions, it may be necessary to prevent loss of volatile reagents while maintaining the reaction mixture at the boiling point, make additions of reagents, keep atmospheric moisture from entering the reaction apparatus, work under inert atmosphere conditions, or prevent noxious vapors from entering the laboratory. Assembly of the apparatus necessary for each of these reaction conditions is described in Techniques 7.1–7.4.

7.1   Refluxing a Reaction Mixture

Most organic reactions do not occur quickly at room temperature but require a period of heating. If the reaction were heated in an open container, the solvent and other liquids in the system would soon evaporate; if the system were closed, pressure could build up and the system could explode when heated. Chemists have developed a simple method of heating a reaction mixture for extended time periods without loss of reagents. This process is called *refluxing,* which simply means boiling a solution while continually condensing the vapor by cooling it and returning the liquid to the reaction flask.

The rate of heating a reflux apparatus is not critical as long as the liquid boils at a moderate rate. With more heat, faster boiling occurs, but the temperature of the liquid in the flask cannot rise above the boiling point of the solvent or solution. If the system is boiling at too rapid a rate, the capacity of the condenser to cool the vapors may be exceeded and reagents (or product!) lost from the top of the condenser.

A condenser mounted vertically above the reaction flask provides the means of cooling the vapor so that it condenses and flows back into the reaction flask. Condensers are available for either water cooling or air cooling. When the boiling point of the reaction mixture is less than 150°C, a water-jacketed condenser is used. For reaction mixtures with boiling points above 150°C, an air condenser is sufficient because the vapor loses heat rapidly enough to condense before it can escape from the top of the condenser.

Miniscale Reflux Apparatus

A funnel keeps the reagents from coating the inside of the ground glass joint.

Begin the assembly of a reflux apparatus by firmly clamping the round-bottomed flask to a ring stand or vertical support rod. Position the clamp holder far enough above the bench top so that a ring or a lab jack can be placed underneath the flask to hold the heating mantle. Add the reagents with the aid of a conical funnel for liquids and a powder funnel for solids. Add a boiling stone or magnetic stirring bar to the flask.

Grease the lower joint of the condenser and fit it into the top of the flask. Attach rubber tubing to the water-jacket outlets as shown in Figure 7.1a. **Water must flow into the water jacket at the**

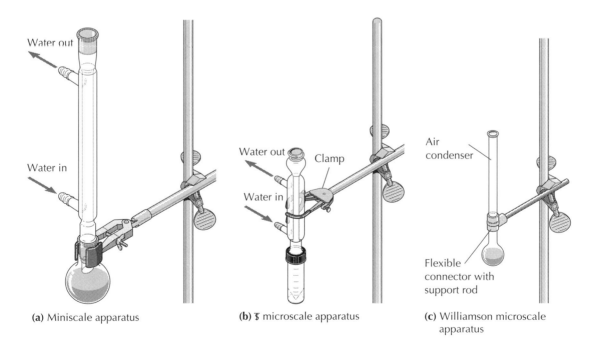

(a) Miniscale apparatus   (b) ꙇ microscale apparatus   (c) Williamson microscale apparatus

FIGURE 7.1 Apparatus for simple reflux.

bottom inlet and out at the top outlet to ensure that a column of water without any air bubbles surrounds the inside tube. Raise the heating mantle, supported on an iron ring or a lab jack, until it is touching the bottom of the round-bottomed flask. At the end of the reflux period, the heating mantle is lowered away from the reaction flask.

Standard Taper Microscale Glassware

Place the reagents for the reaction in a conical vial or 10-mL round-bottomed flask sitting in a small beaker so that it will not tip over. Put a boiling stone or a magnetic spin vane into the reaction vessel. Grease is not used on the joints of microscale glassware except when the reaction mixture contains a strong base such as sodium hydroxide. Fit the condenser to the top of the conical vial or round-bottomed flask with a screw cap and an O-ring as shown in Technique 2, Figure 2.6. Fasten the apparatus to a vertical support rod or a ring stand with a microclamp attached to the condenser. Attach rubber tubing to the water-jacket outlets (Figure 7.1b). **Water must flow into the water jacket at the bottom inlet and out at the top outlet** to ensure that a column of water without any air bubbles surrounds the inside tube. Lower the apparatus into an aluminum heating block, sand bath, or water bath heated on a hot plate or into a sand-filled Thermowell heater. At the end of the reflux period, raise the apparatus out of the heat source.

Williamson Microscale Glassware

Place a 5- or 10-mL round-bottomed flask in a 30-mL beaker and use the plastic funnel to add the reagents to the flask. Add a boiling stone or magnetic stirrer. Attach the air condenser to the flask using the flexible connector with the support rod. Clamp the apparatus to a vertical support rod or a ring stand as shown in Figure 7.1c. Wrap the air condenser with a wet paper towel or wet pipe cleaners to prevent loss of vapor when refluxing reaction mixtures containing solvents or reagents that boil under 120°C. Lower the apparatus into an aluminum heating block, sand bath, or water bath that is heated on a hot plate or into a sand-filled Thermowell heater. At the end of the reflux period, raise the apparatus out of the heat source.

| **7.2** | **Addition of Reagents During a Reaction** |

Miniscale Glassware

When it is necessary to add reagents during the reflux period, a separatory funnel can be used as a dropping funnel. If the round-bottomed flask has only one neck, a Claisen adapter provides a second opening into the flask, as shown in Figure 7.2a. For a three-necked flask, the third neck is closed with a ground glass stopper, as shown in Figure 7.2b. If it is also necessary to maintain anhydrous conditions

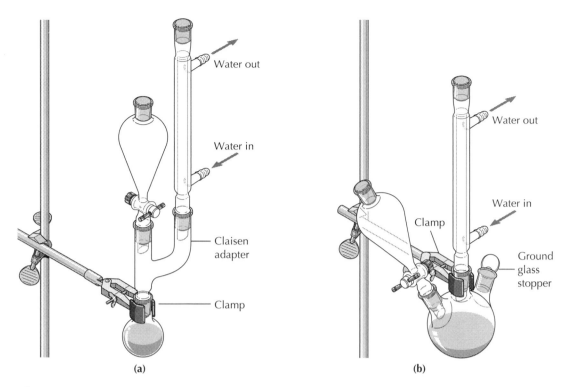

FIGURE 7.2
Assemblies for adding reagents to a reaction heated under reflux in (a) a one-necked reaction flask and (b) a three-necked flask.

[see Technique 7.3] during the reflux period, both the condenser and the separatory funnel can be fitted with drying tubes filled with a suitable drying agent.

Standard Taper Microscale Glassware

The addition of reagents to a microscale reaction is done with a syringe. Figure 7.3a shows a standard taper microscale apparatus assembled for reagent addition using a syringe. The Claisen adapter provides two openings into the system. The opening used for the syringe can be capped either with a screw cap and Teflon septum or with a fold-over rubber septum. The top of the condenser is left open.

Williamson Microscale Glassware

The addition of reagents to a microscale reaction is done with a syringe. For Williamson microscale glassware, the Claisen adapter/distilling head provides two openings in the system. The opening used for the syringe is capped with a fold-over rubber septum (sleeve stopper) and the other opening is left uncovered, as shown in Figure 7.3b.

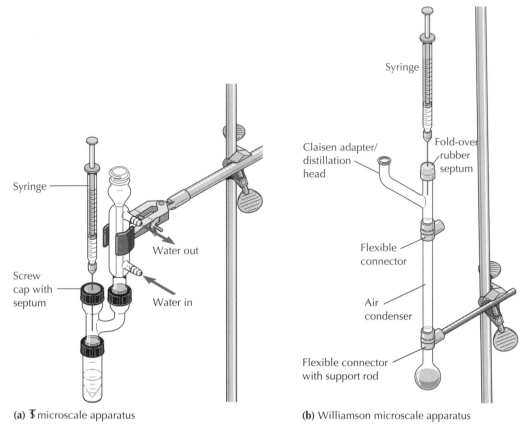

(a) ℥ microscale apparatus   (b) Williamson microscale apparatus

FIGURE 7.3
Using a syringe to add reagents to a microscale reaction.

7.3   Anhydrous and Inert Atmosphere Reaction Conditions

Some reaction procedures need to be carried out in the absence of atmospheric moisture or under inert atmosphere conditions. Anhydrous reaction conditions are discussed in Technique 7.3a and inert atmosphere reaction conditions are discussed in Technique 7.3b.

7.3a Anhydrous Reaction Conditions

Sometimes it is necessary to prevent atmospheric moisture from entering the reaction chamber during the reflux period. In this case, a drying tube filled with a suitable drying agent, often anhydrous calcium chloride, is placed at the top of the condenser.

Miniscale Glassware   For miniscale glassware, a thermometer adapter with a rubber sleeve serves to hold the drying tube (Figure 7.4a). A small piece of

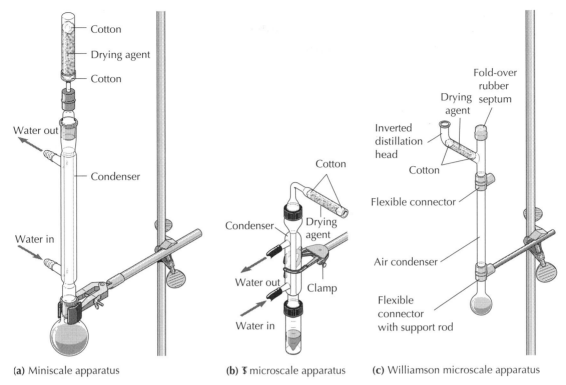

(a) Miniscale apparatus   **(b)** ꟸ microscale apparatus   **(c)** Williamson microscale apparatus

FIGURE 7.4
Refluxing under anhydrous conditions.

cotton is placed at the bottom of the drying tube to prevent drying agent particles from plugging the outlet of the tube; a piece of cotton is also placed over the drying agent at the top of the drying tube to keep the particles from spilling.

Standard Taper
Microscale
Glassware

The L-shaped standard taper microscale drying tube has a ground glass inner joint that fits into the outer ground glass joint at the top of the condenser and is secured with an O-ring and screw cap (Figure 7.4b). A small piece of cotton is pushed into the drying tube to prevent the drying agent particles from falling into the reaction vessel; cotton is also placed near the open end of the drying tube to hold the drying agent in place.

Williamson
Microscale
Glassware

Figure 7.4c shows how the Williamson microscale Claisen adapter/distilling head can be used as a drying tube. A small piece of cotton is pushed to the bottom of the side arm using the tip of a flexible plastic disposable pipet, a suitable drying agent, such as anhydrous calcium chloride, is added, and a second piece of cotton is placed at the top to keep the drying agent from spilling. The other opening is closed with a fold-over rubber septum. The drying tube is fitted to the top of the air condenser with a flexible connector.

7.3b Inert Atmosphere Reaction Conditions

Many useful reagents react quickly and vigorously with oxygen and with moisture. Reactions using these reagents must be conducted in an inert atmosphere with air excluded from the system. Examples of air-sensitive reagents include borane complexes, organoboranes, metal hydrides, and organometallic compounds such as Grignard reagents, organoaluminums, organolithiums, and organozincs. Reactions with these reagents are usually carried out in an atmosphere of nitrogen or argon.

Several special techniques and apparatuses are used for inert atmosphere reactions. Consult your instructor before using any of these specialized techniques:

- Reaction apparatus and its assembly
- Flushing the system with N_2
- Balloon assembly
- Transferring reagents using syringe techniques
- Transferring liquid from a reagent bottle with a syringe
- Transferring liquid from a reagent bottle with a cannula

Reaction Apparatus and Its Assembly

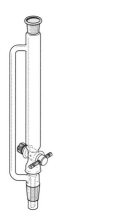

Reactions can be run under inert atmosphere conditions using common standard taper ground glass apparatus. Additional equipment needed includes a bubbler, a source of nitrogen (or argon), rubber septa, and syringes fitted with suitable needles. If the volume of reagent to be added during the reaction is larger than the available syringes will hold, a pressure-equalizing funnel should be included in the reaction apparatus (Figure 7.5). All glassware and syringes used for inert atmosphere reactions need to be oven-dried. Consult your instructor about drying the apparatus for your reaction.

Standard taper joints in an apparatus used for inert atmosphere conditions should have a light coating of grease and Keck clips attached to keep the joints firmly in place. Nitrogen enters the system through a syringe needle placed in a rubber septum fitted over one neck of the reaction apparatus. A bubbler partially filled with mineral oil provides an outlet for nitrogen from the system (Figure 7.6).

FIGURE 7.5
Pressure-equalizing funnel.

FIGURE 7.6
Bubbler. Place in nitrogen exhaust line of inert atmosphere reaction apparatus.

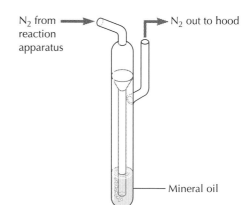

N_2 from reaction apparatus

N_2 out to hood

Mineral oil

Assemble a round-bottomed flask, Claisen adapter, and condenser as shown in Figure 7.7a. Close the tops of the Claisen adapter and the condenser with fold-over rubber septa. Insert a syringe needle into the septum at the top of the condenser and insert the needle attached to the nitrogen source into the septum on the Claisen adapter.

SAFETY PRECAUTION

All reaction assemblies described in the following techniques use syringe needles, which have sharp tips and can cause puncture wounds. Handle the needles with caution.

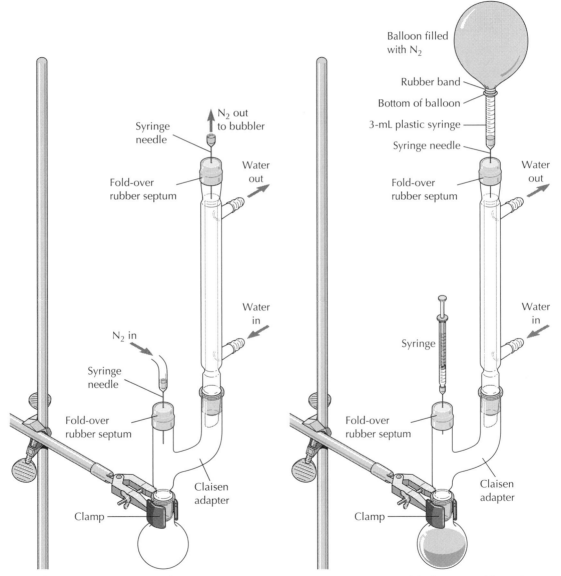

(a) Flushing reaction apparatus with N_2

(b) Reaction apparatus with balloon assembly in place.

FIGURE 7.7
Miniscale reaction apparatus for inert atmosphere conditions using balloon assembly.

Flushing the
Reaction Apparatus
with Nitrogen

The reaction apparatus may be flushed (purged) with nitrogen either before or after the reagents and solvent are placed in the reaction flask, depending on their air and moisture sensitivity. Turn on the nitrogen flow so that a reasonably rapid stream of bubbles passes through the liquid in the bubbler. Flush the apparatus with a gentle flow of nitrogen delivered through the needle in the Claisen adapter; the needle in the top of the condenser serves as the gas exit during purging. When you have finished purging the system, remove both needles from the septa.

Note: If the reaction is heated, increase the nitrogen flow rate as soon as you remove the heat source from the reaction flask. This precaution prevents air from being drawn into the system through the bubbler as the vapors inside the apparatus cool and contract.

Balloon Assembly

For small-scale and microscale reactions, you can often use a balloon assembly to provide an inert atmosphere (Figure 7.7b). Prepare the balloon assembly by removing the plunger and cutting the top off a 3-mL disposable plastic syringe. Fasten a small balloon to the top of the syringe with a small rubber band that is doubled to make a tight seal. Carefully fill the balloon with N_2 through the needle, using plastic tubing to connect to the nitrogen source. When the balloon is inflated, tightly pinch its neck just above the top of the syringe barrel, remove the plastic tubing connected to the gas source from the needle, and immediately push the needle into a solid rubber stopper. The balloon will remain inflated, but it should be used as soon as possible after filling with nitrogen; otherwise, diffusion of oxygen from the atmosphere will contaminate it.

Insert the needle attached to the gas-filled balloon into the septum at the top of the condenser. Add reagent(s) to the reaction with a syringe inserted into the septum on the Claisen adapter.

Figure 7.8a shows a standard taper microscale apparatus and Figure 7.8b shows a Williamson microscale apparatus for inert atmosphere conditions using a balloon assembly.

Transfer of Reagents
Using Syringe
Techniques

Air-sensitive reagents and the dry solvents necessary for their use require special techniques for transferring them from reagent bottles to the reaction apparatus without exposure to atmospheric oxygen and moisture. Small quantities (up to 40 mL) may be transferred from a reagent bottle to the reaction apparatus with a syringe fitted with a long (12–24 in) flexible needle. The all-glass syringe and the needle should be cleaned, dried in an oven, and cooled in a desiccator. Purge the syringe and needle with nitrogen before filling the syringe with reagent (consult your instructor). After purging the syringe and the needle, insert the tip of the needle into a solid rubber stopper unless it will be immediately filled with reagent.

Transferring liquid from a reagent bottle with a syringe. The reagent bottle should be firmly clamped so that it cannot move. Insert a short syringe needle connected to a nitrogen source into the septum, sealing the reagent bottle, and pressurize the bottle to a small degree. Then insert the needle of the transfer syringe so

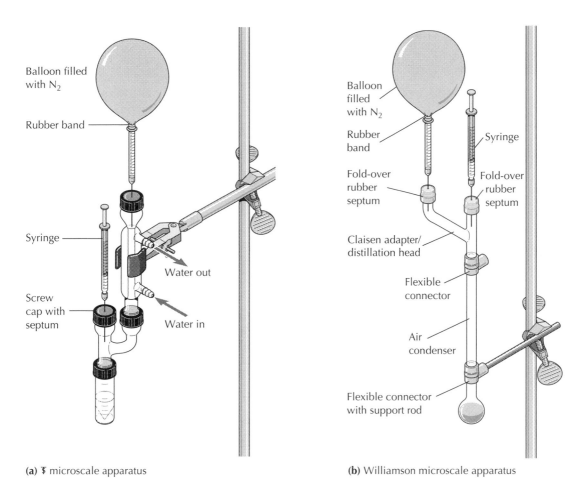

(a) ⌁ microscale apparatus

(b) Williamson microscale apparatus

FIGURE 7.8
Microscale reaction apparatus with balloon assembly for inert atmosphere conditions.

that the open end is below the surface of the liquid in the bottle (Figure 7.9a). Allow the nitrogen pressure in the reagent bottle to assist in filling the syringe until it contains a liquid volume slightly larger than required. Do not pull on the plunger because this may cause leaks or generate gas bubbles. Push the plunger slowly to expel gas bubbles and adjust the volume of reagent to the desired amount (Figure 7.9b). Hold the syringe with one hand. Use the other hand to pull the needle out of the reagent bottle and quickly insert it through the rubber septum on the reaction apparatus.

Transferring liquid from a reagent bottle with a cannula. If larger quantities than will fit in a syringe are needed, the transfer from the reagent bottle to a standard taper graduated cylinder or pressure-equalizing dropping funnel is best accomplished with a *cannula,* a long double-ended needle. The transfer of liquid from a reagent bottle using a cannula is a complex operation for which consultation with your instructor is essential.

The procedure entails two preliminary steps. The reagent bottle should be firmly clamped as shown in Figure 7.9. Second, a gradu-

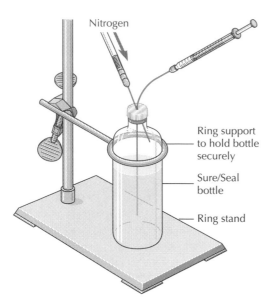

(a) Filling syringe using nitrogen pressure

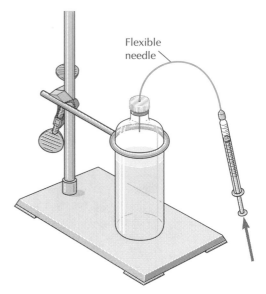

(b) Removing gas bubbles and returning excess reagent to the Sure/Seal bottle

FIGURE 7.9
Filling a syringe with an air-sensitive reagent. (Reprinted with permission from Aldrich Chemical Co., Inc., Milwaukee, WI.)

ated cylinder needs to be purged with N_2. A ground glass adapter fitted with two septa-covered ports is placed in the top of the graduated cylinder (Figure 7.10a) and a Keck clip is positioned over the ground glass joint. The side outlet of the adapter is connected to a bubbler by a short syringe needle. With a syringe needle in the second septum, the graduated cylinder is then flushed with nitrogen.

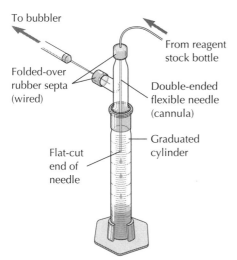

(a) Transfer of liquid from stock bottle to graduated cylinder.

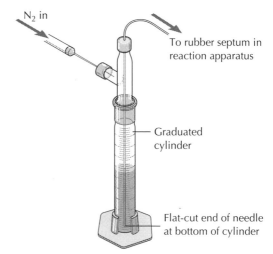

(b) Transfer of liquid from graduated cylinder to reaction apparatus.

FIGURE 7.10
Transfer of a liquid reagent (a) to a graduated cylinder and (b) from a graduated cylinder under inert atmosphere conditions.

After the graduated cylinder has been prepared, insert a short syringe needle connected to a nitrogen source into the septum seal of the reagent bottle and pressurize the bottle. Then flush the cannula with nitrogen and insert one end of it into the reagent bottle so that the needle point is above the level of the liquid. The flow of nitrogen through the cannula will purge it of any remaining air. Insert the other end of the cannula into the septum at the top of the adapter on the graduated cylinder to a depth that is less than the height of liquid to be delivered (Figure 7.10a). Push the end of the cannula that is in the reagent bottle into the liquid to begin the transfer of reagent. When the level of liquid in the graduated cylinder reaches the desired height, immediately pull the cannula out of the reagent bottle and insert it into the reaction apparatus with the tip of the needle above the surface of the reaction mixture.

Transferring liquid to the reaction flask with a cannula.   To transfer the reagent from the graduated cylinder to the reaction apparatus, remove the syringe needle attached to the bubbler from the side arm of the adapter on the graduated cylinder and replace it with a syringe needle attached to a nitrogen source. Push the needle tip of the cannula to the bottom of the graduated cylinder and adjust the nitrogen flow so that the reagent drips slowly into the reaction flask (Figure 7.10b).

Additional Information

References 1–3 provide additional details and information on a wide variety of reaction setups and methods for carrying out reactions under inert atmosphere conditions.

7.4   Removal of Noxious Vapors

Any reaction that emits noxious vapors should be performed in a hood. When a noxious acidic gas such as nitrogen dioxide, sulfur dioxide, or hydrogen chloride forms during a reaction, it must be prevented from escaping into the laboratory. Acidic gases are readily soluble in water, so a gas trap containing either water or dilute aqueous sodium hydroxide solution effectively dissolves them.

Miniscale Apparatus

Attach a U-shaped piece of glass tubing to the top of a reflux condenser by means of a one-hole rubber stopper or a thermometer adapter. Carefully fit the other end of the U tube through a one-hole rubber stopper sized for a 125-mL filter flask. Place about 50 mL of ice water or dilute sodium hydroxide solution in the filter flask and position the open end of the U tube *just above* the surface of the liquid, as shown in Figure 7.11a.

In laboratories equipped with water aspirators, a gas trap can be made by placing a vacuum adapter at the top of a condenser. The side arm of the vacuum adapter is connected by heavy-walled rubber tubing to the side arm of the water aspirator and the water turned on at a moderate flow rate. The noxious gases are pulled from the reaction apparatus and dissolved in the water passing through the aspirator (Figure 7.11b).

Any reaction that emits noxious vapors should be performed in a hood.

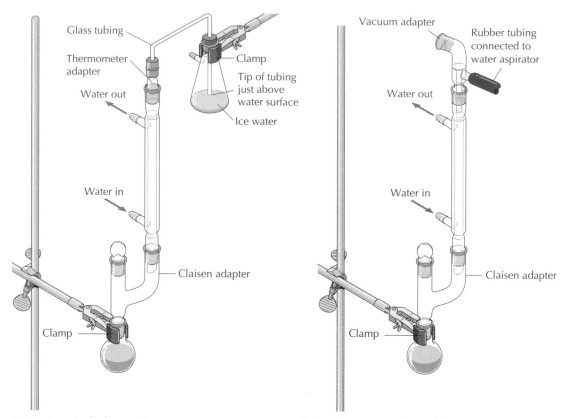

(a) Gas trap attached to reaction apparatus   **(b)** Noxious vapors exhausted through a water aspirator

FIGURE 7.11
Miniscale apparatus to trap water-soluble noxious vapors.

Standard Taper Microscale Apparatus

Any reaction that emits noxious vapors should be performed in a hood.

Pull toothpick and tubing through septum

Round toothpick

Rubber fold-over septum

Teflon tubing

FIGURE 7.12
Threading Teflon tubing through a rubber septum.

A gas trap for microscale reactions can be prepared with fold-over rubber septa, Teflon tubing (1/16 inch in diameter), and a 25-mL filter flask. To insert the Teflon tubing through a rubber septum, carefully punch a hole in the septum with a syringe needle and push a round toothpick through the hole. Fit the tubing over the point of the toothpick and pull the toothpick (with tubing attached) back through the septum, as shown in Figure 7.12. Repeat this process to place a rubber septum on the other end of the tubing.

Half fill a 25-mL filter flask with ice water or dilute sodium hydroxide solution and close the top with one septum. Push the tubing down until the open end is *just above* the surface of the water or sodium hydroxide solution. Attach the other septum to the top of the condenser. The side arm of the filter flask serves as a vent (Figure 7.13a).

In laboratories equipped with water aspirators, a gas trap for standard taper microscale glassware can be made by placing a vacuum adapter at the top of a condenser. The side arm of the vacuum adapter is connected to the side arm of the water aspirator with heavy-walled

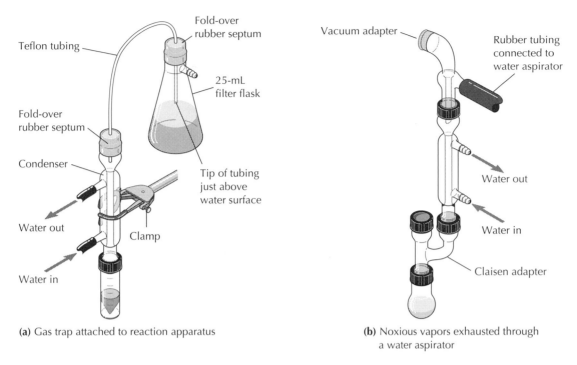

Teflon tubing

Fold-over
rubber septum

25-mL
filter flask

Fold-over
rubber septum

Condenser

Water out

Clamp

Water in

Tip of tubing
just above
water surface

Vacuum adapter

Rubber tubing
connected to
water aspirator

Water out

Water in

Claisen adapter

(a) Gas trap attached to reaction apparatus

(b) Noxious vapors exhausted through
a water aspirator

FIGURE 7.13
Standard taper microscale apparatus to trap water-soluble noxious vapors.

rubber tubing and the water turned on at a moderate flow rate. The noxious gases are pulled from the reaction apparatus and dissolved in the water passing through the aspirator (Figure 7.13b).

*Williamson
Microscale
Glassware*

A gas trap for microscale reactions using Williamson glassware can be prepared with fold-over rubber septa, Teflon tubing (1/16 inch in diameter), and a 25-mL filter flask or a reaction tube. To insert the Teflon tubing through a rubber septum, carefully punch a hole in the septum with a syringe needle and push a round toothpick through the hole. Fit the tubing over the point of the toothpick and pull the toothpick (with tubing attached) back through the septum, as shown in Figure 7.12. Repeat this process to place a rubber septum on the other end of the tubing.

*Any reaction that emits
noxious vapors should
be performed in a hood.*

Half fill a 25-mL filter flask or a Williamson reaction tube with ice water or dilute sodium hydroxide solution and close the top with one septum. Push the tubing down until the open end is *just above* the surface of the water or sodium hydroxide solution. Attach the other septum to the top of the Claisen adapter. If a filter flask serves as the trap, the side arm provides a vent (Figure 7.14a); if the trap is a Williamson reaction tube, then a syringe needle must be inserted into the septum attached to the reaction tube to provide a vent (Figure 7.14b).

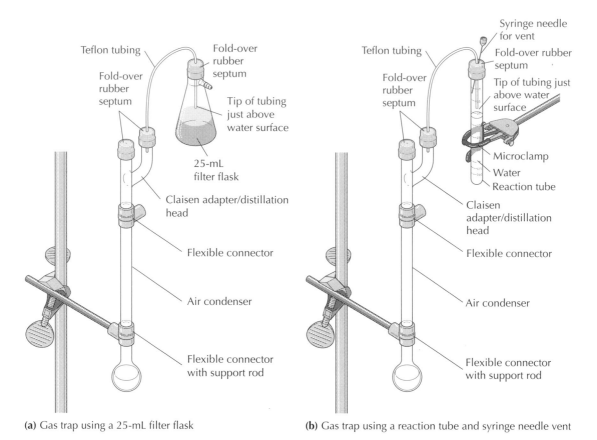

(a) Gas trap using a 25-mL filter flask   (b) Gas trap using a reaction tube and syringe needle vent

FIGURE 7.14
Williamson microscale apparatus to trap water-soluble noxious vapors.

Planning a Chemical Reaction

As you gain more experience in organic chemistry, you may have the opportunity to plan and carry out a chemical reaction where you are not given explicit experimental directions. Open-ended projects, in which you develop your own lab procedures, can be great fun, but they can also be frustrating if you don't plan carefully before beginning the experimental work. Consult with your lab instructor about your planning and your final detailed written procedure before beginning any experimental work.

Often a project focuses on the synthesis of a specific organic compound. Usually, you would begin this kind of project by searching the chemical literature to find a synthesis of the compound. If you cannot find one, you can look for a synthesis of a structurally similar compound to use as a guide. The material presented here is intended to provide you with practical advice for planning a synthesis procedure from literature precedents.

7.5   Library Work

There is a maxim in experimental chemistry: "An hour in the library is worth at least a day in the laboratory." Before attempting any laboratory work, search the chemical literature for examples of the reaction you wish to carry out. You may find several different methods for preparing the desired compound or one similar to it. Compare the various methods critically and carefully in terms of their scale, the availability of starting materials, availability and complexity of equipment, ease of workup, and safety issues.

A good place to start is *Organic Syntheses,* a compilation of carefully checked procedures with full experimental details. The detailed footnotes at the end of each procedure are especially useful. Another good resource is the multivolume series *Reagents for Organic Synthesis* by Fieser and Fieser (Ref. 4). This series provides information on improvements in the preparation and purification of organic compounds. Many newer reagents are safer and easier to handle than older, more traditional reagents.

Other suggestions for information resources can be found in Appendix B, The Literature of Organic Chemistry. There you will find a list of reference books that contain descriptions of techniques and methods, procedures and reactions used in organic synthesis. In an early phase of your library searching, it will be worthwhile to look at *Comprehensive Organic Transformations: A Guide to Functional Group Preparations* by Larock (Ref. 5). It lists ways to carry out specific classes of reactions in the synthesis of specific functional groups; it also gives references to the primary journal literature. Last but by no means least is the invaluable database *Scifinder Scholar,* an excellent search engine. If *Scifinder Scholar* is available on your campus, you will have at your disposal perhaps the most efficient way there is to survey the primary journal literature for the synthesis of particular organic compounds.

7.6   Modifying the Scale of a Reaction and Carrying It Out

Very often, a synthesis procedure found in the literature does not prepare the amount of compound that one wishes to make. The reaction may need to be scaled up or scaled down. Methods from literature published prior to the 1960s and those found in *Organic Syntheses* are usually on a larger scale than most of the reactions carried out in the modern organic chemistry laboratory. These procedures will need to be scaled down. Many of the exemplar synthetic reactions in undergraduate organic chemistry lab manuals are scaled-down procedures from *Organic Syntheses.* Conversely, if a synthesis method is of recent vintage, it may be on the microscale level and need to be scaled up.

At first approximation, the scale-up or scale-down is simply a matter of direct proportionality. If a procedure produces only one-half the material you want, the quantities of all the reagents and solvents should be doubled to produce enough of the product. If the amount you want is only one-tenth the amount produced in the procedure, divide the quantities of all the reagents and solvents by 10. However, when scaling up or down by a large factor, the simple proportionality often needs to be adjusted for some of the reaction components, particularly the solvent volumes.

Also keep in mind that many published synthetic procedures report optimum product yields that are achieved only after a number of iterations. It is likely that your yield on the first attempt will be less than that reported, perhaps only 50–60% as much. If you propose to carry out a synthesis in three steps, lower yields may result in 50% × 50% × 50%, or 13%, of what has been reported. When a reaction looks particularly challenging, it can be useful to try it out on a smaller scale before attempting it on the scale you ultimately need.

Once you have determined the scale of the reaction, you are ready to consider the specific details involved in carrying out the reaction:

- Amount of solvent to use
- Size of reaction apparatus
- How the reagents will be added
- How to determine the reaction time
- Whether the reaction mixture needs stirring
- How you will provide temperature control
- Whether the reaction requires anhydrous or inert atmosphere conditions

Amount of Solvent to Use

In scaling down a very large-scale reaction to miniscale or microscale, reducing the solvent volume by the same factor as the reagents may not provide enough solvent for an effective reflux of the reaction mixture. The volume capacity of the apparatus should probably be much larger proportionately than that of a large-scale model reaction. Otherwise, when the reaction is refluxed, almost all the solvent might vaporize, leaving little to none to dissolve the reaction mixture. In such cases, extra solvent must be used for the scaled-down reaction.

Conversely, when scaling up a microscale reaction by a large factor, the amount of solvent should often be decreased, thus avoiding the use of extremely large volumes of solvent, which can be cumbersome to handle and lead to increased waste disposal costs.

Size of Reaction Apparatus

It is important to use apparatus of a size appropriate for the scale of the reaction. Using large-scale apparatus for a small-scale reaction usually leads to excessive loss of material, which adheres to the surface of the glassware. Large-scale apparatus has a much larger surface area than small-scale equipment, leading to proportionately larger losses of material. With small-scale reactions, flasks with conical bottoms are recommended since they focus the material into a smaller area.

The capacity of the reaction flask should be two to three times the total combined volumes of the reagents and solvent(s). This practice allows for the usual increase in volume as a mixture is heated and allows room for vaporization of the solvent during reflux. If the mixture is known to foam during reflux or if gas is evolved during the reaction, a flask five or more times the volume of the reaction mixture is recommended.

Addition of Reagents

Some reactions give optimal results if one of the reagents is gradually added to the reaction mixture. With large-scale reactions, this slow addition is best accomplished using a dropping funnel for liquid reagents and solutions. For miniscale reactions, the most convenient method is to use a pipet to gradually drip the reagent into the reaction mixture through the reflux condenser attached to the top of the reaction flask. The addition of reagents can be done if the reaction is being heated at reflux or simply being stirred at room temperature. Care must be taken not to lose too much of the reagent on the walls of the condenser. If the reaction system is sealed to isolate it from the atmosphere, a liquid reagent or solution can be added through a rubber septum with a syringe.

Reaction Time

The time required for a scaled-up or scaled-down reaction should be approximately the same as that for the model reaction. That being said, there can be great variation in optimal reaction times due to many variables that cannot be scaled along with the reagents, for example, surface area of the apparatus and heating or cooling efficiency. Miniscale and microscale reactions can take less time than their large-scale counterparts.

The best way to determine when a reaction has reached completion is to monitor it, usually by thin-layer chromatography [see Technique 15] of samples taken from the reaction mixture during the course of the reaction. The reaction is stopped when one of the starting materials is no longer present. Gas chromatography [see Technique 16] can also be used for monitoring reactions. Sometimes other visual clues can be used to decide when to stop a reaction, for example, a color change, disappearance of a solid, or appearance of a solid.

Stirring Reactions

Usually, reaction mixtures are stirred. Magnetic stirring is normally used for miniscale and microscale reactions to ensure that reagents are evenly distributed and available for reaction throughout the entire mixture. Stirring is especially important for mixtures of solids and liquids or immiscible liquids, which are not homogeneous. Even homogeneous mixtures are routinely stirred to avoid concentration gradients and uneven heating.

Temperature
Control

Most organic chemical reactions require heating to drive them to completion in a reasonable amount of time. The exact method of heating—water bath, steam bath, heating mantle, or oil bath—depends on the equipment available in the laboratory. If a thermostat for the heating source is available, it can provide a convenient method for controlling the temperature of the reaction. Alternatively, the temperature of the reaction can be controlled by choice of solvent. In a refluxing reaction mixture, the temperature is close to the boiling point of the solvent; for example, the temperature of a reaction carried out in refluxing ether is close to 35°C and the temperature of a reaction carried out in refluxing hexane is close to 70°C.

Exothermic reactions require external methods for dissipating the generated heat. This heat transfer is often accomplished with a solvent that refluxes into a water-cooled condenser as the reaction heats up. With miniscale and microscale reactions, the large surface area of the apparatus often provides efficient and rapid transfer of heat. Many microscale reactions can be carried out in 10-in test tubes, the sides of the test tube providing the condensing surface for the refluxing solvent. With very exothermic or large-scale exothermic reactions, it is often necessary to use a water or ice-water bath to cool the reaction flask: water is an efficient heat transfer medium.

Another method for controlling the temperature of exothermic reactions is by slow addition of one of the reagents to the stirred reaction mixture. If the reaction becomes too vigorous, addition is stopped or slowed until the reaction rate subsides. Some reactions must be cooled well below 0°C. A 2-propanol/dry ice bath in a low-form Dewar flask works well for reactions that must be carried out in the −30° to −70°C temperature range; Dewar flasks also allow for magnetic stirring [see Technique 6.3].

Using Anhydrous
Reaction and Inert
Atmosphere
Reaction Conditions

The presence of water is deleterious to many organic reactions. It is important to use dry equipment and a drying tube. Even though there may be no visible evidence of water, the surface of glassware can absorb considerable moisture. Glassware should be placed in a 120°C oven to bake off any surface moisture. Because of the relatively large surface area of the glassware relative to the size of the reaction, it is especially important to dry the equipment used for small-scale reactions. Inert atmosphere conditions are discussed in Technique 7.3.

References

1. *Aldrich Technical Bulletin AL-134, Handling Air-Sensitive Reagents,* Aldrich Chemical Co., Inc., Milwaukee, WI.
2. Leonard, J.; Lygo, B.; Procter, G. *Advanced Practical Organic Chemistry;* 2nd ed.; Blackie Academic and Professional: London, 1995.
3. Sharp, J. T.; Gosney, I.; Rowley, A.G. *Practical Organic Chemistry, A Student*

Handbook of Techniques, Chapman and Hall: London, 1989.
4. Fieser, L. F.; Fieser, M. *Reagents for Oganic Synthesis;* 19 vols.; Wiley: New York, 1967–2000.
5. Larock, R. C. *Comprehensive Organic Transformations: A Guide to Functional Group Preparations;* 2nd ed.; Wiley: New York, 1999.

8

EXTRACTION AND DRYING ORGANIC LIQUIDS

Extraction is a technique used for separating a compound from a mixture. For example, a water-insoluble organic compound can be separated from an aqueous mixture by extracting it into a water-insoluble organic solvent. Extractions are frequently part of the workup procedure for isolating and purifying the product of an organic reaction. The theory of extraction and the procedures used for extractions are discussed in the first part of Technique 8.

At the end of an extraction procedure the organic liquid or solution will often contain trace amounts of water, which need to be removed to ensure a dry product. The procedures for drying organic liquids and recovering the desired compound are discussed in the second part of Technique 8.

Extraction

Extractions are frequently used to separate a compound from its reaction mixture and to remove impurities from it. The process of *liquid-liquid extraction* involves the distribution of a compound between two solvents that are *immiscible (insoluble)* in each other. Generally, although not always, one of the solvents in an extraction is water and the other is a much less polar organic solvent, such as diethyl ether, ethyl acetate, hexane, or dichloromethane. By taking advantage of the differing solubilities of a solute in a pair of solvents, compounds can be selectively transported from one liquid phase to the other.

Aqueous Extractions In the extraction procedure, an *aqueous phase (water)* and an immiscible organic solvent, often called the *organic phase,* are gently shaken in a separatory funnel (Figure 8.1). The solutes distribute themselves between the aqueous layer and the organic layer according to their relative solubilities. Inorganic salts generally prefer the aqueous phase, whereas most organic substances dissolve more readily in the organic phase. Two or three extractions of an aqueous mixture often suffice to quantitatively transfer a nonpolar organic compound, such as a hydrocarbon or a halocarbon, to an organic solvent. Separation of low-molecular-weight alcohols or other more polar organic compounds may require additional extractions or a different approach.

FIGURE 8.1
Funnels for extractions.

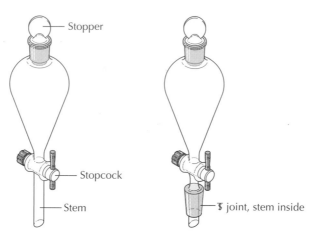

Stopper

Stopcock

Stem

ℑ joint, stem inside

(a) Separatory funnel

(b) Dropping funnel, which can be used as a separatory funnel

Density of Solvent

Before you begin any extraction, look up the density of the organic solvent in Table 8.1 (p. 78) or use a handbook to determine whether the extraction solvent you are using is more dense or less dense than water. **The denser layer is always on the bottom.** Organic solvents that are less dense than water form the upper layer in the separatory funnel (Figure 8.2a), whereas solvents that are denser than water form the lower layer (Figure 8.2b). Occasionally sufficient material is extracted from the aqueous phase to the organic phase or vice versa to change the relative densities of the two phases enough for them to exchange places.

Acid/ Base Extractions

When inorganic acids or bases are present in the organic phase, an extraction with water followed by an extraction with a base or an acid will usually remove the acid or base, respectively. Chemists often use the term *wash* to describe this type of extraction. For

FIGURE 8.2
Solvent densities.

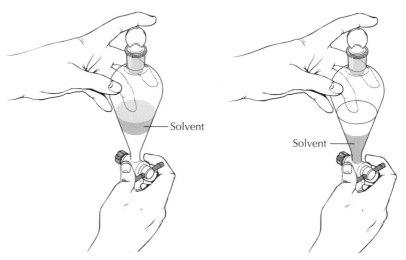

Solvent

Solvent

(a) Organic solvent less dense than water   **(b)** Organic solvent more dense than water

example, if HBr were used in a reaction, the organic phase could be *washed* with water and then with a dilute sodium bicarbonate solution. The reaction between HBr and sodium bicarbonate effectively removes HBr from the organic phase to the aqueous phase by converting it to the ionic salt, sodium bromide.

$$HBr + NaHCO_3 \longrightarrow NaBr + H_2O + CO_2(g)$$

An acid/base extraction can also be used to separate an acidic product from a reaction mixture. For example, in the synthesis of a carboxylic acid (RCO_2H), the product can be purified by first extracting an ether solution of the reaction mixture with a dilute solution of sodium hydroxide. The carboxylic acid is converted to the water-soluble carboxylate anion, which dissolves into the aqueous sodium hydroxide solution, while nonacidic impurities remain in the organic phase.

$$R-C\!\!\overset{O}{\underset{OH}{\bigg|}} + NaOH \longrightarrow R-C\!\!\overset{O}{\underset{O^-Na^+}{\bigg|}} + H_2O$$

Carboxylic acid   Sodium carboxylate

Later, the basic solution of the sodium carboxylate can be acidified, and the purified carboxylic acid can be extracted back into an organic solvent to recover it.

$$R-C\!\!\overset{O}{\underset{O^-Na^+}{\bigg|}} + HCl \longrightarrow R-C\!\!\overset{O}{\underset{OH}{\bigg|}} + NaCl$$

Sodium carboxylate   Carboxylic acid

8.1   Understanding How Extraction Works

Distribution Coefficient

Liquid-liquid extraction involves the distribution of a compound *(solute)*, between two immiscible liquid *solvents (phases)*. When an organic compound is distributed or partitioned between an organic solvent and water, the ratio of solute concentration in the organic solvent, C_1, to its concentration in water, C_2, is equal to the ratio of its solubilities in the two solvents. The distribution of an organic solute, either liquid or solid, can be expressed by

$$K = \frac{C_1\,g}{C_2\,g} = \frac{\text{compound per mL organic solvent}}{\text{compound per mL water}}$$

K is defined as the *distribution coefficient,* or *partition coefficient.*
Any organic compound with a distribution coefficient greater than 1.5 can be separated from water by extraction with a water-insoluble organic solvent. As you will soon see, working through the mathematics of the distribution coefficient shows that a series of

Solvent	Boiling point, °C	Solubility in water, $g \cdot (100 \text{ mL})^{-1}$	Hazard	Density, $g \cdot mL^{-1}$	Fire hazard[a]
Diethyl ether	35	6	Inhalation, fire	0.71	++++
Pentane	36	0.04	Inhalation, fire	0.62	++++
Petroleum ether[b]	40–60	low	Inhalation, fire	0.64	++++
Dichloromethane	40	2	LD_{50},[c] $1.6 \text{ mL} \cdot kg^{-1}$	1.32	+
Hexane	69	0.02	Inhalation, fire	0.66	++++
Ethyl acetate	77	9	Inhalation, fire	0.90	++

TABLE 8.1 Common extraction solvents

a. Scale: extreme fire hazard = ++++.

b. Mixture of hydrocarbons.

c. LD_{50}, lethal dose orally in young rats.

extractions using small volumes of solvents is more efficient than a single large-volume extraction. A volume of solvent about one-third the volume of the aqueous phase is appropriate for most extractions. Commonly used extraction solvents are listed in Table 8.1.

If the distribution coefficient K of a solute between water and an organic solvent is large (>100), a single extraction suffices to extract the compound from water into the organic solvent. Most often, however, the distribution coefficient is less than 10, making multiple extractions necessary.

Using Three Extractions

Consider a simple case of extraction from water into ether, assuming a distribution coefficient of 5.0 for the organic compound being extracted. For illustration, let's use 1.0 g of compound dissolved in 50 mL of water being extracted into 15 mL of ether. C_1 is the final concentration in ether (5 parts), C_2 is the final concentration in water (1 part), and x is the amount of solute remaining in the water layer after extraction:

$$5.0 = K = \frac{C_1}{C_2} = \frac{\dfrac{1.00 \text{ g} - x \text{ g}}{15 \text{ mL ether}}}{\dfrac{x \text{ g}}{50 \text{ mL water}}} = \frac{(1.00 - x)\,50}{15x}$$

$x = 0.40$ g, the amount of the organic compound that remains in the water layer after the first extraction

$1.0 - x \text{ g} = 0.60$ g, the amount transferred to the ether layer

So 0.40 g of solute remains in the water layer after the first extraction, and 0.60 g of solute was extracted into the ether layer.

A second extraction increases the overall efficiency of the extraction procedure. Extracting the remaining water solution with a fresh 15-mL portion of ether removes more of the organic substance that still remains in the water.

Again, C_1 is the final concentration in ether (5 parts), C_2 is the final concentration in water (1 part), and x is the amount of solute remaining in the water layer after the second extraction. As before:

$$5.0 = K = \frac{C_1}{C_2} = \frac{\dfrac{0.40\text{ g} - x\text{ g}}{15\text{ mL ether}}}{\dfrac{x\text{ g}}{50\text{ mL water}}} = \frac{(0.40 - x)\,50}{15x}$$

$x = 0.16$ g, which remains in water layer after the second extraction

$0.40 - x$ g $= 0.24$ g, the amount transferred to ether layer

Thus, 0.16 g of solute remains in the water layer after the second extraction, and 0.24 g has been extracted into the second ether layer.

A third extraction of the residual water layer with a 15-mL portion of ether removes another 0.10 g of the organic compound from the water layer. In three ether extractions, a total of 0.94 g (0.60 g + 0.24 g + 0.10 g) of the organic substance is extracted from the water into the organic solvent. Only 6% of the organic substance remains in the water layer; most of it could be extracted with one more 15-mL portion of ether.

Using Only One Extraction

Suppose that, instead of extracting with three 15-mL portions of ether, we extract the original solution only once with a 45-mL portion of ether. Let x, C_1, and C_2 be defined as before and $K = 5.0$, also as before.

$$5.0 = K = \frac{C_1}{C_2} = \frac{\dfrac{1.00\text{ g} - x\text{ g}}{45\text{ mL ether}}}{\dfrac{x\text{ g}}{50\text{ mL water}}} = \frac{(1.00 - x)\,50}{45x}$$

$x = 0.18$ g, which remains in water layer after one larger extraction

$1.0 - x$ g $= 0.82$ g, the amount transferred to ether layer

After one extraction with the same total volume of ether used for three extractions, 0.18 g of solute remains in the water layer, and 0.82 g of solute has been transferred to the ether layer. Compared with the total of 0.94 g separated by three 15-mL extractions, the single 45-mL extraction is less efficient because 18% of the organic solute remains behind in the water layer versus 6% with the same volume of solvent used in three portions.

In general, the fraction of solute remaining in the original water solvent is given by

$$\frac{C_{2,\text{ final}}}{C_{2,\text{ initial}}} = \left(\frac{V_2}{V_2 + V_1 K}\right)^n$$

where V_1 = volume of ether in each extraction, V_2 = original volume of water, n = number of extractions, and K = distribution coefficient.

| 8.2 | Miniscale Extractions |

Extraction with an Organic Solvent Less Dense Than Water

Place a separatory funnel large enough to hold two to four times the total solution volume in a metal ring firmly clamped to a ring stand or upright support rod (Figure 8.3, step 1). **Make sure that the stopcock fits tightly and is closed.** Pour the cooled aqueous so-

FIGURE 8.3
Using a separatory funnel.

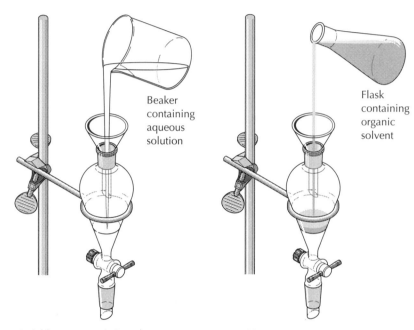

1. Add aqueous solution.   **2.** Add organic solvent.

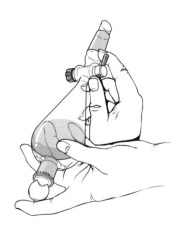

3. Insert stopper and hold stopper with your finger.

4. Invert funnel and immediately open stopcock to release pressure.

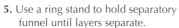

5. Use a ring stand to hold separatory funnel until layers separate.

6. Draw off bottom layer.

7. Pour off top layer.

FIGURE 8.3 *(continued)*

Pour the top layer out of the top of the funnel so that it is not contaminated by the residual bottom layer adhering to the stopcock and tip.

lution to be extracted into the separatory funnel. Add a volume of organic solvent equal to approximately one-third the total volume of the aqueous solution (Figure 8.3, step 2), and put the stopper in place. Remove the funnel from the ring, hold the stopper firmly in place with your index finger (Figure 8.3, step 3), invert the funnel, and **open the stopcock immediately** to release the pressure from solvent vapors (Figure 8.3, step 4). Close the stopcock, and shake the mixture by inverting the separatory funnel four or five times before releasing the pressure by opening the stopcock; repeat this shaking and venting process five or six times to ensure complete mixing of the two phases.

Place the separatory funnel in the ring once more and wait until the layers have completely separated (Figure 8.3, step 5). **Remove the stopper and open the stopcock** to draw off the bottom layer into an Erlenmeyer flask (Figure 8.3, step 6). Pour the remaining organic layer out of the funnel through the top (Figure 8.3, step 7). Do this entire procedure each time an extraction is carried out.

Finally, **label all flasks and do not discard any solution until you have completed the entire extraction procedure and are certain which flask contains the desired product.** If you are in doubt as to which layer is the organic phase and which is the aqueous phase, you can check by adding a few drops of the layer in question to 1–2 mL of water in a test tube and observing whether it dissolves or not.

Extraction with an Organic Solvent Denser Than Water

When extracting an aqueous solution several times with a solvent denser than water, it is not necessary to pour the upper aqueous layer out of the separatory funnel after each extraction. Simply drain the lower organic phase out of the separatory funnel into an Erlenmeyer flask. Then add the next portion of solvent to the aqueous phase remaining in the funnel.

8.3   Additional Information About Extractions

Temperature of the Extraction Mixture

Be sure that the extraction solution is at room temperature or slightly cooler before you add the organic extraction solvent. Most solvents used for extractions have low boiling points and may boil if added to a warm solution. A few pieces of ice can be added to cool the aqueous solution.

Venting the Funnel

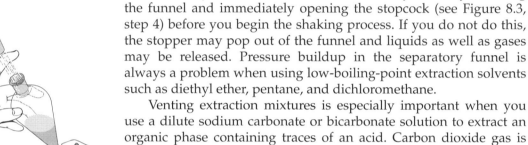

FIGURE 8.4
Failure to vent the separatory funnel when washing with Na_2CO_3 or $NaHCO_3$ solution can cause the stopper to pop out.

Be sure that you vent an extraction mixture by carefully inverting the funnel and immediately opening the stopcock (see Figure 8.3, step 4) before you begin the shaking process. If you do not do this, the stopper may pop out of the funnel and liquids as well as gases may be released. Pressure buildup in the separatory funnel is always a problem when using low-boiling-point extraction solvents such as diethyl ether, pentane, and dichloromethane.

Venting extraction mixtures is especially important when you use a dilute sodium carbonate or bicarbonate solution to extract an organic phase containing traces of an acid. Carbon dioxide gas is given off in the neutralization process. The CO_2 pressure buildup can easily force the stopper out of the funnel, cause losses of solutions (and product!) and possible injury to yourself or your neighbor (as shown in Figure 8.4). When using sodium carbonate or bicarbonate to extract or wash acidic contaminants from an organic solution, **vent the extraction mixture immediately after the first inversion and subsequently after every three or four inversions.**

S A F E T Y   P R E C A U T I O N

Do not point a separatory funnel at yourself or your neighbor.

Washing the Organic Phase

After the extraction is completed and the two immiscible liquids are separated, the organic layer is often *extracted* or *washed* with water or perhaps a dilute aqueous solution of an acid or a base. For example, a chemical reaction involving alkaline (basic) reagents often yields an organic extract that still contains some alkaline material. The alkaline material can be removed by washing the extract (organic phase) with a 5% solution of hydrochloric acid. Similarly, an organic extract obtained from an acidic solution should be washed with a 5% solution of sodium carbonate or sodium bicarbonate (see preceding section on venting). The salts formed in these extractions are very soluble in water but not in typical organic extraction solvents, so they are easily transported into the aqueous phase. If acid or base washes are required, they are done in the same manner as any other extraction and are usually followed by a final water wash.

Salting Out

If the distribution coefficient for a substance to be extracted from water into an organic solvent is close to or lower than 1.0, a simple extraction procedure is not effective. In this case, a salting out procedure can help. Salting out is done by adding a saturated solution of NaCl (sometimes called brine) or Na_2SO_4, or the salt crystals themselves, to the aqueous layer. The presence of a salt in the water layer decreases the solubility of the organic compound in the aqueous phase. Therefore, the distribution coefficient increases because more of the organic compound is transferred from the aqueous phase to the organic layer. Salting out can also help to separate a homogeneous solution of water and a water-soluble organic compound into two phases.

A diethyl ether solution that has been extracted with aqueous solutions is saturated with water, which needs to be removed to recover a dry product. Doing a final extraction of an ether solution with a saturated solution of sodium chloride helps to "dry" (or remove water from) the ether solution. The affinity of the ions in a saturated sodium chloride solution for water molecules draws the water out of the organic phase. The ether solution still needs to be treated with an anhydrous drying agent [see Technique 8.8], but less drying agent is needed because of the sodium chloride extraction.

Emulsions

The formation of an *emulsion*—a suspension of insoluble droplets of one liquid in another liquid—between organic and aqueous layers is sometimes encountered while doing extractions. When an emulsion forms, the entire mixture has a milky appearance, with no clear separation between the immiscible layers, or there may be a third milky layer between the aqueous and the organic phases. Emulsions are not usually formed during diethyl ether extractions, but they frequently occur when aromatic or chlorinated organic solvents are used. An emulsion often disperses if the separatory funnel and its contents are allowed to sit in a ring stand for a few minutes.

Prevention of emulsions. Preventing emulsions is simpler than dealing with them. When using aromatic or chlorinated solvents to extract organic compounds from aqueous solutions, very gentle mixing of the two phases may reduce or eliminate emulsion formation. Instead of shaking the mixture vigorously, invert the separatory funnel and gently swirl the two layers together for 2–3 min. However, use of this swirling technique may mean that you need to extract an aqueous solution with an extra portion of organic solvent for maximum recovery of the product.

What to do if an emulsion forms.   Should an emulsion occur, it can often be dispersed by vacuum filtration through a pad of the filter aid Celite. Prepare the Celite pad by pouring a slurry of Celite and water onto a filter paper in a Buchner funnel (see Figure 9.5). Remove the water from the filter flask before pouring the emulsion through the Buchner funnel. Return the filtrate to the separatory funnel and separate the two phases. Another method, useful when the organic phase is the lower layer, involves filtering the organic phase through a silicone-impregnated filter paper. For microscale extractions [Technique 8.6], centrifugation of the emulsified mixture effectively separates the phases.

Caring for the Separatory Funnel

When the entire extraction is complete, clean up. Cleaning the funnel immediately and regreasing the glass stopcock prevents a "frozen" stopcock later. Grease is not necessary with Teflon stopcocks, but they also may freeze if not loosened during storage.

8.4   Summary of the Miniscale Extraction Procedure

1. With the stopcock closed, pour the aqueous mixture into a separatory funnel with a capacity at least twice as great as the amount of mixture.
2. Add a volume of immiscible organic solvent approximately one-third the volume of the aqueous phase. **You must know the density of the organic solvent.**
3. Invert the funnel with one hand on the stopper and neck of the funnel, and open the stopcock to release any pressure buildup.
4. Close the stopcock, and shake the mixture by inverting the separatory funnel four or five times before releasing the pressure by opening the stopcock; repeat this shaking and venting process five or six times to ensure complete mixing of the two phases (see precautions about emulsions in Technique 8.3). Then vent the funnel.
5. Allow the two phases to separate.
6. *For an organic solvent less dense than water,* draw off the lower aqueous phase into a labeled Erlenmeyer flask; pour the organic phase out of the top of the funnel into a second labeled Erlenmeyer flask; return the aqueous phase to the separatory funnel. *For an organic solvent denser than water,* draw off the lower organic phase into a labeled Erlenmeyer flask; the upper aqueous phase remains in the separatory funnel.
7. Extract the original aqueous mixture twice more with fresh solvent.
8. Combine the organic solvent extracts in the same Erlenmeyer flask and pour this solution into the separatory funnel. Extract this organic solution with dilute acid or base, if necessary, to neutralize any bases or acids remaining from the reaction.
9. Wash the organic phase with water or saturated NaCl.
10. Dry the organic phase with an anhydrous drying agent [see Technique 8.8].

8.5   Using Pasteur Pipets to Transfer Liquids in Microscale Operations

Before we discuss microscale extractions, we must consider methods for transferring the small amounts of liquids involved in microscale operations without significant loss. A liquid should never be poured in any microscale operation; too much will be lost. Losses of even a few drops are significant in microscale work and need to be avoided. A Pasteur pipet serves as the tool of choice for transferring a liquid from one container to another. Volatile organic liquids, however, frequently drip out of a Pasteur pipet during such transfers because solvent vapor pressure increases as the rubber bulb warms from your fingers.

Two methods can be used to prevent volatile liquids from dripping out of a Pasteur pipet. In the first method, a small plug of cotton is pushed into the tip of the Pasteur pipet. In the second method, the Pasteur pipet is fitted with a syringe.

Airborne particles, such as dust and lint, also present problems when working with microscale volumes of liquids. A small cotton plug in the tip of a Pasteur pipet serves to filter the solution each time it is transferred.

SAFETY PRECAUTION

Glass Pasteur pipets are puncture hazards. They should be handled and stored carefully. Dispose of Pasteur pipets in a "sharps" box or in a manner that does not present a hazard to lab personnel or housekeeping staff. Check with your instructor about the proper disposal method in your laboratory.

Pasteur Filter Pipets   Pasteur filter pipets are made by using a piece of wire that has a diameter slightly less than the inside of the capillary portion of the pipet to push a tiny piece of cotton into the tip of a Pasteur pipet (Figure 8.5). A piece of cotton of the appropriate size should offer only slight resistance to being pushed by the wire. If there is so much resistance that the cotton cannot be pushed into the tip of the pipet, then the piece is too large. Remove the wire and insert it through the tip to push the cotton back out of the upper part of the pipet, and tear a bit off the piece of cotton before putting it back into the pipet. The finished cotton plug in the tip of the pipet should be

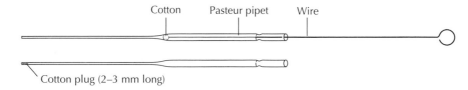

FIGURE 8.5
Preparing a Pasteur filter pipet.

FIGURE 8.6
Syringe fitted to
a Pasteur pipet.

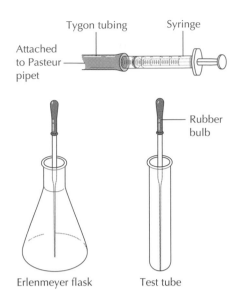

FIGURE 8.7
Correct ways to
temporarily store
Pasteur pipets.

2–3 mm long and should fit snugly but not too tightly. If the cotton is packed too tightly in the tip, liquid will not flow through it; if it fits too loosely, it may be expelled with the liquid. With a little practice, you should be able to prepare a filter pipet easily.

**Pasteur Pipet Fitted
with a Syringe**

When transferring volatile liquids, dripping from the tip of a Pasteur pipet can be minimized by substituting a 3-mL syringe with a Luer-Lok fitting for the rubber bulb.* The syringe is attached to the Pasteur pipet with a short piece of 4-mm-interior-diameter latex or Tygon tubing (Figure 8.6).

**Temporary Storage
of Pasteur Pipets**

When using a Pasteur pipet (or any other pipet with a bulb or syringe attached), set the pipet in a test tube or Erlenmeyer flask to keep it upright. Laying the pipet on the bench top or other horizontal surface allows the rubber bulb or syringe to be contaminated with the liquid being transferred (Figure 8.7).

8.6   Microscale Extractions

*Read the introduction to
Technique 8 and
Techniques 8.1 and 8.3
before undertaking a
microscale extraction for
the first time.*

The small volumes of liquids used in microscale reactions should not be handled in a separatory funnel because most of the material would be lost on the glassware. Instead, use a conical vial or a centrifuge tube to hold the two-phase system and a Pasteur pipet to separate one phase from the other (Figure 8.8). The V-shaped bottom of a conical vial or a centrifuge tube enhances the visibility of the interface between the two phases in the same way that the conical shape of a separatory funnel enhances the visibility of the interface just above the stopcock. Centrifuge tubes are particularly useful for

*Becton-Dickinson 3-mL syringes with a Luer-Lok fitting (B-D product #309585) are available from VWR and Thomas Scientific.

FIGURE 8.8
Equipment for
microscale
extractions.

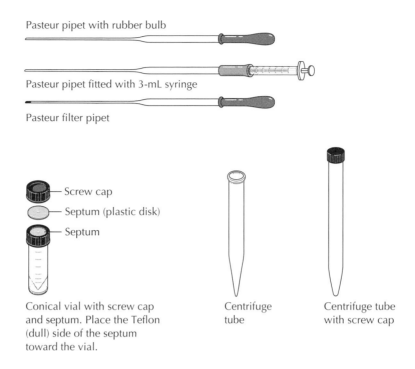

Pasteur pipet with rubber bulb

Pasteur pipet fitted with 3-mL syringe

Pasteur filter pipet

— Screw cap

— Septum (plastic disk)

— Septum

Conical vial with screw cap
and septum. Place the Teflon
(dull) side of the septum
toward the vial.

Centrifuge
tube

Centrifuge tube
with screw cap

extractions with combinations of organic and aqueous phases that form emulsions. The tubes can be spun in a centrifuge to produce a clean separation of the two phases.

8.6a Equipment and Techniques Common to All Microscale Extractions

Before discussing specific types of extractions, we should consider the equipment and techniques common to all microscale extractions.

Conical Vials

Conical vials, with a capacity of 5 mL, work well for extractions in which the total volume of both phases does not exceed 4 mL. Conical vials tip over very easily. **Always place the vial in a small beaker.** The plastic septum used with the screw cap on a conical vial has a chemically inert coating of Teflon on one side. The Teflon looks dull and should be positioned toward the vial. (The shiny side of the septum is not inert to organic solvents.)

Centrifuge Tubes

Centrifuge tubes with a 15-mL capacity and tight-fitting caps serve for extractions involving a total volume of up to 12 mL. Set centrifuge tubes in a test tube rack to keep them upright.

Mixing the Two Phases

As with extractions performed in a separatory funnel, thorough mixing of the two phases is essential for complete transfer of the solute from one phase to the other. Mix the two phases by capping the conical vial or centrifuge tube and shaking it vigorously. Slowly loosen the cap to vent the vial or centrifuge tube. Repeat the shaking and venting process four to six times. Alternatively, or for a centrifuge tube without a screw cap, draw the two phases into a Pasteur pipet (with no cotton plug in the tip) and squirt the mixture back into the centrifuge tube five or six times to mix the two phases thoroughly. The use of a vortex mixer, if available, is another way of mixing the two phases.

Separation of the Phases with a Pasteur Filter Pipet

The liquid in the lower layer is more easily removed from a conical vial or centrifuge tube than the upper layer. A Pasteur filter pipet provides better control for transferring volatile solvents such as dichloromethane or ether during a microscale extraction than does a Pasteur pipet without the cotton plug. **Expel the air from the rubber bulb before inserting the pipet to the bottom of the conical vial or centrifuge tube.** Slowly release the pressure on the bulb and draw the lower layer into the pipet. Maintain a steady pressure on the rubber bulb while transferring the liquid to another container. Hold the receiving container close to the extraction vial or centrifuge tube so that the transfer can be accomplished quickly without any loss of liquid (Figure 8.9).

Separation of the Phases with a Pasteur Pipet and Syringe

A Pasteur pipet fitted with a syringe [see Technique 8.5] can also be used to remove the lower layer. Place the tip of the pipet on the bottom of the V in the conical vial or centrifuge tube. Draw the lower layer into the pipet with a steady pull on the syringe plunger until the interface between the layers reaches the bottom of the vial or tube. **Do not exceed the capacity of the Pasteur pipet (approximately 2 mL); no liquid should be drawn into the syringe.** Remove the Pasteur pipet from the extraction vessel and transfer the contents of the pipet to the receiving container— another conical vial, a centrifuge tube, or a test tube. Hold the receiving container close to the extraction vessel so that the transfer can be accomplished quickly without any loss of liquid (see Figure 8.9). Depress the syringe plunger to empty the pipet.

What to Do if the Upper Phase Is Drawn into the Pasteur Pipet

The interface between the two phases can be difficult to see in some instances, and a small amount of the upper layer may be drawn into the Pasteur pipet. Maintain a steady pressure on the Pasteur pipet with the rubber bulb or syringe and allow the two phases in the pipet to separate. Slowly expel the lower layer into the receiving container until the interface between the phases is at the bottom of the pipet. Then move the pipet to the original container and add the upper layer in the pipet to the remaining upper phase.

Label All Receiving Containers

Extractions always involve the use of several containers. Careful labeling of all containers holding aqueous and organic solutions is essential. **Do not discard any solution until you have completed the entire extraction procedure and know with certainty which vessel contains your product.**

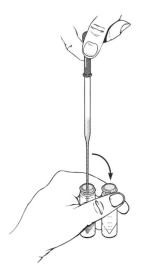

FIGURE 8.9
Holding vials while
transferring solutions.

In any extraction, no
material should be
discarded until you are
certain which container
holds the desired product.

8.6b Microscale Extractions with an Organic Phase Denser Than Water

Extraction of an aqueous solution with a solvent that is denser than water, such as dichloromethane (CH_2Cl_2), and washing a dichloromethane/organic product solution with water are examples of this type of extraction. The dichloromethane solution (lower phase) needs to be removed from the conical vial or centrifuge tube in order to separate the layers.

Place the aqueous solution and the specified amount of organic solvent in a labeled conical vial or centrifuge tube. Tightly cap the vial or tube and shake the mixture thoroughly. Loosen the cap slightly to release the pressure. Alternatively, use the squirt method (five or six squirts) or a vortex mixer to mix the phases. Allow the layers to separate completely.

Put a Pasteur filter pipet or a Pasteur pipet fitted with a syringe [see Technique 8.5] into the conical vial or centrifuge tube with the tip touching the bottom of the cone (Figure 8.10, step 1). Slowly draw the lower layer into the pipet until the interface between the two layers is exactly at the bottom of the V. Transfer the pipet to another centrifuge tube, conical vial, or test tube and expel the dichloromethane solution into the second container (Figure 8.10, step 2). The aqueous layer remains in the extraction tube and can be extracted a second time with another portion of CH_2Cl_2. The second dichloromethane solution is added to the second centrifuge tube after the separation.

FIGURE 8.10
Extracting an aqueous
solution with an
organic solvent
(dichloromethane)
denser than water.

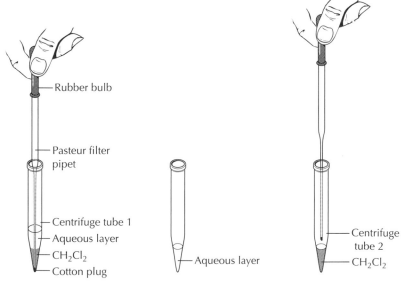

Rubber bulb

Pasteur filter
pipet

Centrifuge tube 1
Aqueous layer
CH_2Cl_2
Cotton plug

Aqueous layer

Centrifuge
tube 2
CH_2Cl_2

1. Expel air from rubber bulb
and insert Pasteur pipet to
bottom of centrifuge tube.
Draw lower layer into pipet.

Aqueous layer remains
in centrifuge tube.

2. Transfer organic layer
to another centrifuge
tube or test tube.

**Washing the
Organic Liquid**

If the organic phase transferred to tube or vial 2 is being washed with an aqueous solution, the aqueous reagent is added to tube 2. Cap the tube or vial, shake it to mix the phases, and then loosen the cap to release any pressure buildup. The lower organic phase is separated and transferred to another centrifuge tube (or conical vial) if more washings are necessary. Otherwise, the organic phase is transferred to a dry test tube for treatment with a drying agent [see Technique 8.8].

8.6c Microscale Extractions with an Organic Phase Less Dense Than Water

The microscale extraction of an aqueous solution with an organic solvent that is less dense than water, such as diethyl ether or hexane, and washing an ether solution with aqueous reagents are examples of this type of extraction.

Two centrifuge tubes or conical vials and a test tube are needed for the extraction of an aqueous solution with a solvent less dense than water. Place the aqueous solution in the first centrifuge tube or conical vial, and add the specified amount of organic solvent— diethyl ether in this example. Cap the tube or vial and shake it to mix the layers. Vent the tube by slowly releasing the cap and allow the phases to separate. Alternatively, use the squirt method (five or six squirts) or a vortex mixer to mix the phases. Allow the layers to separate completely.

In any extraction, no material should be discarded until you are certain which container holds the desired product.

Put a Pasteur filter pipet or a Pasteur pipet fitted with a syringe [see Technique 8.5] into the tube or vial with the tip touching the bottom of the cone (Figure 8.11, step 1). Slowly draw the aqueous layer into the pipet until the interface between the ether and the aqueous solution is at the bottom of the V. Transfer the aqueous solution to the second centrifuge tube or conical vial (Figure 8.11, step 2). The ether solution remains in the first tube. Add a second portion of ether to the aqueous phase in the second tube, cap the tube, and shake it to mix the phases. After the phases separate, again remove the lower aqueous layer and place it in a test tube (Figure 8.11, step 3). Transfer the ether solution in the first tube to the ether solution in the second tube with the Pasteur pipet (Figure 8.11, step 4). Repeat the procedure if a third extraction is necessary.

**Washing the
Organic Liquid**

If the experiment specifies washing an ether or other organic solution that is less dense than water with an aqueous solution, place the organic solution in a centrifuge tube or conical vial. Add the requisite amount of water or aqueous reagent solution, cap the tube (or vial), and shake it to mix the phases. Open the cap to release any built-up vapor pressure and allow the layers to separate. Transfer the lower aqueous layer to a test tube with a Pasteur filter pipet or

FIGURE 8.11
Extracting an aqueous
solution with an
organic solvent (ether)
less dense than water.

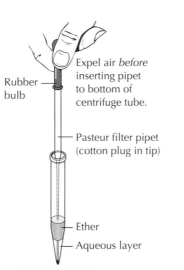

Expel air *before*
inserting pipet
to bottom of
centrifuge tube.

Rubber
bulb

Pasteur filter pipet
(cotton plug in tip)

Ether

Aqueous layer

Centrifuge tube 1

1. Remove lower aqueous phase
with Pasteur pipet.

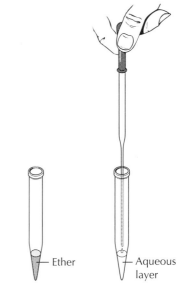

Ether

Aqueous
layer

Centrifuge tube 1   Centrifuge tube 2

2. Transfer aqueous phase to tube 2.

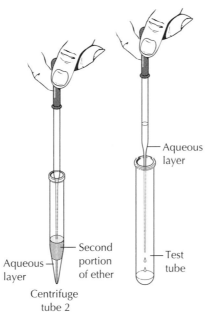

Aqueous
layer

Aqueous
layer

Second
portion
of ether

Test
tube

Centrifuge
tube 2

3. Remove aqueous phase and transfer to
a test tube.

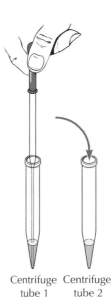

Centrifuge   Centrifuge
tube 1   tube 2

4. Combine ether solution from tube 1
with ether solution in tube 2.

a Pasteur pipet fitted with a syringe [see Technique 8.5]. The upper
organic phase remains in the extraction tube (or conical vial) ready
for the next step, which may be washing with another aqueous
reagent solution or drying with an anhydrous salt (Figure 8.12).

FIGURE 8.12
Washing an organic
phase less dense
than water.

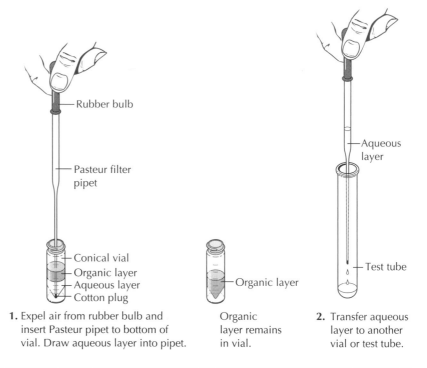

Rubber bulb

Pasteur filter
pipet

Aqueous
layer

Conical vial
Organic layer
Aqueous layer
Cotton plug

Organic layer

Test tube

1. Expel air from rubber bulb and
   insert Pasteur pipet to bottom of
   vial. Draw aqueous layer into pipet.

Organic
layer remains
in vial.

2. Transfer aqueous
   layer to another
   vial or test tube.

8.7   Sources of Confusion in Extractions

Which Layer Is the Organic Phase?

Before beginning any extraction, ascertain the density of the organic solvent (or organic liquid, if a solvent is not being used). If the extraction involves dilute aqueous solutions of inorganic reagents, you can assume their density is close to that of water, $1.0 \ g \cdot mL^{-1}$. If the density of the organic solvent or liquid is less than $1.0 \ g \cdot mL^{-1}$, the organic phase will be the upper layer in the separatory funnel. If the density of the organic solvent or liquid is greater than $1.0 \ g \cdot mL^{-1}$, the organic phase will be the lower layer.

Three Layers Are Present

After mixing the two phases in a separatory funnel, three instead of two layers are visible. The middle layer is an emulsion of the organic and aqueous phases. The section "Emulsions" in Technique 8.3 describes procedures for breaking up emulsions.

No Separation of Phases Is Visible

Several scenarios can lead to no discernible interface between the liquid phases in an extraction.

Solvent added to solvent. This problem occurs in the extraction of an aqueous solution with an organic solvent less dense than water if both the aqueous phase and the upper organic phase are not removed from the separatory funnel (or microscale vial) before the subsequent portion of organic solvent is added. When the second portion of solvent is added to the original solvent remaining in the separatory funnel (or microscale vial), no interface appears because the second solvent is the same as the first one.

The upper layer is too small to be easily visible. Occasionally, the volume of the upper layer in a separatory funnel is too small for the interface to be clearly visible. Draining some of the lower layer will increase the depth of the upper layer as the liquid moves toward the narrower conical portion of the funnel, and the interface will become visible. Another approach to this problem is to add some additional solvent that will become part of the upper layer.

The refractive index of the two solutions is very similar. In rare instances, the refractive index of each solution is so similar that the interface is not visible. Usually adding more water to the aqueous phase will dilute the solution enough to change its refractive index and make the interface visible.

Which Container Holds the Product?   When carrying out a series of extractions, many containers may be used for the various solutions involved. It is **imperative** that all containers be clearly labeled to indicate their contents.

If you are in doubt about the contents of any container, add a few drops of the solution in question to 1–2 mL of water in a small test tube and observe whether it dissolves or not. The organic phase will be insoluble.

A Prudent Practice   **Never discard any solution** during an extraction until you are certain that you know which container holds the product.

Drying Organic Liquids and Recovering the Product

Most organic separations involve extractions from an aqueous solution; no matter how careful you are, some water is usually present in the organic liquid. A small amount of water dissolves in most extraction solvents, and possibly the physical separation of the layers in the extraction process was incomplete. As a result, the organic layer usually needs to be dried before performing additional operations on it.

Following the drying procedure, the organic liquid needs to be separated from the drying agent and the solvent removed to recover the product.

8.8   Drying Agents

The most common way to *dry* (remove the water from) an organic liquid is to add an anhydrous chemical drying agent that binds with water. Chemical drying agents react with water to form crystalline hydrates insoluble in the organic phase:

$$n\text{H}_2\text{O} + \text{drying agent} \longrightarrow \text{drying agent} \cdot n\text{H}_2\text{O}$$

Drying agents for organic liquids are usually anhydrous inorganic salts. Table 8.2 lists common drying agents used for organic liquids.

TABLE 8.2 **Common anhydrous chemical drying agents**

Drying agent	Acid/base properties	Capacity value[a]	Capacity[b]	Efficiency,[c] $mg \cdot L^{-1}$	Speed of drying
$MgSO_4$	neutral	7	high	2.8	fairly rapid
$CaCl_2$	neutral	6	high	1.5	fairly slow
Na_2SO_4[d]	neutral	10	high	25	slow
K_2CO_3	basic	2	low	moderate	fairly rapid
$CaSO_4$	neutral	2	low	0.004	fast
KOH	basic	high	high	0.1	fast
H_2SO_4	acidic	very high	high	0.003	fast

a. Capacity value = (maximum number moles of H_2O)/(moles of drying agent).

b. Capacity = relative amount of water taken up per unit weight of drying agent.

c. Efficiency = measure of equilibrium residual water [mg · (L of air)$^{-1}$] at 25°C.

d. Do not use at temperatures above 32°C.

Selecting a Drying Agent

The factors that need to be considered in selecting a drying agent are its capacity for removing water, its efficiency, the speed with which it works, and its chemical inertness (see Table 8.2). *Capacity* refers to the maximum number of moles of water bound in the hydrated form of the salt and corresponds to the amount of water that can be taken up per unit weight of drying agent. *Efficiency* expresses how much water the drying agent leaves behind in the organic liquid. The higher the efficiency value, the greater amount of water left in the organic liquid, so the poorer job the drying agent does. The *speed* with which the hydrate forms determines how long the drying agent needs to be in contact with the organic solution. Drying agents must be *chemically inert* (unreactive) to both the organic solvent and any organic compound dissolved in the solvent. For example, a base such as K_2CO_3 is not suitable for drying an acidic organic compound because it would undergo a chemical reaction with it. Table 8.3 lists suitable drying agents to use with various classes of organic compounds.

As you can see in Table 8.2, some drying agents have a high capacity but leave quite a bit of water in the organic solution; Na_2SO_4 is a good example. It is particularly useful as a preliminary drying agent, but it is also widely used as a general-purpose drying agent because it is inexpensive and can be used with many types of compounds. $CaSO_4$, on the other hand, leaves little water behind but has a low capacity, which means that it works better after a

---

TABLE 8.3 **Preferred drying agents**

Class of compounds	Recommended drying agents
Alkanes and alkyl halides	$MgSO_4$, $CaCl_2$, $CaSO_4$, H_2SO_4
Hydrocarbons and ethers	$CaCl_2$, $MgSO_4$, $CaSO_4$
Aldehydes, ketones, and esters	Na_2SO_4, $MgSO_4$, $CaSO_4$, K_2CO_3
Alcohols	$MgSO_4$, K_2CO_3, $CaSO_4$
Amines	KOH, K_2CO_3
Acidic compounds	Na_2SO_4, $MgSO_4$, $CaSO_4$

FIGURE 8.13
Adding drying agent
to solution.

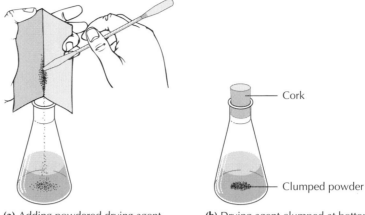

(a) Adding powdered drying agent to solution

(b) Drying agent clumped at bottom of flask

preliminary drying of the liquid with Na_2SO_4 or $MgSO_4$. Some drying agents, such as Na_2SO_4 and $CaCl_2$, do not react quickly with water, so they must be given time to form the hydrate.

Of the drying agents listed in Tables 8.2 and 8.3, magnesium sulfate is used most often because it rarely reacts with solutes or solvents, has a good capacity for water, doesn't leave too much water behind in the solvent, and is reasonably fast at removing water. However, its exothermic reaction with water in the solution being dried sometimes causes the solvent to boil if the drying agent is added too rapidly. Slow addition of the drying agent prevents this problem.

Using a Drying Agent

To remove water from an organic liquid, add a small quantity of the powdered drying agent directly to the solution to be dried (Figure 8.13a). If an organic solution has water in it, the first bit of drying agent you add will clump together (Figure 8.13b). As soon as some of the drying agent moves freely in the mixture while the flask is gently swirled, you have added enough. The solution may be stirred with a magnetic stirring bar or simply swirled occasionally by hand to ensure as much contact with the surface of the drying agent as possible. The best drying efficiency is obtained if the liquid stands over the drying agent for at least 10 min.

Always place the organic liquid being treated with drying agent in an Erlenmeyer flask closed with a cork to prevent evaporation losses.

Often a preliminary drying period of 30–60 s, followed by removal of the drying agent, is useful. Then allowing a second portion of drying agent to stand in the liquid for 10 min removes the water more completely than the use of a single portion.

8.9 Methods for Separating Drying Agents from Organic Liquids

After the drying agent has absorbed the water present in the organic liquid, it must be separated from the liquid. The container receiving the liquid should be clean and dry and have a volume about twice the volume of the organic liquid.

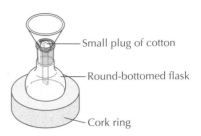

FIGURE 8.15
Filtration of drying agent from
an organic liquid when
no solvent is present.

FIGURE 8.14
Filtration of drying
agent from a solution
when the solvent will
be evaporated. (Replace
the Erlenmeyer flask
with a round-bottomed
flask if the solvent is to
be removed by distilla-
tion or with a rotary
evaporator.)

Miniscale
Separation of
Drying Agents

The product is dissolved in solvent.   If the solvent needs to be evaporated to recover the product, place fluted filter paper in a small
funnel and set the funnel in an Erlenmeyer flask (Figure 8.14). If the
solvent will be distilled from the compound, use a round-bottomed
flask as the receiving container and set it on a cork ring. Decant the
solution slowly into the filter paper, leaving most of the drying
agent in the flask. Rinse the drying agent in the flask with 2–3 mL
of dry solvent and also pour this rinse into the filter paper. The
extraction and drying procedures are now complete, and the
organic liquid is ready for the removal of the solvent.

A liquid product is not dissolved in solvent.   In some extraction procedures, the organic liquid is *neat*, not dissolved in a solvent. In this
situation, you must minimize the loss of liquid product during the
removal of the drying agent. Instead of filter paper, put a tightly
packed cotton plug about 5–6 mm in diameter into the outlet of the
funnel. The plug traps the drying agent and absorbs only a small
amount of the organic liquid (Figure 8.15). Slowly pour the liquid
from the drying agent.

The drying agent is granular or chunky.   If the drying agent is
granular or chunky, the cotton plug can be omitted and the liquid
carefully decanted into the funnel, keeping all drying agent in the
original flask. The drying agent may or may not be rinsed with a
few milliliters of solvent in this procedure. The organic liquid is
ready for the final distillation or evaporation of the solvent.

Microscale
Separation of
Drying Agents

In microscale extractions, the organic liquid is usually dried in a centrifuge tube or a test tube. A Pasteur filter pipet [see Technique 8.5] is
used to remove the liquid from the drying agent and to transfer it to
a clean, dry container (Figures 8.16a and c). In an alternative procedure, useful for a powdered drying agent such as magnesium sulfate, a Pasteur pipet is used as a filtering funnel and a second
Pasteur pipet is used to transfer the liquid. The cotton is packed
tightly at the top of the constriction (Figures 8.16b and d).

| 8.10 | **Recovery of the Organic Product**
from the Dried Extraction Solution |

Once the extraction solution has been dried, the solvent must be
removed to recover the desired organic product. In experiments in
which the amount of solvent is small (less than 25 mL), it can be

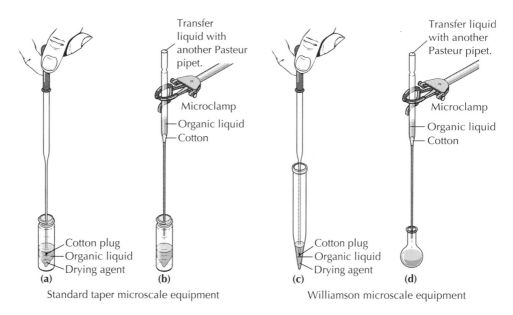

FIGURE 8.16
Using microscale equipment and a Pasteur pipet (a, c) with a cotton plug in the tip to separate an organic liquid from the drying agent and (b, d) as a filtering funnel.

removed by evaporation on a steam bath in a hood or by blowing it off with a stream of nitrogen or air in a hood. Your instructor will advise you whether these methods for removing solvents are followed in your laboratory. Concern for the environment and environmental laws now limit and sometimes prohibit using them. Removing solvents by distillation or with a rotary evaporator are alternatives to evaporation; both methods allow the solvents to be recovered.

Evaporation

Place a boiling stick or boiling stone in the Erlenmeyer flask containing the solution to be evaporated and heat the flask on a steam bath in a hood. The product will be the liquid or solid residue left in the flask when the boiling ceases. The last of the solvent can be blown off in a hood with a stream of nitrogen or air.

Distillation

Assemble the simple distillation apparatus shown in Technique 11, Figure 11.6. If the solvent is ether, pentane, or hexane, use a steam bath as a heat source to eliminate the fire hazard an electric heating mantle poses with the very flammable vapors from these solvents. Continue the distillation until the solvent has completely distilled, an endpoint indicated by a drop in the temperature reading on the thermometer. The drop in temperature occurs because there is no longer any hot vapor surrounding the thermometer bulb. The product and a small amount of solvent will remain in the boiling flask.

Using a Rotary Evaporator

Read Technique 11.6 on vacuum distillation before using a rotary evaporator.

A rotary evaporator is an apparatus for removing solvents rapidly in a vacuum (Figure 8.17). No boiling stones or sticks are necessary because the rotation of the flask minimizes bumping. Rotary evaporation is usually done in a round-bottomed flask that is no more than half filled with the solution being evaporated. A receiving flask (also called a trap) is placed between the round-bottomed flask and the vacuum source so that the evaporated solvent can be recovered.

The following protocol is a generalized outline of the steps in using a rotary evaporator, but you should consult your instructor about the exact operation of the rotary evaporators in your laboratory. Select a round-bottomed flask of a size that will be only half full or less with the solution undergoing evaporation. Connect the flask to the rotary evaporator with a joint clip. Use an empty trap and be sure that it is also clipped tightly to the rotary evaporator housing. Position a room-temperature water bath under the flask containing the solution so that the flask is approximately one-third submerged in the water bath. Turn on the water to the condenser and then turn on the vacuum source. Make sure the stopcock is closed. As the vacuum develops, turn on the motor that rotates the evaporating flask. When the vacuum stabilizes at 20–30 torr or lower, begin to heat the water bath. A temperature of 50°–60°C will quickly evaporate solvents with boiling points under 100°C.

When solvent no longer condenses in the trap (receiving flask), the evaporation is complete. Stop the rotation of the flask and remove the water bath. Open the stopcock slowly to release the vacuum and allow air to bleed slowly into the system. Hold the flask with one hand, take off the clip holding it to the evaporator, and remove the flask from the rotary evaporator. Turn off the vacuum source and the condenser water. Disconnect the trap from the rotary evaporator housing and empty the solvent in the trap into the appropriate waste or recovered solvent container.

FIGURE 8.17
Schematic diagram of a rotary evaporator.

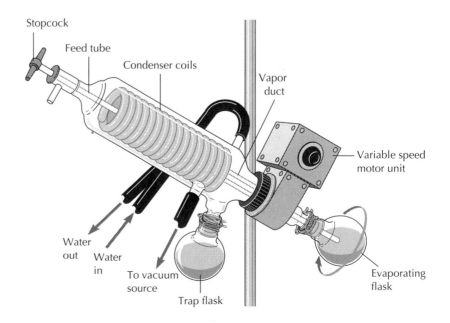

Stopcock

Feed tube

Condenser coils

Vapor duct

Variable speed motor unit

Water out

Water in

To vacuum source

Trap flask

Evaporating flask

8.11   Sources of Confusion in Drying Liquids

Amount of Drying Agent to Use

The amount of drying agent necessary to remove residual water from the organic liquid cannot be specified; it depends on how much water is present in the liquid. You need to learn to judge when enough drying agent has been added. If all the drying agent particles are clumped together, not enough has been used. Continue adding small amounts from the tip of a spatula until there is a thin layer of particles that look very similar to the original particles of the drying agent and that move freely in the flask.

Remember that the use of too much drying agent can cause a loss of product by its adsorption on the drying agent. If you have to add quite a bit of drying agent to reach the clumping point, you must have had a large amount of water present initially. In this case you may wish to add more organic solvent to minimize the loss of product.

Is the Organic Liquid Dry?

Drying agents do not absorb water instantaneously. Allow a minimum of 10 min for the drying agent to become hydrated. When an organic liquid is dry, it will be clear and at least a portion of the drying agent will still have the particle size and appearance of the anhydrous form. If all the drying agent has become clumped or the organic liquid is still cloudy after 10 min, decant the organic liquid into a clean Erlenmeyer flask and add another portion of drying agent. Allow the mixture to stand for another 10 min.

A White Liquid Surrounds the Drying Agent

When the drying agent is added to the organic liquid, a milky white liquid may appear around the drying agent particles, particularly when anhydrous calcium chloride in pellet form is being used. The pellets do not provide as much surface area for reaction with water as powder or granules do. The white liquid is a saturated water solution of calcium chloride. Continue adding pellets until the liquid is absorbed and some of the pellets move freely in the organic liquid. Allow at least 15 min for the drying agent to be effective.

Questions

1. An extraction procedure specifies that an aqueous solution containing dissolved organic material be extracted twice with 10-mL portions of diethyl ether. A student removes the lower layer after the first extraction and adds the second 10-mL portion of ether to the upper layer remaining in the separatory funnel. After shaking the funnel, the student observes only one liquid phase with no interface. Explain.

2. A crude nonacidic product mixture dissolved in diethyl ether contains acetic acid. Describe an extraction procedure that could be used to remove the acetic acid.

3. What precautions should be observed when an aqueous sodium carbonate solution is used to extract an organic solution containing traces of acid?

4. When the two layers form during an ether/water extraction, what would be

an easy, convenient way to tell which layer is which if the densities were not available?

5. You have 100 mL of a solution of benzoic acid in water; the amount of benzoic acid in the solution is estimated to be about 0.30 g. The distribution coefficient of benzoic acid in diethyl ether and water is approximately 10. Calculate the amount of acid that would be left in the water solution after four 20-mL extractions with ether. Do the same calculation using one 80-mL extraction with ether to determine which method is more efficient.

6. K_2CO_3 is an excellent drying agent for some classes of organic compounds. Would it be a better choice for an acid (RCO_2H) or an amine (RNH_2)? Why?

TECHNIQUE

9

RECRYSTALLIZATION

A pure organic compound is one in which there are no detectable impurities. Because experimental work requires an immense number of molecules (Avogadro's number per mole), it is not true that 100% of the molecules in a "pure" compound are identical to one another. Seldom is a pure compound purer than 99.99%. Even if it were that pure, one mole would still contain more than 10^{19} molecules of other compounds. Nevertheless, we want to work with compounds that are as pure as possible; therefore, we must have ways to purify impure materials. Recrystallization is one of the major techniques for purifying solid compounds.

9.1   Theory of Recrystallization

Recrystallization

Recrystallization is a process in which a crystalline material (solute) dissolves in a hot solvent, then returns to a solid again by crystallizing in the cooled solvent. The technique of recrystallization depends on the increased solubility of a compound in a hot solvent. A saturated solution at a higher temperature normally contains more solute than the same solute/solvent pair at a lower temperature. Therefore, the solute precipitates when a warm saturated solution cools.

Impurities in the solid being recrystallized are usually significantly lower in concentration than the concentration of the substance being purified. As the mixture cools, the impurities remain in solution while the highly concentrated product crystallizes.

Crystal Formation

Crystal formation of a solute from a solution is a selective process. When a solid crystallizes at the right speed under the appropriate conditions of concentration and solvent, an almost perfect crystalline material can result because only molecules of the right shape fit into the crystal lattice. In recrystallization, dissolution of the impure solid in a suitable hot solvent destroys the impure crystal lattice, and recrystallization from the cold solvent selectively produces a new, more perfect (purer) crystal lattice. Slow cooling of the saturated solution promotes formation of pure crystals because the molecules of the impurities, which do not fit as well into the newly forming crystal lattice, have time to return to the solution. Crystals that form slowly are larger and purer than ones that form quickly. Indeed, rapid crystal formation traps the impurities because the lattice grows so quickly that the impurities are simply surrounded by the crystallizing solute as the crystal forms.

9.2 Choice of Recrystallization Solvent

The most crucial aspect of a recrystallization procedure is the choice of solvent, because **the solute should have a maximum solubility in the hot solvent and a minimum solubility in the cold solvent.** Table 9.1 lists common recrystallization solvents.

Solvent Properties

In general, a solvent with a structure similar to that of the solute will dissolve more of the solvent than will solvents with dissimilar structures. Although the appropriate choice of a recrystallization solvent is a trial-and-error process, a relationship exists between the solvent's structure and the solubility of the solute. This relationship is simply described as **like dissolves like.** Nonionic compounds generally dissolve in water only when they can associate with the water molecules through hydrogen bonding. Thus, all hydrocarbons and alkyl halides are virtually insoluble in water, whereas carboxylic acids, which readily form hydrogen bonds, are often recrystallized from water solution. Molecules that associate with water through hydrogen bonds include carboxylic acids, alcohols, and amines (Figure 9.1). Carboxylic acids

TABLE 9.1 Common recrystallization solvents

Solvent	Boiling point, °C	Miscibility[a] in water	Solvent polarity	Dielectric constant (ε)	Fire[b] hazard
Diethyl ether	35	−	low	4.3	++++
Acetone	56	+	intermediate	20.7	+++
Petroleum ether[c]	60–80	−	nonpolar	ca. 2	++++
Methanol	65	+	polar	32.6	++
Hexane	69	−	nonpolar	1.9	++++
Ethyl acetate	77	−	low	6.0	++
Ethanol	78.5	+	polar	24.3	++
Water	100		very polar	80	0
Toluene	111	−	nonpolar	2.4	++

a. Infinite solubility = +.

b. Scale: Extreme fire hazard = ++++.

c. Petroleum ether (or ligroin) is a mixture of isomeric alkanes. The term *ether* refers to volatility, not the presence of an ether functional group.

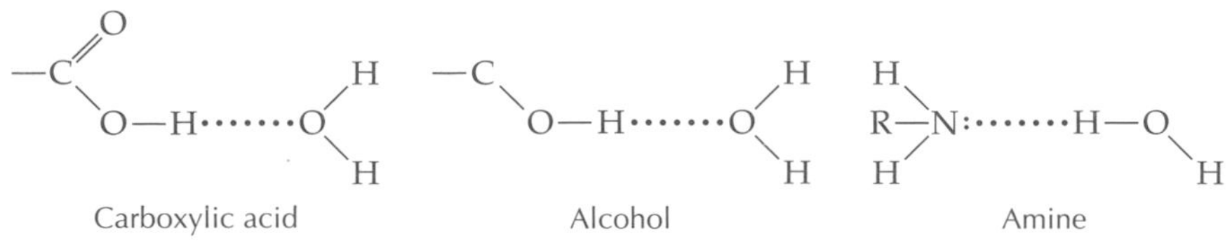

FIGURE 9.1
Compounds that associate with water by hydrogen bonding (dotted lines).

hydrogen bond to a lone pair of electrons of water through the acidic proton; alcohols do likewise. Amines hydrogen bond through the lone pair on nitrogen and a hydrogen atom of water.

Polarity of Solvents

The polarity of the solvent is a crucial factor in solubility phenomena. An important measure of solvent polarity is the *dielectric constant, ε*. Table 9.1 shows the range of dielectric constants of common recrystallization solvents, from highly polar water ($\varepsilon = 80$) to nonpolar hexane ($\varepsilon = 1.9$). As a rough estimate, solvents with higher dielectric constants should more readily dissolve polar compounds than would solvents with lower dielectric constants. The members of homologous series of carboxylic acids, alcohols, and amines are water soluble up to $C_4—C_5$. But as the molecular weight increases, the water solubility of a compound decreases because the hydrocarbon, or nonpolar, portion of the molecule begins to dominate its physical behavior.

The salts of low-molecular-weight carboxylic acids are quite water soluble. For example, potassium acetate, $CH_3CO_2^-K^+$, and sodium propionate, $CH_3CH_2CO_2^-Na^+$, are ionic compounds with solubility characteristics similar to sodium chloride or potassium bromide.

Meth eth diss ←

High-polarity solvents. Among the more polar organic solvents, both methanol and ethanol are commonly used for recrystallization because they dissolve a wide range of both polar and nonpolar compounds to the appropriate degree. Ethanol and methanol also evaporate easily and possess water solubility, which allows recrystallization from an alcohol/water mixture.

Low-polarity solvents. Organic solvents of low polarity also dissolve many nonionic organic compounds with ease. Even polar organic compounds can dissolve in solvents of low polarity if the ratio of polar functional groups per carbon atom is not too high and if hydrogen bonding can occur between the solute and the solvent.

Among the low-polarity solvents, diethyl ether and ethyl acetate appear to provide the best solvent properties, although the extreme flammability of diethyl ethyl and its low boiling point (35°C) require careful attention to safety. Ether in combination with hexane or methanol has excellent solvent properties for recrystallizations.

SAFETY PRECAUTION

Both ether and hexane are very flammable and should be heated with a steam or hot-water bath. They should **never** be heated with a flame or on a hot plate.

9.3 How to Select a Proper Recrystallization Solvent

An effective recrystallization solvent is one in which the solid compound dissolves in the hot solvent but doesn't dissolve in the cold solvent. For recrystallization to work effectively, the solubility of the organic solid should not be too large or too small in the recrystallization solvent. If the solubility is too large, it is difficult to recover the compound; if the solubility is too small, a very large volume of solvent will be needed to dissolve the compound.

Testing a Single Solvent

To select a suitable solvent for recrystallization, take a small sample (20–30 mg) of the compound to be recrystallized, place it in a test tube, and add 5–10 drops of a trial solvent. Shake the tube to mix the materials. If the compound dissolves immediately, it is probably too soluble in the solvent for a recrystallization to be effective. If no solubility is observed, heat the solvent to its boiling point. If complete solubility is observed, cool the solution to induce crystallization. The formation of crystals in 10–20 min suggests that you have a good recrystallization solvent.

Careful measurements and observation are essential when testing potential solvents.

Testing Mixed Solvent Pairs

When no single solvent seems to work, a pair of *miscible solvents,* solvents that are very soluble in one another, can sometimes be used. Mixed solvent pairs usually include one solvent in which a particular solute is very soluble and another in which its solubility is marginal. Typical mixed solvent pairs are listed in Table 9.2.

Record the exact amounts of each solvent used for the tests.

Mixed solvent tests. To select a suitable mixed solvent pair, place 20–30 mg of the solute in a test tube and add 5–10 drops of the solvent in which it is more soluble. Warm the solution nearly to its boiling point. When the solid dissolves completely, add the other solvent drop by drop until a slight cloudiness appears and persists as mixing continues, indicating that the hot solution is saturated with the solute. If no cloudiness appears, the compound is too soluble for an effective recrystallization.

If cloudiness appears, add the first solvent again drop by drop until the cloudiness just disappears and then add a few drops more to ensure an excess. Let the solution cool slowly. The formation of crystals in 10–20 minutes suggests that you have found a good solvent pair.

Scaling up a mixed solvent method. If one of the tests for a mixed solvent is more successful than those using a single solvent, you

T A B L E   9 . 2   Solvent pairs for mixed solvent recrystallizations[a]

Solvent 1	Solvent 2	Solvent 1	Solvent 2
Ethanol	Acetone	Ethyl acetate	Hexane
Ethanol	Petroleum ether	Methanol	Diethyl ether
Ethanol	Water	Methanol	Water
Acetone	Water	Diethyl ether	Hexane (or petroleum ether)

a. Properties of these solvents are given in Table 9.1.

If you are working with less than 1.0 g of compound, the solid can be recovered by evaporating the solvent.

can scale it up for the recrystallization of your compound. Follow the procedures of either Technique 9.5 or Technique 9.7, depending on the amount of the sample. Use approximately the same proportions of the two solvents in the scaled-up procedure. If the solid is more soluble in the solvent with the lower boiling point, any excess solvent can simply be boiled away in the hood until cloudiness is reached. Then the solution can be allowed to cool.

9.4   Planning and Carrying Out Successful Recrystallizations

In many ways recrystallization is as much an art as a science. A recrystallization is straightforward if you are told what solvent to use and given explicit directions about the ratio of solvent to solute. But when you have to determine these factors yourself, recrystallization can be more challenging. To be successful, you must consider the choices and then pay careful attention to your experimental observations and what they tell you.

Begin by carefully selecting what seems to be a good recrystallization solvent using the procedure in Technique 9.3. However, when you scale up a recrystallization from the test quantities, you need to be flexible enough to question your solvent choice if the recrystallization does not seem to be working. For example, if most of the crystals dissolve immediately in a small volume of solvent, you may have to boil away the solvent you are using and start again with a different choice.

The following factors are also important to consider and apply to both miniscale and microscale recrystallizations. When you are recrystallizing a product, attention to these details will make the process proceed more smoothly and successfully.

Scale of the Recrystallization

The amount of solid to be recrystallized will determine the size of the flask used for the recrystallization and the volume of solvent needed.

Size of flask.   The size of the Erlenmeyer flask you use depends on the scale. For miniscale recrystallizations you will probably never use a flask of smaller capacity than 50 mL. A 125- or 250-mL flask is usually more appropriate for recrystallizations of 1–10 g.

A good rule of thumb is to use a flask 2–3 times larger than the amount of solvent you think you will need.

Volume of solvent. After making your solvent choice, you need to decide the approximate volume of solvent needed for the recrystallization. That amount will naturally differ if you are purifying 400 mg or 4.0 g of a compound. You would not want to recrystallize 4.0 g of compound in 10 mL of solvent. It would be difficult to achieve much purification that way. You will probably want to use 20–40 times the amount of solvent as compound being recrystallized. This ratio was used in testing solvent choices in Technique 9.3.

Incremental additions of solvent. If you are using approximately 20 mL of solvent, it works best to make incremental additions of warm solvent with a Pasteur pipet. If you are using a larger amount of solvent, pour small portions of warm solvent directly from the flask holding the solvent into the recrystallization flask.

Insoluble Impurities

Consider a situation where you have added 40 mL of warm solvent to your compound. When you heat the mixture to just under the boiling point of the solvent, most of the solid dissolves immediately. With the addition of another 5 mL of solvent, more of the solid dissolves. But after you add another 10 mL of solvent and heat the mixture again to the boiling point, no more solid has gone into solution. Now it is time to consider that your compound contains an insoluble impurity, which needs to be removed by filtration of the hot solution [see Technique 9.5, step 2]. In this situation you have to make accurate experimental observations and then act on them if necessary.

Seed Crystals

Always set aside a small amount of the crude crystalline product to use as seeds for catalyzing the formation of crystals in the event that recrystallization does not occur. If no crystals appear in the cooled solution, it could mean that the solution is not saturated with your compound. But it could also mean that the solution is supersaturated and won't form crystals until an appropriate surface is present on which crystal growth can occur. Deciding which situation pertains can be difficult, but adding two or three small crystals of the compound will tell you.

Maximum Recovery of Product

Many students recover a smaller amount of product from a recrystallization than they should because of mechanical losses on the walls of oversized flasks or during the filtration step. Losses also occur because (1) too much solvent is added, (2) too much charcoal is added to decolorize colored solutions, (3) premature crystallization occurs during a gravity filtration, or (4) the crystals are filtered or centrifuged before recrystallization is complete.

Ensuring Dry Crystals

When a higher boiling-point solvent, such as ethyl alcohol, water, or toluene, is used as the recrystallization solvent, the recrystallized product dries slowly and should be allowed to dry at least overnight before determining its mass and melting point.

9.5 Miniscale Procedure for Recrystallizing a Solid

The procedure for recrystallizing a solid involves several steps:

- Dissolving the solid
- Removing impurities from the recrystallization solution
- Cooling the solution to allow for crystal growth
- Collecting the recrystallized solid by vacuum filtration

SAFETY PRECAUTIONS

1. Most organic solvents used for recrystallizations are volatile and flammable. Therefore, they should be heated on a steam bath or in a hot-water bath, not on a hot plate or with an open flame.
2. Lift a hot Erlenmeyer flask with flask tongs. **Note:** Test tube holders are not designed to hold Erlenmeyer flasks securely and the flask may fall onto the bench top.

Step 1. Dissolving the Solid

Always set aside a small amount of the crude product to use as seeds in the event that recrystallization does not occur.

Place the solid to be recrystallized on a creased weighing paper and carefully pour it into an Erlenmeyer flask (Figure 9.2a). Alternatively, a plastic powder funnel may be set in the neck of the Erlenmeyer flask to prevent spillage (Figure 9.2b). Add one or two boiling chips or a boiling stick. Heat an appropriate volume of the solvent in another Erlenmeyer flask. Then add small portions of hot (just below boiling) solvent to the solid being recrystallized. Begin heating the solid/solvent mixture, allowing it to boil briefly between additions, until the solid dissolves; then add a little excess solvent. Remember that some impurities may be completely insoluble, so do not add too much solvent in trying to dissolve the last bit of solid. Bring the solution to a boil on a steam bath (Figure 9.3).

FIGURE 9.2
Two ways to add a solid to an Erlenmeyer flask for recrystallization.

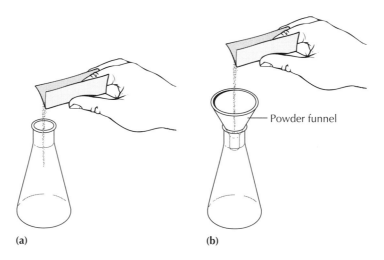

Powder funnel

(a)   (b)

Boiling a mixed solvent [Technique 9.3] can preferentially remove the lower boiling poing solvent and affect the solubility of the solute.

With particularly volatile organic solvents, such as ether or hexane, it is often easier to add a small amount of cold solvent and then heat the mixture nearly to boiling. Slowly add more cold solvent to the heated mixture until the solid just dissolves when the solution is boiling; then add a slight excess of solvent.

If you have no insoluble material or colored impurities in your hot recrystallization solution, proceed to step 3; however, if you have insoluble material or colored impurities, they need to be removed before cooling the recrystallization solution. Proceed to step 2.

Step 2. Removing Impurities from the Recrystallization Solution

Impurities in a recrystallization solution may be solid particles—dust, bits of filter paper, or other insoluble impurities—or colored impurities, which have dissolved in the recrystallization solvent. They need to be removed before cooling the solution to recrystallize the product.

Filtering solid impurities. If the solution contains solid impurities, carry out a gravity filtration using a prefolded (fluted) filter paper. Fluted filter paper provides a larger surface area than the usual filter paper cone, so filtration is faster. Suction filtration under reduced pressure does not work well because the solution cools too rapidly under suction and premature crystallization can occur. Also, small particles may pass through the filter paper when suction is used.

Place the fluted filter paper in a clean, short-stemmed funnel (a plastic powder funnel works well) and put the funnel in a second clean Erlenmeyer flask. Add a small amount (1–3 mL) of the recrystallization solvent to this receiving flask and heat the flask, funnel, and solvent on the steam bath (Figure 9.4, step 1). The boiling

Boiling stick

Steam in

Steam out

FIGURE 9.3
Heating a solution on a steam bath.

FIGURE 9.4
Filtering a recrystallization solution.

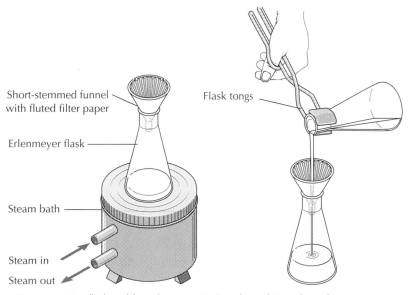

Short-stemmed funnel with fluted filter paper

Erlenmeyer flask

Steam bath

Steam in

Steam out

Flask tongs

1. Heat receiving flask and funnel.

2. Pour hot solution through fluted filter paper.

solvent warms the funnel and helps prevent premature crystallization of the solute during filtration. If your steam bath is large enough,
keep both flasks hot during the filtration process; if it is too small for
both, keep the unfiltered solution hot and set the receiving flask on
the bench top. Next pour the hot recrystallization solution through
the fluted filter paper (Figure 9.4, step 2).

SAFETY PRECAUTION

Lift a hot Erlenmeyer flask with flask tongs.

Be sure that the hot solution is added in small quantities to the
fluted filter paper, because cooling at this stage may cause crystallization. Keep the unfiltered solution hot during this step so that all
the solid remains in solution. If you have difficulty keeping the
solution from crystallizing on the filter paper, add extra hot solvent
to the flask containing the unfiltered solution and reheat it to the
boiling point before continuing the filtration. Then, after the filtration is complete, add a boiling stone or stick and boil away the
extra solvent you added.

When all the hot solution has drained through the filter paper,
check to see whether any crystallization occurred in the Erlenmeyer
receiving flask due to rapid cooling during the filtration step. If it
has, reheat the mixture to dissolve the solid. Then allow the solution to cool slowly. While the solution is cooling, the Erlenmeyer
flask should be loosely stoppered or covered with a small watch
glass. Proceed to step 3.

Using activated charcoal to remove colored impurities. If the
compound you are recrystallizing is colorless and if the recrystallization solution is deeply colored after the compound dissolves,
treatment with activated charcoal (Norit, for example) may remove
the colored impurities. Activated charcoal has a large surface area
and a strong affinity for highly conjugated colored compounds,
allowing it to readily adsorb colored impurities and effectively
remove them from the recrystallization solution. Using too much
charcoal, however, may cause some of the compound you are purifying to be adsorbed by the charcoal and reduce your yield.

Add 40–50 mg of Norite activated carbon decolorizing pellets
to the hot but not boiling recrystallization solution.

SAFETY PRECAUTION

Adding charcoal to a boiling solution can cause the solution to
foam out of the flask. **Cool the hot solution briefly before adding
the charcoal.**

Then heat the solution to just under boiling for a few minutes. (Boiling actually hinders decolorization, but heating to keep the compound in solution is necessary.)

While the solution is still very hot, gravity-filter it according to the procedure described in the previous section on filtering solid impurities. Proceed to step 3.

Step 3. Cooling the Solution

If there are no insoluble or highly colored substances in the hot recrystallization mixture, you can proceed directly to the cooling step. The size of the crystals obtained will depend on the rate at which the solution cools: the slower the cooling, the larger the crystals. Cork the Erlenmeyer flask while the solution cools. Allow the hot solution to stand on the bench top until crystal formation begins and the flask reaches room temperature, followed by final cooling for 10–15 min in an ice-water bath. Occasionally, it may take 30 min or more before crystals appear. This slow cooling usually produces crystals of a reasonable purity and intermediate size. The cooling process will take at least 20 min.

What to do if no crystals appear in the cooled solution. If no crystals appear in the solution after at least 15 min of cooling in an ice-water bath, add one or two seed crystals. If you do not have any seed crystals, scratch the bottom of the flask vigorously with a glass stirring rod. Tiny particles of glass scratched from the flask can initiate crystallization. If crystallization still does not occur, there is probably too much solvent. Boil off a small portion of the solvent and cool the solution again.

Attention to detail. Careful attention to detail and slow cooling of the hot solution often results in the formation of beautiful, pure crystals. Beautiful crystals are to an organic chemist what a home run is to a baseball player!

Step 4. Collecting the Recrystallized Solid

To complete the recrystallization procedure after all the crystals appear to have formed, collect the crystals by vacuum filtration, using a Buchner funnel, neoprene adapter, filter flask, and trap bottle or flask (Figure 9.5). The trap flask avoids backflow of water from a water aspirator coming into contact with your remaining recrystallization solution; with a house vacuum system, the trap flask keeps any overflow from the filter flask out of the vacuum line.

Choose the correct size of filter paper, one that will fit flat on the bottom of the Buchner funnel and just cover all the holes. Turn on the vacuum source and wet the paper with the recrystallizing solvent to pull it tightly over the holes in the funnel. Pour a slurry of crystals and solvent into the funnel.

Wash the crystals on the Buchner funnel with a small amount of cold recrystallization solvent (1–5 mL, depending on the amount of crystals) to remove any supernatant liquid adhering to them. To wash the crystals, allow air to enter the filtration system by removing the rubber tubing from the water aspirator nipple. Then turn off the water (to prevent backup of water into the system), or

FIGURE 9.5
Apparatus for vacuum
filtration. The second
filter flask serves as
a backflow trap.

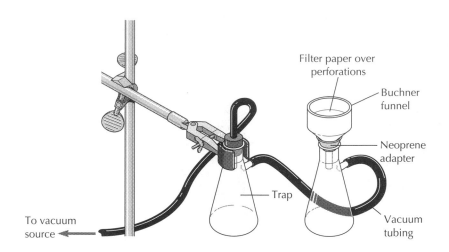

turn off the vacuum line and loosen the neoprene adapter connecting the Buchner funnel to the filter flask. Cover the crystals with the
cold solvent, reconnect the vacuum, and draw the liquid off the
crystals. Initiate the crystal drying process by pulling air through
the crystals for a few minutes. Again disconnect the vacuum as
described earlier. Place the crystals on a *tared (preweighed)* watch
glass. You will probably need to leave the crystals open to the air
in your desk for a time to dry them completely. Remove any boiling
chips or sticks before you weigh the crystals.

*A Second "Crop" of
Crystals*

A second "crop" of crystals can sometimes be obtained by evaporating about half the solvent from the filtrate and again cooling
the solution. This crop of crystals should be kept separate from the
first crop of crystals until the melting points of both crops [see
Technique 10] have been determined. If the two melting points
are the same, indicating that the purity is the same, the crops may
be combined. Usually the second crop has a slightly lower melting
point and a larger melting range, indicating that some impurities
crystallized with the desired product.

9.6   Summary of the Miniscale Recrystallization Procedure

1. Dissolve the solid sample in a minimum volume of hot solvent
   with a boiling stone or boiling stick present.
2. If the color of the solution reveals impurities, add a small
   amount of activated charcoal to the hot but not boiling solution
   (optional).
3. If insoluble impurities are present or charcoal treatment is used,
   gravity filter the hot solution through a fluted filter paper.
4. Cool the solution slowly to room temperature and then in an
   ice-water bath to induce crystallization.

5. Recover the crystals from the solvent by vacuum filtration.
6. Wash the crystals with a small amount of cold solvent.
7. Allow the crystals to air-dry completely on a watch glass before weighing them and determining their melting point.

9.7 Microscale Recrystallization Methods

Read Techniques 9.1 – 9.4 before you undertake your first microscale recrystallization.

Microscale methods are used for recrystallizations of less than 300 mg of solid. They generally follow the steps outlined in Technique 9.5, using smaller equipment. For recrystallization of less than 150 mg of material, a specialized apparatus called a Craig tube may be used.

9.7a Recrystallizing Less Than 300 Milligrams of a Solid

For microscale recrystallizations of less than 300 mg of a solid, a 25-mL Erlenmeyer flask holds the recrystallization solution and a Hirsch funnel replaces the Buchner funnel for collecting the crystals. If the amount of solid being recrystallized is less than 150 mg, a 10-mL Erlenmeyer flask or a test tube can be used. The following steps outline the procedure for a microscale recrystallization.

Always save a few crude crystals to use as seeds in the event that recrystallization does not occur.

1. Place the solid in a 25-mL or 10-mL Erlenmeyer flask, depending on the mass of crude product to be recrystallized; add a boiling stick or boiling stone. With a Pasteur pipet, add only enough solvent to just cover the crystals.
2. Heat the contents of the test tube or flask to the boiling point, then add additional solvent by drops, allowing the mixture to boil briefly after each addition. Continue this process until just enough solvent has been added to dissolve the solid. Be aware that some impurities may not dissolve.
3. If colored impurities are present, cool the mixture slightly and add 10 mg of Norit carbon-decolorizing pellets (about 10 pellets). Keep the mixture heated just under the boiling point. If the color is not removed after 1–2 min, add a few more Norit pellets and heat briefly. Prepare a Pasteur filter pipet [see Technique 8.5]. Warm the Pasteur pipet by immersing it in a test tube of hot solvent and drawing the hot solvent into it several times; use the heated pipet to separate the hot recrystallization solution from the Norit pellets and transfer it to another test tube or flask. If crystallization begins in the solution with the carbon pellets during this process, add a few more drops of solvent and warm the mixture to boiling to redissolve the crystals before completing the transfer.
4. If the recrystallization mixture contains insoluble impurities, use a Pasteur filter pipet as outlined in step 3 to separate the solution from the insoluble impurities.

FIGURE 9.6
Vacuum filtration using a Hirsch funnel.

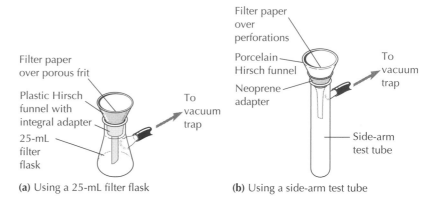

Filter paper over porous frit

Plastic Hirsch funnel with integral adapter

25-mL filter flask

To vacuum trap

(a) Using a 25-mL filter flask

Filter paper over perforations

Porcelain Hirsch funnel

Neoprene adapter

To vacuum trap

Side-arm test tube

(b) Using a side-arm test tube

5. Allow the solution to cool slowly to room temperature, and then chill it in an ice-water bath to complete crystallization.
6. Collect the crystals by vacuum filtration, using a Hirsch funnel (Figure 9.6).
7. Allow the crystals to air-dry completely on a watch glass before weighing them and determining their melting point.

9.7b Recrystallizing Up to 150 Milligrams of a Solid in a Craig Tube

Recrystallizations of up to 150 mg may be done in a Craig tube, a small device that serves as both recrystallization vessel and filtration apparatus. The Craig tube eliminates the losses associated with separation of the crystals from the solution by vacuum filtration. It also allows several recrystallizations to be performed without removing the crystals from the tube, thus preventing significant losses of material on the glassware.

Always save a few crude crystals to use as seeds in the event that recrystallization does not occur.

The Craig tube consists of a glass tube that has a band with a rough surface at the point where the tube widens and a separate plug made of Teflon or glass (Figure 9.7). When the plug is inserted into the tube and the apparatus inverted as described later, the rough glass surface of the tube forms an incomplete seal with the plug. This imperfect seal allows the crystallization solution (mother liquor) to flow out during centrifugation, leaving the crystals in the tube. This process separates the crystals from the recrystallization solution.

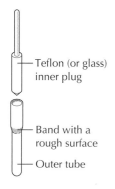

Teflon (or glass) inner plug

Band with a rough surface

Outer tube

FIGURE 9.7
Craig tube.

Heat for a Craig tube recrystallization is supplied by an aluminum block heated to 10°–20°C above the boiling point of the solvent; one of the holes in the block is sized for the Craig tube. Bumping and boiling over occur very quickly with a Craig tube; prevent this problem by using a boiling stick or agitating the hot recrystallization solution by rolling a microspatula inserted in the Craig tube between your fingers. Boiling chips should not be used with a Craig tube because their subsequent removal is difficult.

Craig Tube Recrystallization Procedure

Boiling a mixed solvent [Technique 9.3] can preferentially remove the lower boiling point solvent and affect the solubility of the solute.

Slow cooling promotes the formation of the larger and purer crystals essential for recovery of a pure product.

During centrifugation, the recrystallization solvent flows through the small pores in the rough etched ring of the Craig tube, while the crystals remain on the end of the inner plug.

Place the material to be recrystallized in the outer tube of the Craig assembly. If the amount of solvent is not specified, cover the crystals with a few drops of solvent and begin heating the mixture in the aluminum block. Continue adding solvent by drops and boiling the mixture gently between additions until the solid dissolves completely. When dissolution is complete, insert the plug into the tube and set the assembly inside a 25-mL Erlenmeyer flask to cool slowly. When crystallization appears to be complete, cool the tube in an ice-water bath for 5 min.

Recover the crystals in the following manner. Obtain a centrifuge tube. If the centrifuge tube and the Craig tube plug are glass, place a small piece of cotton in the bottom of the centrifuge tube; cotton is unnecessary with either a plastic centrifuge tube or a Teflon plug. Slip a loop of thin copper wire over the stem of the plug (Figure 9.8, step 1) and invert the centrifuge tube over the Craig apparatus (Figure 9.8, step 2). Then turn the centrifuge tube upright and bend the end of the wire over the lip of the centrifuge tube (Figure 9.8, step 3).

Set the tube in a centrifuge and balance its weight in the opposite hole with another centrifuge tube that is two-thirds to three-fourths full of water; centrifuge for 1–2 min. If you are having difficulty balancing the centrifuge, weigh the centrifuge tube containing the Craig apparatus in a small beaker, then adjust the amount of water in the other centrifuge tube until the mass approximately equals that of the tube with the Craig apparatus.

Remove the tube from the centrifuge and carefully remove the Craig tube from the centrifuge tube by pulling on the copper wire. Turn the Craig apparatus upright. Remove the plug from the outer

FIGURE 9.8
Removing solvent from a Craig tube by centrifugation.

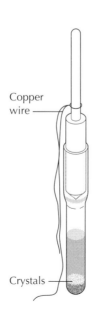

Copper wire

Crystals

1. Place loop of copper wire over plug.

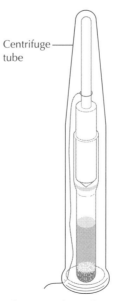

Centrifuge tube

2. Place centrifuge tube over Craig apparatus, invert, and put in centrifuge.

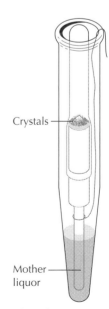

Crystals

Mother liquor

3. Centrifuge tube/Craig apparatus after centrifugation.

tube and set the tube upright in your desk to allow the crystals to dry. Carefully scrape any crystals adhering to the plug onto a small watch glass and store them until the crystals in the Craig tube are dry. The loss of crystals adhering to the inside of the tube is less if the crystals have dried before they are removed from the tube.

9.8	## Summary of Microscale Recrystallization Procedures

Recrystallizing Less than 300 Milligrams of a Solid

1. Dissolve the solid in a minimum volume of hot solvent in a 25-mL Erlenmeyer flask; use a boiling stick or boiling chip to prevent bumping. If the amount of solid is less than 150 mg, a 10-mL Erlenmeyer flask or test tube can be used.
2. If colored impurities are present, heat the mixture briefly with 8–10 Norit pellets.
3. If insoluble impurities are present or Norit pellets were used, transfer the hot recrystallization solution to another test tube or flask, using a warm Pasteur filter pipet.
4. Cool the solution slowly to room temperature to induce crystallization, then complete cooling in an ice-water bath.
5. Collect the crystals by vacuum filtration on a Hirsch funnel.
6. Allow the crystals to air-dry completely on a watch glass before weighing them.

Recrystallizing Up to 150 Milligrams of a Solid in a Craig Tube

1. Dissolve the solid in a minimum volume of hot solvent in the outer tube of the Craig apparatus; use a boiling stick or continuously roll a stirring rod between your fingers to prevent bumping.
2. Insert the plug into the tube and set the assembly inside a 25-mL Erlenmeyer flask to cool undisturbed. When crystallization is complete, cool the tube in an ice-water bath.
3. Place an inverted centrifuge tube over the Craig apparatus, then turn the centrifuge tube upright. Separate the solvent from the crystals by spinning the centrifuge tube in a centrifuge.
4. Allow the crystals to dry in the outer tube before removing them for weighing.

9.9	## Sources of Confusion

As you read this section, recall Technique 9.4, which discusses the importance of scale, volume of solvent, insoluble impurities, the use of seed crystals, maximum recovery factors, and ensuring dry crystals. Technique 9.4 also pointed out the need for flexibility in doing a recrystallization, to make good observations, form hypotheses from them, and be willing to test your hypotheses.

Did I Use the Proper Recrystallization Solvent?

Probably the most confusing part of recrystallization is deciding whether you have used the best recrystallization solvent. The best time to ask this question is while you are choosing the solvent by the methods of Technique 9.3. If you are unsure, repeat the procedure at that stage. It can save a great deal of time in the long run. If loss of the product that you use for the solubility tests would be substantial without its recovery, you can always recover it by evaporation of the solvents.

How Much Solvent Should I Use?

The answer to this question depends on the solubilities of the compound in the hot and cold solvent and the amount of material you are recrystallizing. General recrystallization guidelines are always somewhat ambiguous because they cannot be applied in a straightforward manner for every one of the many thousands of organic compounds you might be recrystallizing.

A recrystallization is often started with only enough solvent to cover the crystals in the recrystallization flask. After heating the solvent to boiling, solvent is added in small increments, 1–5 mL for miniscale recrystallizations and a few drops for microscale recrystallizations, heating the recrystallization container after the addition of each solvent increment. Only enough solvent is added to just dissolve the crystals when the solvent is boiling, plus another increment to provide a modest excess of solvent.

I Added Solvent but the Volume Did Not Change

If the rate of heating is too rapid, solvent may be evaporating from the recrystallization flask as fast as you are adding it. Evaporation is a particular problem when working with a mixed solvent recrystallization. Rapid heating in this instance probably results in preferential loss of the lower-boiling solvent. Lessen the heat to a setting that just maintains the solvent at its boiling point.

No Crystallization Occurred in the Cooled Solution

In many instances, recrystallization fails because too much solvent is used in the process. In these cases, you need to boil off a small portion of the solvent and try the recrystallization again. If crystallization still does not occur from the supersaturated solution, the best approach is to add one or two seed crystals. If you do not have a few seed crystals available, it may be possible to promote crystal formation by scratching the inside of the bottom of the flask vigorously with a glass stirring rod. Tiny particles of glass scratched from the flask can serve as centers for crystallization.

Formation of Oils

The formation of oils may be the most frustrating outcome of an attempted recrystallization. The presence of impurities lowers the melting point, making "oiling out" especially prevalent during recrystallization of a solute with a melting point below the boiling point of the solvent. Oiling out also occurs if too little recrystallization solvent has been used so that the compound becomes insoluble at a high temperature, even if it is below the boiling point of the solvent. The presence of insoluble liquid oil allows impurities to distribute themselves between the solvent and the oil before crystallization can occur. This means that impurities are trapped in the oil when it cools; it often hardens into a viscous glasslike substance.

If you have oil rather than crystals, you can add more solvent so that the compound does not come out of solution at so high a temperature. It may also help to switch to a solvent with a lower boiling point (consult Table 9.1, page 101).

Some oils can be crystallized by dissolving them in a small amount of diethyl ether or hexane and allowing the solvent to evaporate slowly in a hood. Crystallization often occurs as the solution slowly becomes more concentrated. Once crystals form, seed crystals are available to assist further purification by recrystallization.

Questions

1. Describe the characteristics of a good recrystallization solvent.
2. The solubility of a compound is 59 g per 100 mL in boiling methanol and 30 g per 100 mL in cold methanol, whereas its solubility in water is 7.2 g per 100 mL at 95°C and 0.22 g per 100 mL at 2°C. Which solvent would be better for recrystallization of the compound? Explain.
3. Explain how the rate of crystal growth can affect the purity of a recrystallized compound.
4. In what circumstances is it necessary to filter a hot recrystallization solution?
5. Why should a hot recrystallization solution be filtered by gravity rather than by vacuum filtration?
6. Low-melting solids often "oil out" of a recrystallization solution rather than crystallizing. If this were to happen, how would you change the recrystallization procedure to ensure good crystals?
7. An organic compound is quite polar and is thus much more soluble in methanol than in pentane (bp 36°C). Why would methanol and pentane be an awkward solvent pair for recrystallization? Consult Table 9.1 to assist you in deciding how to change the solvent pair so that recrystallization would proceed smoothly.
8. For a variety of reasons, N,N-dimethylformamide (DMF) is not usually utilized as a recrystallization solvent even though it has a moderate dielectric constant ($\varepsilon = 38$). Included among its properties are an unpleasant odor, a high boiling point (153°C), and the fact that it is highly hygroscopic (absorbs water). Refer to Table 9.1 and discuss what other solvent(s) might serve as an alternative.

10

MELTING POINTS AND MELTING RANGES

Molecules in a crystal are arranged in a regular pattern. Melting occurs when the fixed array of molecules in the crystalline solid rearranges to the more random, freely moving liquid state. The transition from solid to liquid requires energy in the form of heat to break down the crystal lattice. The temperature at which this transition occurs is the solid's *melting point*, an important physical property of any solid compound. The melting point of a compound is useful in establishing its identity and as a criterion of its purity. Relatively pure compounds normally melt over a temperature range of 0.5°–2.0°C, whereas impure substances often melt over a much larger range. Until the advent of modern chromatography and spectroscopy, the melting point was the primary index of purity for an organic solid. Melting points are still used as an effective preliminary indication of purity.

10.1 Melting-Point Theory

A solid at any temperature has a finite vapor pressure. As the temperature of a solid increases, its vapor pressure increases as well. Both the solid and the liquid phases are always in equilibrium with the vapor and, at the melting point, are also in equilibrium with each other (Figure 10.1). A vapor pressure/temperature diagram for both the solid and liquid phases of a chemical compound is shown in Figure 10.2.

Melting Behavior

The temperature at which a compound melts is a physical characteristic of the substance, and for pure compounds it is generally reproducible. However, the presence of even small amounts of impurities usually depresses the melting point a few degrees and causes melting to occur over a relatively wide temperature range. Because the melting point is the temperature at which the vapor pressures of the pure liquid and the pure solid are equal, the presence of an impurity that is soluble in the liquid can change the temperature at which this equilibrium state occurs.

The phase diagram depicted in Figure 10.3 shows the observed melting curve for various mol % mixtures of compounds A and B ranging from 100 mol % of A with 0 mol % B to 0 mol % A with 100 mol % B. A pure sample of compound A melts at temperature T_a whereas pure compound B melts at temperature T_b. At T_a and T_b, pure samples of A and B melt sharply over a very narrow temperature range of 1° or less.

Mixtures of A and B exhibit different melting behavior. Consider the behavior of a solid consisting of 80% of compound A and 20% of compound B. As compound A, with the lower melting point, begins to melt, compound B, still a solid, starts to dissolve in the liquid of compound A. The vapor pressure of the liquid

FIGURE 10.1
Solid and liquid in equilibrium.

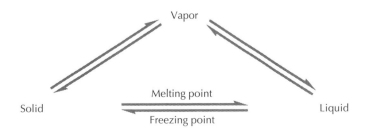

FIGURE 10.2
Vapor pressure/
temperature diagram
for solid and liquid
phases of a pure
compound.

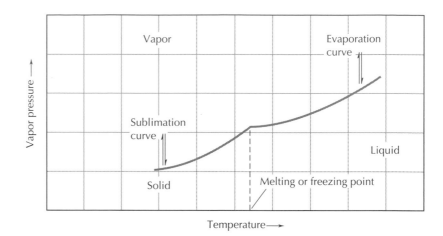

solution of A and B is lower than that of pure liquid A at the melting point, whereas the vapor pressure of solid A at a particular temperature is virtually unchanged by the impurity B because the solids do not mix together intimately. Thus, the vapor pressures of the liquid and the solid are equal at a lower temperature, and the temperature at which solid A melts is lower when B is present. One practical application of this behavior is salting roads to melt ice at a temperature lower than 0°C.

*Eutectic
Composition*

There is a limit, however, to how far the melting point can be lowered. This limit is reached when the liquid solution becomes saturated in B, a condition causing some of solid B to remain

FIGURE 10.3
Melting-point compo-
sition diagram for the
binary mixture A + B.
T_a is the melting point
of pure solid A, T_b of
pure solid B, and T_e
of eutectic mixture E.
The temperature range
T_e–T_m is the melting
range of a solid con-
taining 80 mol % A
and 20 mol % B.

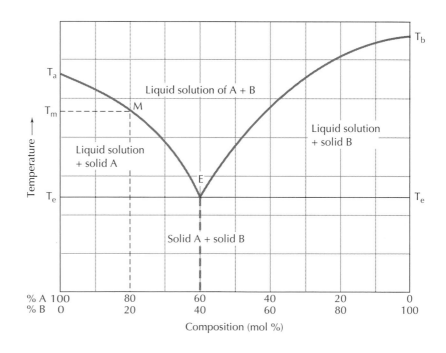

after all of A has melted. Until point E is reached in Figure 10.3, all of B dissolves in the melting A. After point E, when all of A is melted, a portion of solid B remains. Point E defines the composition of a saturated solution of B in liquid A and is called the *eutectic composition.* A solid mixture with the eutectic composition (60% A and 40% B) will melt sharply at the eutectic temperature, T_e. The eutectic point E can be thought of as the point at which increasing the percentage of A means that B acts as the impurity, and increasing the percent of B means that A acts as the impurity.

For another perspective on the melting points of mixtures, again consider the melting of a solid mixture composed of 80% A and 20% B. As heat is applied, the temperature of the solid mixture rises. When it reaches temperature T_e, A and B will melt together at a constant ratio (the eutectic composition) and the temperature will remain constant. When the minor component, B, is completely liquefied and more heat is applied, solid A, in equilibrium with the eutectic composition of liquid A and B, continues to melt as the temperature increases. Because the vapor pressure of liquid A increases as the mole fraction of A in the liquid increases, the temperature required to melt A also rises. Melting occurs along curve EM in Figure 10.3, giving an observed temperature range of $T_e - T_m$.

It is possible for a binary mixture like A + B to have a more complex melting-point composition diagram than the one shown in Figure 10.3. There can be two eutectic points when the two components interact to form a molecular compound of definite composition. Discussions of more complex melting phenomena are found in the references given at the end of the chapter.

10.2   Apparatus for Determining Melting Ranges

Two types of electrically heated melting-point devices are commonly used in introductory organic chemistry laboratories—the Mel-Temp apparatus and the Fisher-Johns hot-stage apparatus.

Mel-Temp Apparatus   A Mel-Temp apparatus is shown in Figure 10.4. A thin-walled glass capillary tube holds the sample. The capillary tube fits into one of three sample chambers in the heating block; multiple chambers allow simultaneous determinations of three melting points. A cylindrical cavity in the top of the heating block holds the thermometer, a light illuminates the sample chamber, and an eyepiece containing a small magnifying lens facilitates observation of the sample. A digital thermometer can also be used with a Mel-Temp apparatus.

A rheostat controls the rate of heating by allowing continuous adjustment of the voltage. The higher the rheostat setting, the faster the rate of heating. Figure 10.5 shows graphically how the rate of heating changes at different rheostat settings. As the

FIGURE 10.4
Mel-Temp apparatus (shown without the heat shield around the light and heating block). (Courtesy of Laboratory Devices, Inc., Holliston, MA.)

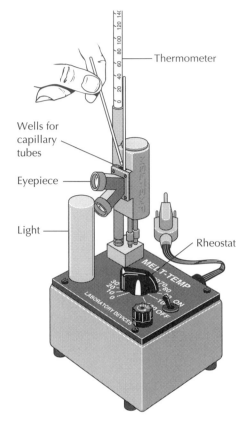

heating curves indicate, the rate of heating at any particular setting increases more rapidly at the start, then slows as the temperature increases. The decreasing rate of heating at the higher temperatures allows for the slower rate of heating needed as the melting point is approached.

FIGURE 10.5
Heating rate curves for a Mel-Temp apparatus at various rheostat settings (indicated in volts). (Courtesy of Laboratory Devices, Inc., Holliston, MA.)

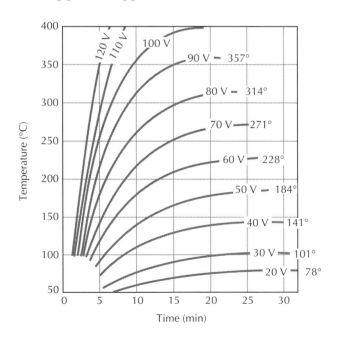

FIGURE 10.6
Fisher-Johns hot-stage melting-point apparatus. (Courtesy of Fisher Scientific, Pittsburgh, PA.)

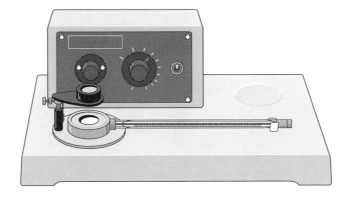

Fisher-Johns Apparatus

The Fisher-Johns hot-stage apparatus represents a second device for the determination of melting points (Figure 10.6). The crushed sample is placed between thin, circular, microscope coverslips rather than in a capillary tube. The coverslips fit in a depression in the metal block surface. A rheostat controls the rate of heating, and the lighted sample area is viewed through a small magnifying glass.

SAFETY PRECAUTIONS

1. If the heater on a Fisher-Johns apparatus is not turned off after the sample melts, the high heat may ruin the thermometer calibration or even break the thermometer. The latter event may lead to a spill of toxic mercury in the laboratory.
2. Never use ice to cool the hot stage. The sudden decrease in temperature may break the thermometer and cause a spill of toxic mercury.

10.3   Determining Melting Ranges

Sample Preparation

The melting range of an organic solid can be determined by introducing a small amount of the substance between two cover slips or into a capillary tube with one sealed end. Such capillary tubes, which are approximately 1–2 mm in diameter, are commercially available.

Filling a capillary tube. Place a few milligrams of the dry solid on a piece of smooth-surfaced paper and crush it to a fine powder with a spatula. Introduce the solid into the capillary tube by tapping the open end of the tube in the powdered substance. A small amount of material will stick in the open end. Invert the capillary tube so that the sealed end is down, and holding it very near the sealed end, tap it lightly with quick motions against the bench top. The solid will fall to the bottom of the tube. Repeat this operation until the amount of solid in the tube is 1–2 mm in height. A small sample is essential for accurate melting points. Melting-point determinations

The ideal sample for a melting point is only 1–2 mm in height in the capillary tube.

made with too much sample lead to a broad melting range because more time is required to melt the complete sample as the temperature continues to rise.

SAFETY PRECAUTION

Care must be taken while tapping the capillary tube against the bench top; it could break and cause a cut.

An alternative method for getting the solid to the bottom of a capillary tube is to drop the tube down a piece of glass tubing about 1 m in length or down the inside tube of a condenser, the bottom end of which is resting on the lab bench. After a few trips down the glass tubing, the solid will usually have fallen to the bottom of the capillary tube.

Samples for the Fisher-Johns apparatus. Samples for the Fisher-Johns apparatus also need to be finely powdered. Put a few grains of the powdered sample on one coverslip and set it in the metal heating block. Place a second coverslip over the sample and gently flatten the powder until the two glass surfaces just touch each other; contact between the two coverslips ensures good heat transfer to the sample.

Wet samples. If the solid is still wet from a recrystallization, it will not fall to the bottom of a capillary tube but will stick to the capillary wall. This failure to behave properly is probably a good thing, because melting points of wet solids are always low and thus nearly worthless. If your sample is still wet, allow it to dry completely before continuing with the melting-range determination.

Heating the Sample to the Melting Point

The melting-point apparatus can be heated rapidly until the temperature is about 20°C below the expected melting point. Then decrease the rate of heating so that the temperature rises only 1°–2° per minute and the sample has time to melt before the temperature rises above the true melting point. When you are taking successive melting points, remember that the apparatus needs to cool to at least 20° below the expected melting point before it can be used for the next determination.

Approximate melting point. If you do not know the melting point of a solid sample, you can make a quick preliminary determination by heating the sample rapidly and watching for the temperature at which melting begins. In a more accurate second determination, you can then carefully control the temperature rise to 1°–2° per minute when you get within 15°–20° of the expected melting point.

Use a fresh sample for each determination. Always prepare a fresh sample for each melting-point determination; many organic compounds decompose at the melting point, making reuse of the solidified sample a poor idea. Moreover, many low-melting-point compounds (mp 30°–80°C) do not easily resolidify with cooling.

Digital thermometers. Digital thermometers have a metal probe that responds more rapidly than a mercury-filled glass thermometer to temperature changes. The rate of heating near the melting point must be 1°–2° per minute or else the observed melting-point range will very probably be above the true melting point. Consult your instructor before using a digital thermometer.

Reporting the Melting Range

Unless you have an extraordinarily pure compound in hand, you will always observe and report a melting range—from the temperature at which the first drop of liquid appears to the temperature at which the solid is completely melted and only a clear liquid is present. This melting range is usually 1°–2° or slightly more. For example, salicylic acid usually gives a melting range of 157°–159°C. An extremely pure sample of salicylic acid, however, melts over less than a 1° range (for example, 160.0°–160.5°C) and it may have 160°C listed as its melting point. Published melting points are usually the highest values obtained after several recrystallizations; the values you observe will probably be slightly lower.

10.4   Mixture Melting Point

We have already discussed how impurities can lower the melting point of a compound. This behavior can be useful not only in evaluating a compound's purity but also in helping to identify the compound. Assume that two compounds have virtually identical melting ranges. Are the compounds identical? Possibly, but not necessarily, because the identical melting ranges may be a coincidence.

Sample Preparation

If roughly equal amounts of the two compounds are finely ground together with a spatula, the melting range of the resulting mixture can provide useful information. If there is a melting-point depression or if the melting range is expanded by a number of degrees, it is reasonably safe to conclude that the two compounds are not identical. One compound has acted as an impurity toward the other by lowering the melting range. If there is no lowering of the mixture's melting range relative to that for each separate compound, the two are probably the same compound.

Sometimes only a modest melting-point depression is observed. To know whether this change is significant, the mixture melting point and the melting point of one of the two compounds should be determined simultaneously in separate capillary tubes. This experiment allows simultaneous identity and purity checks. Infrequently, a eutectic point (point E in Figure 10.3) can be equal to the melting point of the pure compound of interest. In a case where you have accidentally used the eutectic mixture, a mixture melting point would not be a good indication of purity or identity. Errors of this type can be discerned by testing various mixtures other than a 1:1 composition. The subsequent use of 1:2 and 2:1 mixtures can avoid eutectic-point-induced misinterpretation.

Other Ways of Determining Identity   Other ways of determining the identity of a solid organic compound involve spectroscopic methods [see Techniques 18–20] and thin-layer chromatography [see Technique 15].

10.5   Thermometer Calibration

The accuracy of melting-point determinations can be no better than the accuracy of the thermometer. It cannot be assumed that a thermometer has been accurately calibrated. Although frequently this is the case, it is not always true. Thermometers can give high or low temperature readings of 1°–2° or more.

A thermometer can be calibrated with a series of pure compounds that are readily available and whose melting points are relatively easy to reproduce. A useful series of compounds is given Table 10.1.

You may want to record the temperature deviation of your thermometer using the melting points of a number of the compounds suggested in Table 10.1 and make a graph of the thermometer corrections. Plot the observed temperatures against the necessary temperature corrections and interpolate to correct future melting-point determinations with your thermometer. Usually these plots are linear.

TABLE 10.1	Compounds suitable for thermometer calibration*
Compound	**Melting point, °C**
Benzophenone	48
Acetamide	81
Benzil	95
Benzoic acid	122
Phenacetin	135
Salicylic acid	160
Succinic acid	189
4-Fluorocinnamic acid	210
Anthraquinone	285

*A kit of compounds for melting-point standards for Mel-Temp calibration is available from the Aldrich Chemical Co.

10.6   Summary of Mel-Temp Melting-Point Determinations

1. Introduce the powdered, dry solid sample to a height of 1–2 mm into a capillary tube that is sealed at one end.
2. Place the capillary tube in the melting-point apparatus.
3. Adjust the rate of heating so that the temperature rises at a moderate rate. The rate can be faster if, for example, the melting point is 170°C rather than 70°C.
4. When a temperature 15°–20° below the expected melting point is reached, decrease the rate of heating so that the temperature rises only 1°–2° per minute. **Note:** There will be a time lag before the rate of heating changes.
5. If the temperature is rising more than 1°–2° per minute at the time of melting, determine the melting point again using a new sample.
6. Record the melting range as the range of temperatures between the temperature at the onset of melting and the temperature at which only liquid remains in the tube.

10.7   Sources of Confusion

When you heat a sample for a melting-point determination, you may see some strange and wonderful things happen before the first drop of liquid actually appears. The compound may soften and shrivel up as a result of changes in its crystal structure. It may "sweat out" some solvent of crystallization. It may decompose, changing color as it does so. None of these changes should be called melting. **Only the appearance of liquid indicates the onset of true melting.** Even so, it can be difficult to distinguish exactly when melting starts. In fact, even with careful heating, two people may disagree on the melting point by as much as 1°–2°.

Rate of Heating

Heating faster than 1°–2° per minute may lead to an observed melting range that is higher than the correct one, particularly when using a digital thermometer with a metal probe. And if the rate of heating is extremely rapid (>10°C per minute), you may also observe thermometer lag with a mercury thermometer, a condition caused by failure of the mercury temperature to increase as rapidly as the temperature of the metal heating block. This error causes the observed melting range to be lower than it actually is. Determining accurate melting points requires patience.

Sublimation

Another possible complication in melting-point determinations occurs if the sample sublimes. Sublimation is the change that occurs when a solid is transformed directly to a gas, without passing through the liquid phase [see Technique 12]. If the sample in the capillary tube sublimes, it can simply disappear as it is heated.

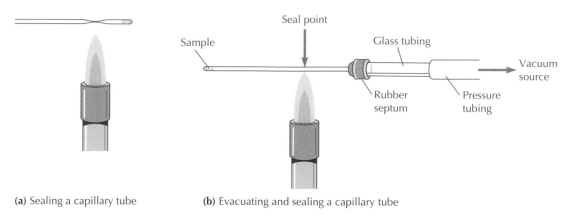

(a) Sealing a capillary tube   (b) Evacuating and sealing a capillary tube

FIGURE 10.7
Methods for sealing a capillary tube with a Bunsen burner.

Many common substances sublime, for example, camphor and caffeine. You can determine their melting points by sealing the open end of the capillary tube in a Bunsen burner flame before it is placed in the melting-point apparatus (Figure 10.7a).

Decomposition

Some compounds decompose as they melt, a behavior usually indicated by a change in color of the sample to dark red or brown. The melting point of such a compound is reported in the literature with the letter d after the temperature, for example, 186°C d, meaning that the compound melts at 186°C with decomposition.

Sometimes decomposition occurs as a result of a reaction between the compound and oxygen in the air. If this is the case, when the air is evacuated from the capillary tube and the tube is sealed, the melting point can be determined without decomposition (Figure 10.7b).

Place the sample in the capillary tube as directed earlier. Punch a hole in a rubber septum, and insert the sealed end of the capillary tube through the inside of the septum. Fit the septum over a piece of glass tubing that is connected to a vacuum line. Turn on the vacuum source, and while heating the upper portion of the capillary tube in a Bunsen burner flame, hold and pull on the sample end of the capillary tube until it seals.

SAFETY PRECAUTION

Be sure no flammable solvents are in the vicinity when you are using a Bunsen burner.

References

1. Skau, E. L.; Arthur, J. C. Jr. In *Physical Methods of Chemistry*, A. Weissberger and B. W. Rossiter (Eds.); Wiley-Interscience: New York, 1971, vol. 1, Part V.

2. Everett, T. S. "Eliminating Mercury Thermometers from the Lab," *J. Chem. Educ.* **1997,** *74,* 1204.

Questions

1. A student performs two melting-point determinations on a crystalline product. In one determination, the capillary tube contains a sample about 1–2 mm in height and the melting range is found to be 141°–142°C. In the other determination, the sample height is 4–5 mm and the melting range is found to be 141°–145°C. Explain the broader melting-point range observed for the second sample. The reported melting point for the compound is 143°C.

2. Another student reports a melting point of 136°–138°C for the melting range of the unknown in question 1 and mentions in her notebook that the rate of heating was about 12° per minute. NMR analysis of this student's product does not reveal any impurities. Explain the low melting point.

3. A compound melts at 120°–122°C on one apparatus and at 128°–129°C on another. Unfortunately, neither apparatus is calibrated. How might you check the identity of your sample without calibrating either apparatus?

4. Why does sealing the open end of a melting-point capillary tube allow you to measure the melting point of a compound that sublimes?

5. A white crystalline compound melts at 111°–112°C and the melting-point capillary is set aside to cool. Repeating the melting-point analysis, using the same capillary, reveals a much higher melting point of 140°C. Yet repeated recrystallization of the original sample yields sharp melting points no higher than 114°C. Explain the behavior of the sample that was cooled and then remelted.

TECHNIQUE

11

BOILING POINTS AND DISTILLATION

Distillation is a method for separating two or more liquid compounds on the basis of boiling-point differences. Unlike the liquid-liquid and solid-liquid separation techniques of extraction and crystallization, distillation is a liquid-gas separation in which vapor pressure differences are used to separate materials.

A liquid at any temperature exerts a pressure on its environment. This pressure, called the *vapor pressure,* results from molecules leaving the surface of the liquid to become vapor.

$$\text{molecules}_{\text{liquid}} \rightleftharpoons \text{molecules}_{\text{vapor}}$$

As a liquid is heated, the kinetic energy of its molecules increases. The equilibrium shifts to the right and more molecules move into the gaseous state, increasing the vapor pressure. Figure 11.1 shows the relationship between vapor pressure and temperature for benzene, water, and *tert*-butylbenzene.

FIGURE 11.1
Examples of the
dependence of
vapor pressure on
temperature.

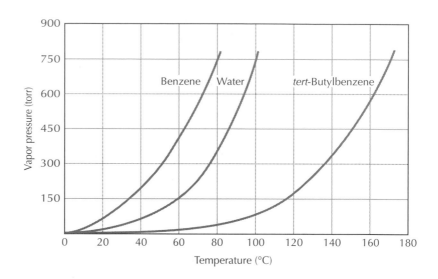

11.1   Determination of Boiling Points

Boiling Point

The boiling point of a pure liquid is defined as the temperature at which the vapor pressure of the liquid exactly equals the pressure exerted on it by the atmosphere. At an external pressure of 1.0 atm (760 torr), the boiling point is reached when the vapor pressure equals 1.0 atm; thus the boiling point of a liquid depends on the atmospheric pressure. Table 11.1 gives boiling points of several common solvents in Laramie, Wyoming (elevation 7520 ft), at the experimental station of the U.S. Department of Agriculture in Death Valley, California (elevation –285 ft), and in New York City (elevation 0 ft). When the boiling point of a substance is determined, both the atmospheric pressure and the boiling point need to be recorded.

Every pure and thermally stable organic compound has a characteristic boiling point at 1 atm. The boiling point reflects its molecular structure, specifically the types of weak intermolecular interactions that bind the molecules together in the liquid state. Intermolecular interactions must be overcome for molecules to leave the liquid state.

**TABLE 11.1   Boiling points of common compounds
at different elevations (pressures)**

Compound	Death Valley, CA Elevation = −285 ft P = 1.01 atm	New York City Elevation = 0 ft P = 1.00 atm	Laramie, WY Elevation = 7520 ft P = 0.75 atm
Water	100.3	100.0	93
Diethyl ether	35.0	34.6	27
Benzene	80.8	80.2	73
Acetic acid	119.0	118.8	110

Polar compounds have higher boiling points than nonpolar compounds of similar molecular weight. Intermolecular hydrogen bonding and dipole-dipole interactions always produce a higher boiling point. Increased molecular weight usually produces a larger molecular surface area and also leads to a higher boiling point.

Miniscale Determination of Boiling Points

The boiling point of 5 mL or more of a pure liquid compound can be determined by a simple distillation using miniscale standard taper glassware. The procedure for setting up a simple miniscale distillation is described in Technique 11.3. When distillate is condensing steadily and the temperature stabilizes, the boiling point of the substance has been reached.

Rather than using the entire sample for a distillation, the microscale procedure described next can be used to determine the boiling point of any pure liquid and requires only 0.3 mL of liquid.

Microscale Determination of Boiling Points

Place 0.3 mL of the liquid and a boiling stone in a Craig tube or a reaction tube. Set the tube in the appropriate-size hole of an aluminum heating block [see Technique 6.2]. Alternatively, heat may be supplied by a sand bath [see Technique 6.2], in which case the tube and the thermometer need to be held by separate clamps. Clamp the thermometer so that the bottom of the bulb is about 0.5 cm above the surface of the liquid; be sure that the thermometer does not touch the wall of the tube (Figure 11.2).

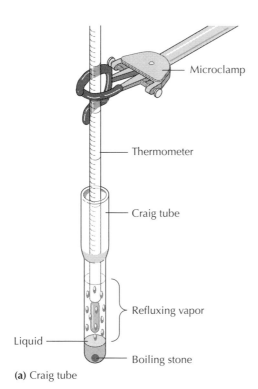

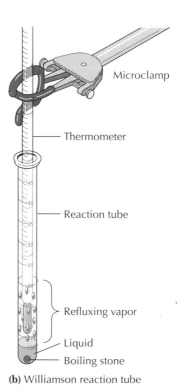

(a) Craig tube   (b) Williamson reaction tube

FIGURE 11.2
Apparatus for microscale boiling-point determinations.

Gradually heat the sample to boiling and continue to increase the rate of heating *slowly* until the ring of condensate is 1–2 cm above the top of the thermometer bulb. When the temperature reaches a maximum and stabilizes for at least 1 min, you have reached the boiling point of the liquid. Rapid or excessive heating of the tube can lead to superheating of the vapor and can also radiate heat from the tube to the thermometer bulb, causing the observed boiling point to be too high.

| 11.2 | **Distillation and Separation of Mixtures** |

The boiling point of a mixture depends on the vapor pressures of its components. Impurities can either raise or lower the observed boiling point. Consider, for example, the boiling characteristics of a mixture of pentane and hexane. Pentane and hexane are mutually soluble, and their molecules interact with one another only by van der Waals forces, which are the weakest intermolecular interactions. A solution composed of both pentane and hexane boils at temperatures between the boiling points of pentane (36°C) and hexane (69°C).

Raoult's and Dalton's Laws

If pentane alone were present, the vapor pressure above the liquid would be due only to pentane. However, when pentane is only a fraction of the solution, the partial pressure (P) exerted by pentane is equal to only a fraction of the vapor pressure of pure pentane ($P°$). The fraction is determined by $X_{pentane}$, the **mole fraction** of pentane, which is the ratio of moles of pentane to the total number of moles of pentane and hexane in the solution.

$$\text{Mole fraction of pentane: } X_{pentane} = \frac{moles_{pentane}}{moles_{pentane} + moles_{hexane}}$$

$$\text{Partial pressure of pentane: } P_{pentane} = P°_{pentane} \, X_{pentane} \tag{1}$$

The hexane present in the solution also exerts its own independent partial vapor pressure.

$$\text{Mole fraction of hexane: } X_{hexane} = \frac{moles_{hexane}}{moles_{pentane} + moles_{hexane}}$$

$$\text{Partial pressure of hexane: } P_{hexane} = P°_{hexane} \, X_{hexane} \tag{2}$$

The vapor pressure/mole fraction relationships expressed in equations 1 and 2 are valid only for ideal liquids in the same way that the ideal gas law strictly applies only to ideal gases. Equations 1 and 2 are applications of Raoult's law, named after the French chemist François Raoult, who studied the vapor pressures of solutions in the late nineteenth century.

Using Dalton's law of partial pressures, we can now calculate the total vapor pressure of the solution, which is the sum of the partial vapor pressures of the individual components:

$$P_{total} = P_{pentane} + P_{hexane} \qquad (3)$$

Figure 11.3 shows the partial-pressure curves for pentane and hexane at 25°C using Raoult's law and the total vapor pressure of the solution using Dalton's law. The boiling point of a pentane/hexane mixture is the temperature at which the individual vapor pressures of both pentane and hexane add up to the total pressure exerted on the liquid by its surroundings.

Composition of the Vapor Above the Solution

Being able to calculate the total vapor pressure of a solution can be extremely useful; knowing the composition of the vapor above a solution can be just as important. Qualitatively, it is not hard to see that the vapor above a 1:1 molar pentane/hexane solution will be richer in pentane as a result of its greater volatility and vapor pressure. Quantitatively, we can predict the composition of the vapor above a solution, for which Raoult's law is valid, simply by knowing the vapor pressures of its volatile components and the composition of the liquid solution.

Here is an illustration of how it is done. Applying the ideal gas law to the mixture of gases above a solution of pentane and hexane, we have equation 4. The quantity $Y_{pentane}$ is the fraction of pentane molecules in the vapor above the solution.

$$Y_{pentane} = \frac{P_{pentane}}{P_{total}} \qquad (4)$$

A single expression for the total vapor pressure (equation 5) can be derived easily from equations 1, 2, and 3, because $X_{hexane} = 1.0 - X_{pentane}$

$$P_{total} = X_{pentane}(P°_{pentane} - P°_{hexane}) + P°_{hexane} \qquad (5)$$

FIGURE 11.3
Vapor pressure/mole fraction diagram for pentane/hexane solutions at 25°C.

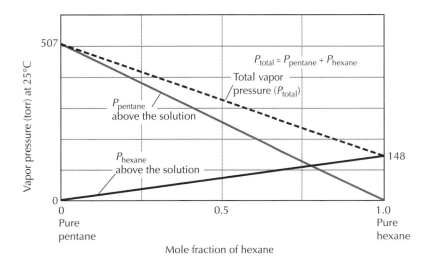

Finally, the combination of equations 1, 4, and 5 allows the calculation of the mole fraction of pentane in the vapor state.

$$Y_{pentane} = \frac{P°_{pentane} X_{pentane}}{X_{pentane}(P°_{pentane} - P°_{hexane}) + P°_{hexane}} \tag{6}$$

Temperature/ Composition Diagrams

If the vapor pressures of pure pentane and pure hexane at various temperatures and the composition of the liquid are known, the fraction of pentane in the vapor above the solution can be calculated. This kind of calculation can be used to construct a temperature/ composition diagram (sometimes called a phase diagram) like the one shown in Figure 11.4. A similar diagram can also be constructed directly from experimental data.

It is useful to follow the dashed line in Figure 11.4, moving from initial liquid composition L_1 to initial vapor composition V_1 to L_2, and so on. Point L_1 indicates a boiling point of 44°C at atmospheric pressure for a solution containing a 1:1 molar ratio of pentane to hexane. Analysis of the vapor composition at V_1 reveals a molar composition of 87% pentane and 13% hexane. The mole fraction of the component with the lower boiling point is greater in the vapor than in the liquid. Now, if the vapor at V_1 condenses, the liquid that collects (L_2) will have the same composition as the vapor (V_1). If the condensed liquid (L_2) is vaporized, the new vapor will be even richer in pentane, point V_2. Repeating the boiling and condensing processes several more times allows us to obtain essentially pure pentane.

Fractional and Simple Distillation

As pentane-enriched vapor is removed, the remaining liquid contains a decreasing proportion of pentane. The liquid, originally at L_1, now is richer in hexane (the component with the higher boiling point). As the mole fraction of hexane in the liquid increases, the boiling point of the liquid also increases until the boiling point of pure hexane, 69°C, is reached. In this way pure hexane can also be collected. The process of repeated vaporizations and condensations, called *fractional distillation,* allows us to separate components of a mixture by exploiting the vapor pressure differences of the components [see Technique 11.4].

FIGURE 11.4
Calculated temperature/composition diagram for pentane/ hexane solutions at 1.0 atm pressure.

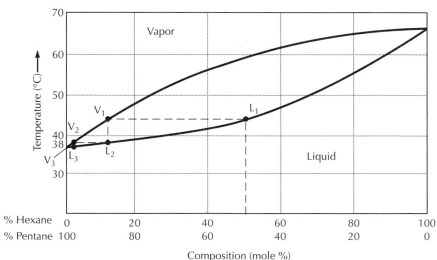

FIGURE 11.5
Distillation curve for
simple distillation of a
1:1 molar solution of
pentane and hexane.

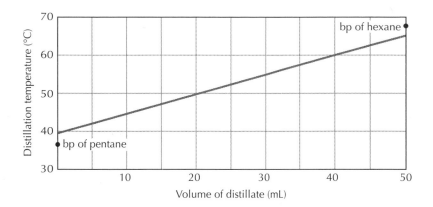

In a *simple distillation*, only one or perhaps two vaporizations and condensations occur, corresponding to points L_1 and V_1 in Figure 11.4. The condensed liquid is called the *distillate* or *condensate*. Simple distillation would not effectively separate a mixture such as a 1:1 molar solution of pentane and hexane. As the distillation proceeds, the remaining pentane/hexane solution becomes increasingly more concentrated in hexane and less concentrated in pentane. Consequently, the boiling point of the solution continues to increase. Figure 11.5 shows a distillation curve of vapor temperature versus volume of distillate for the simple distillation of a 1:1 pentane/hexane solution, which produces an incomplete separation of the two compounds. The initial distillate is collected at a temperature above the boiling point of pure pentane and the final distillate never reaches the boiling point of pure hexane.

11.3 Simple Distillation

Even though simple distillation does not effectively separate a mixture of liquids whose boiling points differ by less than 60°–70°C, organic chemists use simple distillations in two commonly encountered situations: (1) when the last step in the purification of a liquid compound involves a simple distillation to obtain the pure product and determine its boiling point; (2) when simple distillation is used to remove a low-boiling-point solvent from a dissolved organic compound with a high boiling point.

In a simple distillation, **the distilling flask should be filled only one-third to one-half full of the liquid being distilled.** With a flask that is too full, liquid can easily bump over into the condenser. If the flask is nearly empty, a substantial fraction of the material will be needed just to fill the flask and distilling head with vapor, along with a thin liquid film on the glass surfaces. When the desired liquid is dissolved in a large quantity of a solvent with a lower boiling point, the distillation should be interrupted after the solvent has distilled, and the liquids with higher boiling points should be poured into a smaller flask before continuing the distillation. Figure 11.6 shows the miniscale apparatus for a simple distillation.

FIGURE 11.6
Simple distillation apparatus. The enlargement shows the correct placement of the thermometer bulb for accurate measurement of the boiling point.

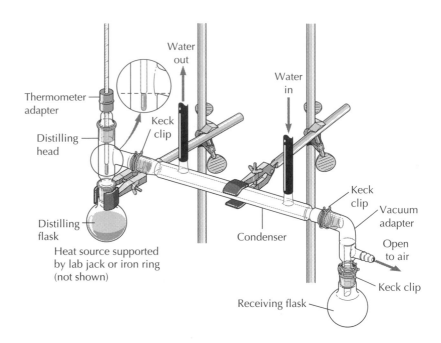

Temperature Measurement

Position of the thermometer. Correct positioning of the thermometer in the distilling head is essential for obtaining accurate temperature readings during a distillation. The enlargement in Figure 11.6 shows the correct placement of the thermometer bulb. **The thermometer is positioned correctly when the top of the thermometer bulb is aligned with the bottom of the distilling-head side arm.** This position allows the bulb to be completely surrounded by vapor and the thermometer to register the correct temperature of the vapor just before it exits into the condenser.

Accuracy of the thermometer. The accuracy of a boiling-point determination is no better than the accuracy of the thermometer. It cannot be assumed that the thermometer has been accurately calibrated. Although frequently this is the case, it is not always true. Thermometers may give high or low temperature readings of $1°–2°$ or more. Thermometer calibration is discussed in Technique 10.5, page 124.

Types of thermometers. Several types of thermometers are available. The *mercury thermometer* was the traditional type of thermometer found in a chemistry laboratory. Concern for the environment, the hazards of cleaning up a mercury spill from a broken thermometer, and the cost of waste disposal for mercury have led to the use of other types of temperature measuring devices, namely nonmercury thermometers and digital thermometers with temperature probes. The *nonmercury thermometers* (spirit or alcohol filled) found in undergraduate laboratories do not always give temperature readings sufficiently accurate for a distillation; we do not recommend nonmercury thermometers for distillations. The use of a *Teflon-coated metal temperature probe* with a digital thermometer is another alternative to a mercury thermometer. Uncoated metal probes can react with hot vapor and are not recommended for distillations. Consult your instructor before using a digital thermometer.

The length of the probe that is placed below the side arm needs to be determined experimentally by a series of distillations using pure compounds. Consult your instructor about the placement of a temperature probe in the distilling head.

Steps in Assembling a Miniscale Apparatus for Simple Distillation

A funnel keeps the ground glass joint from becoming coated with the liquid and prevents loss of product.

1. Select a round-bottomed flask of a size that will be one-third to one-half filled with the liquid being distilled. Place a clamp firmly on the neck of the flask and attach the clamp to a ring stand or support rod. Using a conical funnel, pour the liquid into the flask. Add one or two boiling stones.

SAFETY PRECAUTION

Boiling stones should **never** be added to a hot liquid because this may cause a superheated liquid to boil violently out of the flask.

2. *Lightly* grease the bottom joint and the side-arm joint [see Technique 2.2] on the distilling head. Fit the distilling head to the round-bottomed flask and twist the joint to achieve a tight seal. Finish assembling the rest of the apparatus before inserting the thermometer adapter and thermometer. **Note: The distilling flask and distilling head need to be in a completely vertical position and the condenser should be positioned with a downward slant.**

3. Attach rubber tubing to the outlets on the condenser jacket. Wire hose clamps are often used to prevent water hoses from being blown off the outlets by a surge in water pressure. Grease the inner joint at the bottom of the condenser, attach the vacuum adapter, and while the pieces are lying on the desktop, place a Keck clip over the joint. Clamp the condenser to another ring stand or upright support rod, as shown in Figure 11.6. If the clamp used to support the condenser has a stationary and a movable jaw, position it with the stationary jaw underneath the condenser and the movable jaw above the condenser. Fit the upper joint of the condenser to the distilling head, twist to spread the grease, and place a Keck clip over the joint.

The use of Keck clips ensures that ground glass joints do not come apart.

4. Figure 11.6 shows a round-bottomed flask serving as the receiving vessel. Depending on the particular procedure being carried out, an Erlenmeyer flask or a graduated cylinder may be substituted for the round-bottomed flask. Position the Erlenmeyer flask or graduated cylinder so that the outlet of the vacuum adapter is slightly inside the mouth of the receiving vessel. A beaker should not be used as the receiving vessel because its wide opening readily allows vapor (product) to escape. It is usually necessary to have at least two receiving vessels at hand; the first container is for collecting the initial distillate that consists of impurities with low boiling points before the expected boiling point of the desired fraction is attained.

5. Gently push the thermometer through the rubber sleeve on the thermometer adapter. Alternatively, a thermometer with a

standard taper fitting may be used instead of the thermometer and rubber-sleeved adapter.

SAFETY PRECAUTION

Grasp the thermometer close to the bulb and push it gently 1–2 cm into the adapter. Move your hand several centimeters up the thermometer stem and repeat the pushing motion. Continue this process until the thermometer is properly positioned. Holding the thermometer by the upper part of the stem while inserting it through the rubber sleeve of the thermometer adapter could break the thermometer and force a piece of broken glass into your hand.

The position of the thermometer bulb is crucial to obtaining an accurate boiling point.

A slow to moderate flow rate for the condenser water is usually sufficient and lessens the chance of blowing the rubber tubing off the condenser.

6. Grease the joint on the thermometer adapter and fit it into the top joint of the distilling head. Adjust the position of the thermometer to **align the top of the thermometer bulb with the bottom of the side arm** on the distilling head (see detail in Figure 11.6).
7. Check to ensure that the rubber tubing is tightly attached to the condenser and that **water flows in at the bottom and out at the top.** Slowly turn on the water.
8. Place a heating mantle under the distillation flask, using an iron ring or lab jack to support the mantle, and begin heating the flask.

Carrying Out the Distillation

The expected boiling point of the liquid being distilled determines the heat input, controlled by a variable transformer [see Technique 6.2]; vaporization of a liquid with a high boiling point requires more heat than does a liquid with a low boiling point. Heat the liquid slowly to a gentle boil. A ring of condensate will begin to move up the inside of the flask and then up the distilling head. The temperature observed on the thermometer will not rise appreciably until the ring of vapor reaches the thermometer bulb because it is measuring the vapor temperature, not the temperature of the boiling liquid. If the ring of vapor stops moving before it reaches the thermometer, increase the setting on the variable transformer.

When the vapor reaches the thermometer, the temperature reading should increase rapidly. Collect any liquid that condenses below the expected boiling point as the first fraction, or forerun—which is usually discarded—then change to a second receiving vessel to collect the desired fraction when the temperature stabilizes at or slightly below the expected boiling point of the liquid. Record the temperature at which you begin to collect the desired fraction. Adjust the heat input to maintain a distillate collection rate of 1 drop every 1–2 s. It may be necessary to increase the heat input during the distillation if the rate of distillate collection slows.

It is essential to stop the distillation by lowering the heat source before the distillation flask reaches dryness or when the temperature either begins to climb above the expected boiling range or begins to drop. Record the temperature at which the last drop of distillate is collected; the initial and final temperatures for the main fraction are the boiling range.

SAFETY PRECAUTION

A distillation flask should never be allowed to reach dryness. By leaving a small residue of liquid in the boiling flask, you will not overheat the flask and break it, nor will you char the last drops of residue, which causes cleaning difficulty. Moreover, some compounds, such as ethers, secondary alcohols, and alkenes, form peroxides by air oxidation. If a distillation involving one of these compounds is carried to dryness, the peroxides could explode.

11.3a Miniscale Short-Path Distillation

When only 4–6 mL of liquid are distilled, a simple distillation apparatus can be modified to a short path by omitting the condenser, as shown in Figure 11.7. The short path reduces the *holdup volume,* the amount of space that must be filled with vapor, and also prevents distillate (product) from being lost on the walls of the condenser. A beaker or crystallizing dish of water surrounding the receiving flask replaces the condenser. If the liquid boils between 50°C and 100°C, the beaker should contain an ice/water mixture; in this case, it may be necessary to attach a drying tube to the side arm of the vacuum adapter to prevent moisture from condensing inside the receiving flask. If the liquid boils above 100°C, tap water provides sufficient cooling. For liquids that boil above 150°C, air cooling of the receiving flask suffices.

FIGURE 11.7
Short path distillation apparatus.

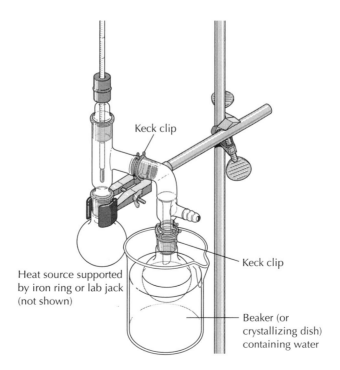

Keck clip

Keck clip

Heat source supported by iron ring or lab jack (not shown)

Beaker (or crystallizing dish) containing water

Carry out the distillation as described in Technique 11.3 for a simple distillation, but do the short-path distillation at a rate of less than 1 drop per second. If the receiving flask is being cooled by a water bath, it may be necessary to stop the distillation by removing the heat source while changing receiving flasks.

11.3b Microscale Distillation Using Standard Taper Apparatus

**Distillation Using a
ᵀ 14/10 Apparatus**

Microscale apparatus is required when the volume of a liquid to be distilled is only 1–3 mL. Standard taper microscale glassware can be assembled into a short-path distillation apparatus with a 14/10 distillation head, a thermometer adapter (Figure 11.8), and a bent vacuum adapter, as shown in Figure 11.9. For the distilling vessel, use a conical vial for 1–3 mL of liquid or a 10-mL round-bottomed flask for 4–5 mL of liquid. For distillation of very volatile liquids, a water-jacketed condenser can be inserted between the distilling head and the vacuum adapter in the same manner as it is used in a miniscale distillation apparatus.

**Using a Hickman
Distilling Head**

Another type of standard taper microscale distillation apparatus consists of a Hickman distilling head (Figure 11.10) and a 3-mL or 5-mL conical vial or a 10-mL round-bottomed flask. The Hickman distilling head also serves as the receiving vessel, an arrangement that considerably reduces the holdup volume. Vapors condense on the upper portion of the Hickman still and drain into the bulbous collection well. One version of the Hickman still has a port at the side for easy removal of the condensate (Figure 11.10a).

Setting up the apparatus. To carry out a microscale distillation, select a conical vial or 10-mL round-bottomed flask appropriate for the volume of liquid to be distilled; the vessel should be no more than two-thirds full. Use a Pasteur pipet to place the liquid in the vial and add a magnetic spin vane or a boiling stone. Attach the Hickman distilling head to the vial with a screw cap and O-ring. Usually an air

FIGURE 11.8
Thermometer adapters for 14/10 microscale glassware.

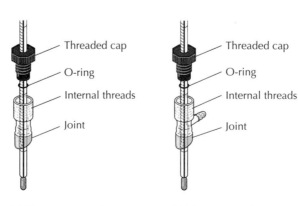

(a) Thermometer adapter   (b) Thermometer/vacuum

FIGURE 11.9
Short-path standard taper microscale distillation apparatus.

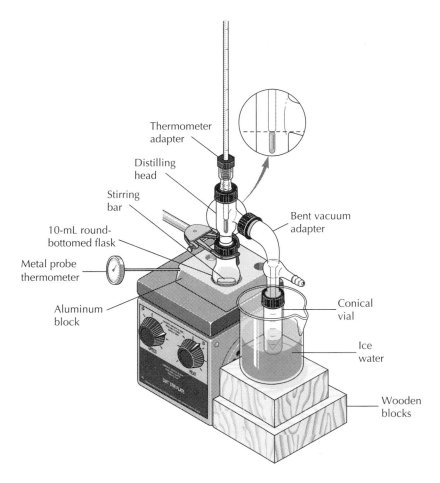

Thermometer adapter

Distilling head

Stirring bar

10-mL round-bottomed flask

Metal probe thermometer

Aluminum block

Bent vacuum adapter

Conical vial

Ice water

Wooden blocks

Grease is not used on ground glass joints of microscale glassware because its presence could contaminate the product.

It may be necessary to wrap the distillation vial loosely with glass wool to prevent rapid heat loss, but do not wrap the well of the Hickman distilling head.

condenser or a water-cooled condenser (for particularly volatile liquids) is placed above the Hickman distilling head to minimize the loss of vapor (Figure 11.11). Clamp the assembled apparatus at the Hickman distilling head, and place the vial in an aluminum heating block. If you are using a spin vane, turn on the magnetic stirrer.

Carrying out the distillation. Begin heating the aluminum block slowly to a temperature 20°–30°C above the boiling point of the liquid being distilled. Position a thermometer inside the condenser and the Hickman distilling head, with the top of the thermometer bulb aligned with the bottom of the head's collection well, as shown in Figure 11.11. Clamp the thermometer firmly above the condenser.

FIGURE 11.10
Hickman distilling heads. The condensate collects in the well at the bottom of the head in both versions.

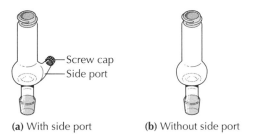

Screw cap
Side port

(a) With side port   **(b)** Without side port

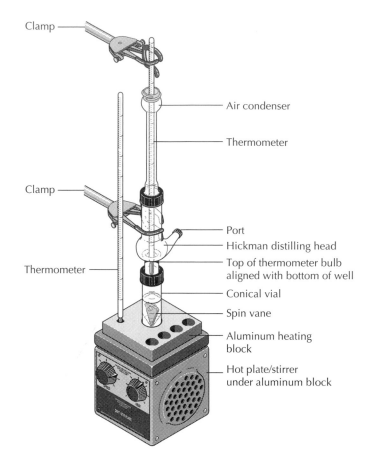

FIGURE 11.11
Standard taper apparatus for a microscale distillation using a Hickman distilling head with a side port.

Clamp

Air condenser

Thermometer

Clamp

Port

Hickman distilling head

Thermometer

Top of thermometer bulb aligned with bottom of well

Conical vial

Spin vane

Aluminum heating block

Hot plate/stirrer under aluminum block

Removing the distillate. After the liquid in the vial boils, you should notice a ring of condensate slowly moving up the vial and into the Hickman distilling head. The temperature observed on the thermometer rises as the vapor reaches the thermometer bulb. You may also see the upper neck of the Hickman distilling head become wet and shiny as the vapor condenses and begins to fill the well. The distillation should be done at a rate slow enough to allow the vapor to condense and not evaporate out of the condenser.

The collection well has a capacity of about 1 mL, so the distillate may need to be removed once or twice during a distillation. Open the port and quickly remove the distillate with a clean Pasteur pipet. Alternatively, withdraw the distillate using a syringe inserted through the plastic septum in the screw cap of the port.

11.3c Microscale Distillation Using Williamson Apparatus

Microscale apparatus is required when the volume of liquid to be distilled is only a few milliliters. The Williamson microscale distillation apparatus is essentially a miniature version of the standard taper short-path distillation apparatus [see Technique 11.3]. The apparatus consists of a 5-mL or 10-mL round-bottomed flask and a distillation head connected by a flexible connector with a support rod.

The thermometer is held in place by the flexible thermometer adapter, as shown in Figure 11.12. The distillate is collected in a small vial that is at least three-fourths submerged in a 50-mL beaker of ice and water.

Assembling the Apparatus

Select a 5-mL or 10-mL round-bottomed flask appropriate for the volume of liquid to be distilled; the flask should be no more than one-half full. Using a Pasteur pipet, transfer the liquid to the flask and add a magnetic stirring bar or a boiling stone. Attach the flexible connector with support rod to the flask and clamp the rod to a vertical support rod or ring stand.

Fit the flexible thermometer adapter to the top of the distilling head and carefully push a thermometer through the adapter.

SAFETY PRECAUTION

Grasp the thermometer close to the bulb and push it gently 1–2 cm into the adapter. Move your hand several centimeters up the thermometer stem and repeat the pushing motion; continue this process until the thermometer is properly positioned. Holding the thermometer by the upper part of the stem while inserting it through the rubber sleeve of the thermometer adapter could break the thermometer and force a piece of broken glass into your hand.

Place the top of the thermometer bulb just below the side arm, as shown by the dashed line drawn across the distillation head in Figure 11.12. Fit the distillation head into the flexible connector holding the distillation flask. Place the receiving vial in a 50-mL beaker of ice and water, and position the vial under the outlet of the distillation head as far as it will go. Put a sand bath or an aluminum heating block with a flask depression under the round-bottomed flask. The temperature of the sand bath or aluminum block needs to be 20°–50°C above the boiling point of the liquid being distilled.

FIGURE 11.12
Williamson microscale distillation apparatus.

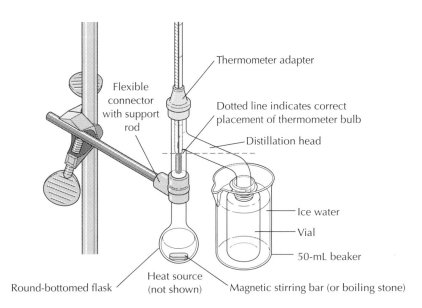

Thermometer adapter

Flexible connector with support rod

Dotted line indicates correct placement of thermometer bulb

Distillation head

Ice water

Vial

50-mL beaker

Round-bottomed flask

Heat source (not shown)

Magnetic stirring bar (or boiling stone)

Carrying Out the
Distillation

After the liquid in the flask boils, you should notice a ring of condensate slowly moving up the flask and into the distillation head. The temperature observed on the thermometer rises as the vapor reaches the thermometer bulb. The distillation should be done at a rate slow enough for the vapor to condense and not evaporate out of the system. It may be necessary to wrap a wet pipe cleaner or a wet paper towel around the side arm of the distillation head to increase its cooling efficiency, particularly for the distillation of compounds boiling below 100°C.

11.4   Fractional Distillation

Fractionating
Columns

In a *fractional distillation,* repeated vaporizations and condensations separate the components of a mixture based on their vapor pressure differences. A *fractionating column* is inserted between the distillation flask and the distilling head of a simple distillation apparatus. The use of a fractionating column in the distillation apparatus provides a large surface area over which a number of separate liquid-vapor equilibria can occur. As vapor travels up a column, it cools, condenses into a liquid, then vaporizes again after it comes into contact with hotter vapor rising from below. The process can be repeated many times. If the fractionating column is efficient, the vapor that finally reaches the distilling head at the top of the column is composed entirely of the component with the lower boiling point.

Efficiency of a fractionating column.   The efficiency of a fractionating column is expressed as its number of *theoretical plates*—a term best defined with the help of Figure 11.4 (page 132). Assume that the original solution being distilled has a 1:1 molar ratio of pentane to hexane. A fractionating column would have one theoretical plate if the liquid that distills from the top of the column has the composition L_2. In other words, a column has one theoretical plate if one complete vaporization of the original solution, followed by condensation of the vapor, occurs in the column.

The column would have two theoretical plates if the liquid that distills has the composition L_3; notice that L_3 is already 98% pentane and only 2% hexane. Figure 11.4 indicates that a column with three theoretical plates would seem sufficient to obtain essentially pure pentane from the 1:1 pentane/hexane mixture present at the start of the distillation. However, as the distillation progresses, the residue in the boiling flask becomes richer in hexane, so a few more theoretical plates are required for complete separation of the two compounds.

Types of fractionating columns.   Fractionating columns that can be used to separate two liquids boiling at least 25°C apart are shown in Figure 11.13. The larger the column surface area on which liquid-vapor equilibria can occur, the more efficient the column will be. The fractionating columns shown in Figure 11.13 have from two to eight theoretical plates. A fractionating column with two theoretical plates can be used to separate liquids with boiling

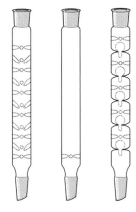

FIGURE 11.13
Examples of fractionating columns.

points differing by about 70°C; an eight-theoretical-plate column can separate liquids boiling only 25°C apart.

Increasing the efficiency of fractionating columns.   A more efficient column can be made by packing a simple fractionating column with a wire spiral, glass helices, metal sponge, or thin metal strips. These packings provide additional surface area on which liquid-vapor equilibria can occur. Care must be used in selecting packing materials to ensure that the packing does not undergo chemical reaction with the hot liquids in the fractionating column. Among the most efficient fractionating columns are those with helical bands of Teflon mesh that spin at thousands of rotations per minute. Spinning-band columns can have more than 150 theoretical plates and can be used to separate liquids that have a boiling-point difference of only a degree or two.

Composition of the Vapor

In a fractional distillation the composition of the vapor phase at the top of the column, as it encounters the thermometer bulb and moves into the side arm, determines the composition of the liquid that forms in the condenser and collects in the receiving flask. For example, consider the fractional distillation of the 1:1 molar solution of pentane and hexane described in Technique 11.2. If the fractionating column has enough theoretical plates to completely separate the two compounds, the initial condensate will appear when the temperature is very close to 36°C, the boiling point of pure pentane. The observed boiling point will remain essentially constant at 36°C while all the pentane distills into the receiving vessel. Then the boiling point will rise rapidly to 69°C, the boiling point of hexane. Figure 11.14 shows a distillation curve for the fractional distillation of pentane and hexane. The abrupt temperature increase in boiling point at approximately 22–24 mL of distillate demonstrates an efficient fractional distillation. On the contrary, a steady increase in the distillation temperature during a simple distillation indicates that incomplete separation of pentane and hexane has occurred.

Miniscale Fractional Distillation Apparatus

The distilling flask capacity should be two to three times the volume of liquid being distilled. When the desired material is contained in a large quantity of a solvent with a lower boiling point, the distillation should be interrupted after the solvent has distilled, and the liquids with higher

FIGURE 11.14
Distillation curve for the fractional distillation of a 1:1 molar solution of pentane and hexane. The dashed line represents the distillation curve for a simple distillation of the same solution.

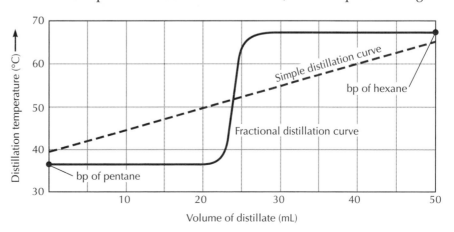

FIGURE 11.15
Miniscale fractional
distillation apparatus.
The fractionating
column is inserted
between the distilling
flask and the distilling
head.

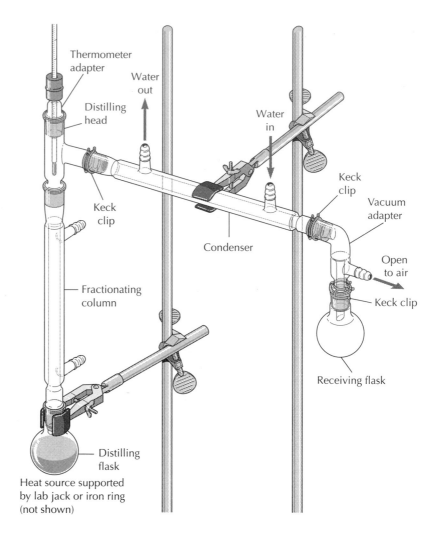

boiling points (the solution that remains in the boiling flask) should
be transferred to a smaller flask before continuing the distillation.

Figure 11.15 shows the apparatus for a fractional distillation.
Follow the steps listed in Technique 11.3 for assembling a simple
distillation apparatus with the addition of the fractionating column
between the distilling flask and the distilling head. Be sure to add
one or two boiling stones to the distilling flask and that the thermometer is placed correctly, as shown in the detail in Figure 11.6.

*Carrying Out
a Fractional
Distillation*

Rate of distillation.   The rate of distillation is always a compromise
between the speed of the distillation and the efficiency of the frac-
tionation. For an easy separation, 1–2 drops per second can be collected. Generally a slow, steady distillation where 1 drop is collected
every 2–3 s is a more reasonable rate. Difficult separations (when the
boiling points of the distilling compounds are close together) require
a slower distillation rate as well as a more efficient fractionating
column—one with more theoretical plates. The distillation rate can
be increased during collection of the last fraction, when all the compounds with lower boiling points have already been distilled.

Rate of heating.   Heat the distilling flask slowly. Control of heating in a fractional distillation is extremely important; the rate of heating needs to be increased gradually as the distillation proceeds. However, applying too much heat causes the distillation to occur so quickly that the repeated liquid-vapor equilibria required to bring about maximum separation on the surfaces of the fractionating column cannot occur. On the other hand, if too little heat is applied, the column may lose heat faster than it can be warmed by the vapor, thus preventing the vapor from reaching the top of the column. Therefore, too little heat during the distillation causes the thermometer reading to drop below the boiling point of the liquid, simply because vapor is no longer reaching the thermometer bulb. The thermometer temperature may also drop during a fractional distillation when a compound with a lower boiling point has completely distilled and not enough heat is being supplied to force the vapor of the compound with the next-higher boiling point up to the top of the column. The addition of more heat corrects this situation.

Collecting the Fractions

You will need a labeled receiving container (round-bottomed flask, vial, or Erlenmeyer flask) for each fraction you plan to collect. The cutoff points for the fractions are the boiling points (at atmospheric pressure) of the substances being separated. For example, in a fractional distillation of the 1:1 molar solution of pentane (bp 36°C) and hexane (bp 69°C) described in Figure 11.14, the first fraction would be collected when the temperature at the distilling head reached 35°–36°C. The temperature would stay at 36°C for a period of time while the pentane distilled.

Eventually the temperature either rises or drops several degrees; the latter change indicates that there is no longer enough pentane vapor to maintain the boiling-point temperature at the thermometer bulb. At this point, increase the heat input and change to the second receiving flask. Liquid then begins to distill again. Leave the second receiver in place until the temperature reaches 69°C, the boiling point of hexane; then change to the third receiving flask. The second receiver should contain only a small amount of distillate. Continue collecting fraction 3 (hexane) until only 1–2 mL of liquid remain in the boiling flask.

S A F E T Y   P R E C A U T I O N

A distillation flask should **never** be allowed to boil dry.

Summary of a Miniscale Fractional Distillation Procedure

1.   Use a round-bottomed flask that has a capacity two or three times the volume of the liquid mixture you wish to distill. Clamp the flask to a ring stand or upright support rod. Pour the liquid into the flask and add one or two boiling chips.
2.   Set up the rest of the apparatus as shown in Figure 11.15.
3.   Heat the mixture to boiling and collect the distillate in fractions based on the boiling points of the individual components in the mixture. Use a separate receiving container for each fraction.

11.5 Azeotropic Distillation

The systems described up to this point are solutions whose compounds interact only slightly with one another and thus approximate the behavior of an ideal solution. As described in Technique 11.2, the behavior of such solutions follows Raoult's law:

$$P = P° X$$

where P = partial pressure of a compound, $P°$ = its vapor pressure, and X = mole fraction of the compound in the liquid phase.

Most liquid solutions, however, deviate from ideality. The deviations result from intermolecular interactions (for example, hydrogen bonding) in the liquid state. In the distillation of some solutions, mixtures that boil at a constant temperature are produced. Such constant-boiling mixtures cannot be further purified by distillation and are called *azeotropes,* or *azeotropic mixtures.*

One of the best-known binary mixtures that forms an azeotropic mixture during distillation is the ethanol/water system, shown in Figure 11.16. The azeotrope boils at 78.2°C and consists of 95.6% ethanol and 4.4% water by weight. Liquid that has the azeotropic composition will vaporize to a gas that has exactly the same composition because the liquid and vapor curves intersect at this point (see Figure 11.16). No matter how many more liquid-vapor equilibria take place as the materials travel up the column, no further separation will occur. Continued distillation never yields a liquid that contains more than 95.6% ethanol. Pure ethanol must be obtained by other means.

More detailed discussion about the formation of azeotropic distillation mixtures from nonideal solutions can be found in the references at the end of the chapter. Extensive tables of azeotropic data are available in references such as the *CRC Handbook of Chemistry and Physics.* Table 11.2 lists a few azeotropes formed by common solvents.

FIGURE 11.16
Temperature/composition diagram for ethanol/water solutions at 1.0 atm pressure. The mixture of 95.6% ethanol and 4.4% water is an azeotrope with a boiling-point minimum.

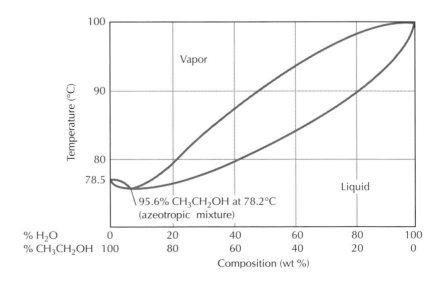

Vapor

Temperature (°C)

95.6% CH_3CH_2OH at 78.2°C
(azeotropic mixture)

Liquid

| % H_2O | 0 | 20 | 40 | 60 | 80 | 100 |
| % CH_3CH_2OH | 100 | 80 | 60 | 40 | 20 | 0 |

Composition (wt %)

T A B L E 1 1 . 2	Azeotropes formed by common solvents			
Component X (bp)	% by wt	Component Y (bp)	% by wt	Azeotrope bp
Water (100)	13.5	Toluene (110.7)	86.5	84.1
Water (100)	1.4	Pentane (36.1)	98.6	34.6
Methanol (64.7)	12.1	Acetone (56.1)	87.9	55.5
Methanol (64.7)	72.5	Toluene (110.7)	27.5	63.5
Ethanol (78.3)	68	Toluene (110.7)	32	76.7
Water (100)	1.3	Diethyl ether (34.5)	98.7	34.2

11.6 Vacuum Distillation

Many organic compounds decompose at temperatures below their atmospheric boiling points. These compounds can be distilled at temperatures lower than their atmospheric boiling points when a partial vacuum is applied to the distillation apparatus. Distillation at reduced pressure, called *vacuum distillation,* takes advantage of the fact that the boiling point of a liquid is a function of the pressure under which the liquid is contained [see Technique 11.1]. Although vacuum distillation is inherently less efficient than fractional distillation at atmospheric pressure, it is often the only feasible way to distill compounds with boiling points above 200°C.

A partial vacuum can be obtained in the laboratory with either a vacuum pump or a water aspirator. Vacuum pumps can easily produce pressures of less than 0.5 torr. The pressure obtained with a water aspirator can be no lower than the vapor pressure of water, which is 13 torr at 15°C and sea level. In practice, an efficient water aspirator produces a partial vacuum of 15–25 torr.

The boiling point of a compound at any given pressure other than 760 torr is difficult to calculate exactly. As a rough estimate, a 50% drop in pressure lowers the boiling point of an organic liquid 15°–20°C. Below 25 torr, reducing the pressure by one-half lowers the boiling point approximately 10°C (Table 11.3).

A nomograph provides a better way of estimating the boiling points of relatively nonpolar compounds at either reduced or atmospheric pressure (Figure 11.17). For example, if the boiling point of a compound at 760 torr is 200°C and the vacuum distillation is

T A B L E 1 1 . 3	Boiling points (°C) at reduced pressures		
Pressure (torr)	Water	Benzaldehyde	Diphenylether
760	100	179	258
100	51	112	179
40	34	90	150
20	22	75	131

being done at 20 torr, the approximate boiling point is found by aligning a straightedge on 200 in column B with 20 in column C; the straightedge intersects column A at 90°C, the approximate boiling point of the compound at 20 torr, as shown by the line on Figure 11.17. Similarly, the boiling point at atmospheric pressure can be estimated if the boiling point at a reduced pressure is known. By aligning the boiling point in column A with the pressure in column C, a straightedge intersects column B at the approximate atmospheric boiling point. The graph gives a less than accurate estimate of boiling points for polar compounds that associate strongly in the liquid phase.

Apparatus for Miniscale Vacuum Distillation

The vacuum distillation apparatus shown in Figure 11.18 works adequately for most vacuum distillations, although a fractionating column may be needed to provide satisfactory separation of some mixtures. Because liquids often boil violently at reduced pressures, a Claisen connecting adapter is always used in a vacuum distillation to lessen the possibility of liquid bumping up into the condenser. If undistilled material jumps through the Claisen adapter into the condenser, you must begin the distillation again.

If a satisfactory vacuum is to be maintained, each connecting surface must be completely greased with high-vacuum silicone grease, and the rubber tubing to the aspirator or vacuum pump must be thick-walled so that it does not collapse. If the partial vacuum is not as low as expected, carefully check all connections for possible leaks.

FIGURE 11.17
Nomograph for estimating boiling points at different pressures.

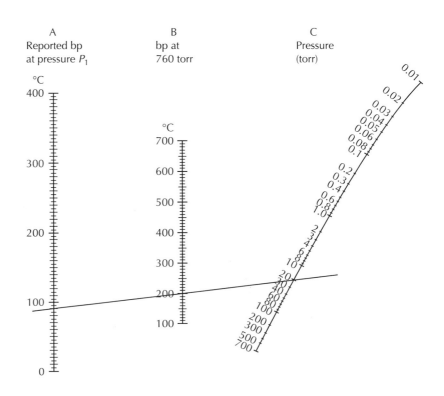

FIGURE 11.18
Vacuum distillation
apparatus.

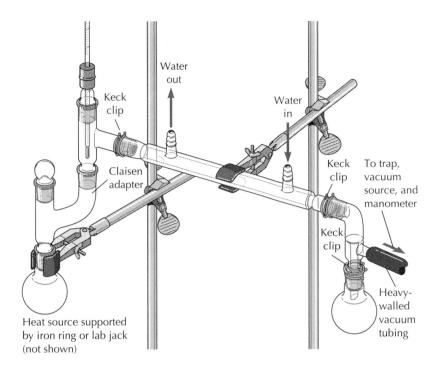

Water out

Keck clip

Water in

Claisen adapter

Keck clip

To trap, vacuum source, and manometer

Keck clip

Heavy-walled vacuum tubing

Heat source supported
by iron ring or lab jack
(not shown)

Uncontrolled bumping during a vacuum distillation can be lessened by using a large distillation flask, by adding small pieces of wood splints in place of boiling stones, or by magnetic stirring. Instead of using these methods, a very finely drawn-out Pasteur pipet capillary tip can provide a steady stream of very small bubbles (Figure 11.19). The bottom of the capillary tube

FIGURE 11.19
Vacuum distillation
apparatus fitted with
a capillary bubbler.

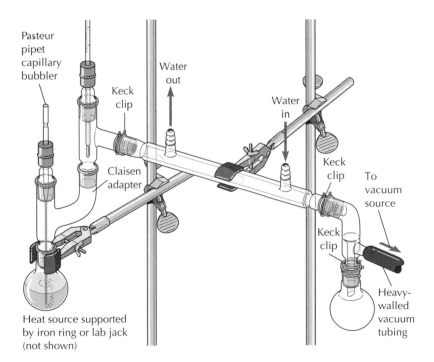

Pasteur pipet capillary bubbler

Keck clip

Water out

Water in

Claisen adapter

Keck clip

To vacuum source

Keck clip

Heavy-walled vacuum tubing

Heat source supported
by iron ring or lab jack
(not shown)

bubbler should be just above the bottom surface of the distilling flask and must always be below the liquid's surface. Do not use wood splints or boiling stones when you use a capillary bubbler; their violent motions may break the fragile tip of the bubbler, making it useless. A capillary bubbler should not be used with air-sensitive compounds unless an inert gas is fed into the top of the Pasteur pipet.

Standard Taper Microscale Apparatus for Vacuum Distillation

The well in a Hickman distilling head has a capacity of only 1 mL.

For a volume of 2–5 mL of liquid, a 10-mL round-bottomed flask and the microscale 14/10 apparatus shown in Figure 11.20 can be used for a vacuum distillation. If the volume of liquid to be distilled is less than 2 mL, the microscale apparatus shown in Figure 11.21 can be used for a vacuum distillation.

The ground glass joints of microscale glassware should not be greased. Usually clean standard taper joints are completely sealed by compression of the O-ring when the cap is screwed down tightly. Only if the requisite reduced pressure cannot be obtained should microscale joints be greased with high-vacuum silicone grease. Care must be exercised to use a very thin film of grease applied only at the top of the inner joints. No grease should be allowed to seep from the bottom of any joint because the grease might contaminate the liquid being distilled.

FIGURE 11.20
Short-path standard taper microscale apparatus for vacuum distillation.

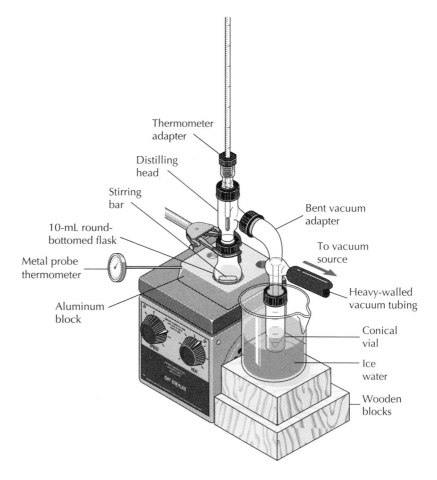

Thermometer adapter

Distilling head

Stirring bar

10-mL round-bottomed flask

Metal probe thermometer

Aluminum block

Bent vacuum adapter

To vacuum source

Heavy-walled vacuum tubing

Conical vial

Ice water

Wooden blocks

FIGURE 11.21
Standard taper
microscale apparatus
for distillation with
a Hickman distilling
head.

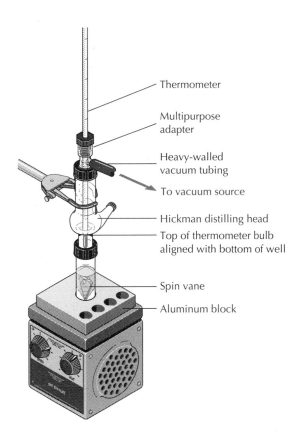

Thermometer

Multipurpose
adapter

Heavy-walled
vacuum tubing

To vacuum source

Hickman distilling head
Top of thermometer bulb
aligned with bottom of well

Spin vane
Aluminum block

*Monitoring the
Pressure During a
Vacuum Distillation*

The pressure can be continuously monitored with a manometer
(Figure 11.22) or read periodically with a McLeod gauge
(Figure 11.23). If a water aspirator is used as the source of the
vacuum, a trap bottle or flask must be used to prevent any backflow of water from entering the distillation apparatus. When a
vacuum pump is used, a cold trap, kept at the temperature of

FIGURE 11.22
Two types of closed-
end manometers used
in vacuum distillation.

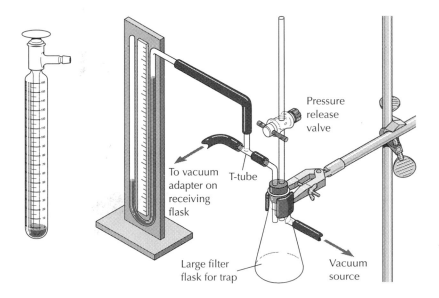

Pressure
release
valve

To vacuum
adapter on
receiving
flask

T-tube

Large filter
flask for trap

Vacuum
source

isopropyl alcohol/dry ice (−77°C) or liquid nitrogen (−196°C), must be placed between the distillation system and the pump. The trap collects any volatile materials that could otherwise get into the pump oil and cause a rise in the vapor pressure of the oil, which would decrease the efficiency of and possibly damage the pump. A pressure relief valve serves to close the system from the atmosphere and to release the vacuum after the system has cooled following the distillation. Consult your instructor before you do a distillation using a vacuum pump.

A McLeod gauge is used to measure pressures below 5 torr. It works by compressing the gas inside the gauge into a closed capillary tube with a pressure great enough to be measured with a mercury column. Initially the gauge must be in the horizontal, resting position with the mercury in the reservoir (Figure 11.23a). When the pressure inside the distillation apparatus has stabilized, the gauge is slowly rotated until the open-ended reference capillary tube is in the vertical position. The pressure is indicated by the scale on the closed-end capillary tube when the mercury level in the reference capillary tube reaches the calibration mark (Figure 11.23b). After the pressure has been read, the gauge must be returned to the horizontal, resting position.

Steps in a Miniscale Vacuum Distillation

S A F E T Y   P R E C A U T I O N

Safety glasses must be worn at all times while carrying out a vacuum distillation because of the danger of an implosion, which can shatter the glassware.

FIGURE 11.23
McLeod gauge used in vacuum distillation.

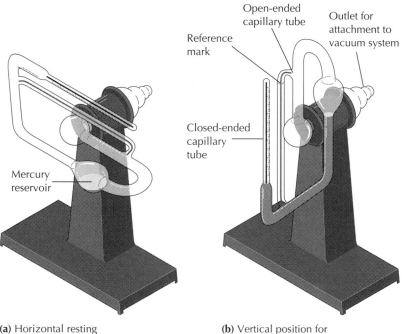

(a) Horizontal resting position

(b) Vertical position for reading pressure

1. Add the liquid to be distilled to a round-bottomed flask sized so that it will be less than half filled. Add some wood splints or a magnetic stirring bar and set up the apparatus as shown in Figure 11.18, or use a capillary bubbler, as shown in Figure 11.19.
2. Attach a trap and a manometer (Figure 11.22) or a McLeod gauge (Figure 11.23) to the system and connect the apparatus to the vacuum source.
3. Close the pressure release valve and turn on the vacuum.
4. When the vacuum has reached an appropriate level, heat the distilling flask cautiously to obtain a moderate distillation rate. Periodically monitor the pressure during the distillation.
5. When the distillation is complete, remove the heat source and allow the apparatus to cool nearly to room temperature before allowing air into the apparatus. Turn off the aspirator or vacuum pump only after the vacuum is broken. If you have used a cold trap, empty its contents immediately.

11.7   Steam Distillation

Codistillation with water, called *steam distillation*, allows distillation of relatively nonvolatile organic compounds without complex vacuum systems. Steam distillation can be thought of as a special kind of azeotropic distillation; it is especially useful for separating volatile organic compounds from nonvolatile inorganic salts or from the leaves and seeds of plants. Indeed, the process has found wide application in the flavor and fragrance industries as a means of separating essences or flavor oils from plant material.

Mutual Insolubility and Vapor Pressure

Steam distillation depends on the mutual insolubility or immiscibility of many organic compounds with water. In such two-phase systems at any given temperature, each of the two components exerts its own full vapor pressure. The total vapor pressure above the two-phase mixture is equal to the sum of the vapor pressures of the pure components independent of their relative amounts.

Consider the codistillation of iodobenzene (bp 188°C) and water (bp 100°C). The vapor pressures ($P°$) of both substances increase with temperature, but the vapor pressure of water will always be higher than that of iodobenzene because water is more volatile. At 98°C,

$$P°_{\text{iodobenzene}} = 46 \text{ torr}$$

$$P°_{\text{water}} = 714 \text{ torr}$$

$$P°_{\text{iodobenzene}} + P°_{\text{water}} = 760 \text{ torr}$$

Therefore, a mixture of iodobenzene and water codistills at 98°C.

An ideal gas law calculation shows that the mole fraction of iodobenzene in the vapor at the distilling head is 0.06 (46 torr/760 torr), and the mole fraction of water in the vapor is 0.94. However, because iodobenzene has a much higher molecular weight than water (204 g·mol^{-1} versus 18 g·mol^{-1}), its weight percentage in the vapor is much larger than 0.06, as the following calculation shows:

$$\frac{\text{moles}_{\text{iodobenzene}}}{\text{moles}_{\text{water}}} = \frac{P^{\circ}_{\text{iodobenzene}}}{P^{\circ}_{\text{water}}}$$

$$\frac{g_{\text{iodobenzene}}/\text{MW}_{\text{iodobenzene}}}{g_{\text{water}}/\text{MW}_{\text{water}}} = \frac{P^{\circ}_{\text{iodobenzene}}}{P^{\circ}_{\text{water}}}$$

Rearranging the preceding expression and substituting the molecular weights and vapor pressures allow us to calculate the weight ratio of iodobenzene to water in the distillate from the steam distillation:

$$\frac{g_{\text{iodobenzene}}}{g_{\text{water}}} = \frac{0.73\ g_{\text{iodobenzene}}}{1.0\ g_{\text{water}}}$$

In other words, the distilling liquid contains 42% iodobenzene and 58% water by weight. In any steam distillation, a large excess of water is used in the distilling flask so that virtually all the organic compound (iodobenzene in this example) can be distilled from the mixture at a temperature well below the boiling point of the pure compound.

The temperature in any steam distillation of a reasonably volatile organic compound will never rise above 100°C, the boiling point of water, unless your laboratory is below sea level. The steam distillation of most compounds occurs between 80°C and 100°C. For example, at 1.0 atm, octane (bp 126°C) steam distills at 90°C, and 1-octanol (bp 195°C) steam distills at 99°C. The lower distillation temperature has the added advantage of preventing decomposition of organic compounds during distillation.

Apparatus for Steam Distillation

Use more water than the amount of organic mixture being distilled and select a distilling flask that will be no more than half filled with this organic/water mixture. Add one or two boiling stones to the flask. For a steam distillation, modify a simple distillation apparatus by adding a Claisen connecting tube or adapter between the boiling flask and the distilling head. This adapter provides a second opening into the system to accommodate a source of steam or the addition of water.

Steam can be generated simply by boiling a large amount of water with the mixture in the distillation flask. If the codistillation of an organic compound with low volatility requires a large volume of steam, a separatory funnel placed in the second opening of the Claisen adapter provides a way of adding more water to the system without stopping the distillation (Figure 11.24). The apparatus for a steam distillation that uses externally generated steam, such as a steam line or a flask of boiling water, is shown in Figure 11.25.

Steps in a Steam Distillation

1. Set up the distillation apparatus.
2. Add the organic mixture and an excess of water to a firmly clamped distilling flask at least twice as large as the combined organic/water volume. Add one or two boiling stones.

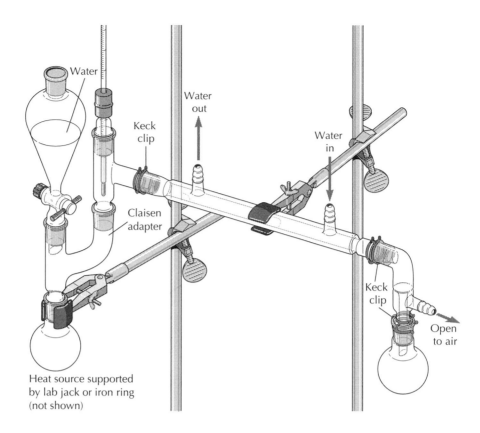

FIGURE 11.24
Steam distillation apparatus for internal generation of steam. During the distillation, additional water can be added to the distilling flask from the dropping funnel.

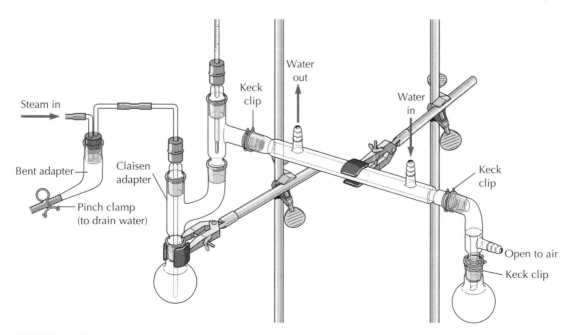

FIGURE 11.25
Steam distillation apparatus for use with an external steam source.

3. Heat the mixture until the entire top organic layer has distilled into the receiving flask. Sometimes it is worthwhile to collect an additional 10–15 mL of water after organic material is no longer apparent in the distilling flask.
4. Separate the organic phase of the distillate from the aqueous phase in a separatory funnel.

11.8 Sources of Confusion

Distillation is an important method for separating and purifying organic liquids. However, successful distillations require careful attention to a number of factors.

What Type of Distillation Should I Use?

Simple distillation. Simple distillation is used in two commonly encountered situations: (1) to remove a low-boiling solvent from an organic compound with a high boiling point; (2) as the last step in the purification of a liquid compound to obtain a pure product and determine its boiling point.

Fractional distillation. Fractional distillation is used for the separation of a mixture of two or more liquid compounds whose boiling points differ by less than 60°–70°C.

Vacuum distillation. When the boiling point of a liquid compound is over 200°C, the compound may decompose thermally before its atmospheric boiling point is reached. The reduced atmospheric pressure of a vacuum distillation allows the compound to boil at a lower temperature and thus distill without decomposition.

Steam distillation. Steam distillation is used to separate volatile compounds from a complex mixture. For example, limonene (oil of orange) can be separated from ground orange peels by steam distillation. It can also used to separate an organic product from an aqueous reaction mixture containing inorganic salts.

The Thermometer Reading Seems Too Low

If the liquid in the distilling flask is boiling but the temperature recorded on the thermometer in the distilling head is still 25°–30°C, it is likely that the vapor has not yet reached the thermometer bulb. The space between the boiling liquid and the thermometer bulb in the distilling head must become filled with vapor before a temperature increase is observed. Filling the space above the boiling liquid with vapor may require several minutes, depending on the rate of heating.

If the distillation is well underway and liquid is collecting in the receiving flask, yet the thermometer reading is still near room temperature, it is likely that the thermometer bulb is positioned improperly above the side arm (see Figure 11.6).

The Temperature Drops Suddenly During a Fractional Distillation

A sudden drop in temperature before all the liquid has distilled indicates a break between fractions. There is not enough vapor of the higher-boiling compound reaching the thermometer bulb to register on the thermometer. Increase the rate of heating until vapor again envelops the thermometer bulb.

When Do I Change Receiving Flasks?

Simple distillation. If you are conducting a simple distillation of a liquid that previously was dissolved in a low-boiling solvent, any liquid that distills at a temperature less than 5°C below the product's reported boiling point should be collected in a separate receiving flask. At 5°C or less from the reported boiling point of the liquid, the receiving flask should be changed to the tared (weighed) flask.

Fractional distillation. In a fractional distillation, the receiving flasks are changed soon after a sudden increase in temperature is noted, after a wait only long enough to allow the lower-boiling fraction to be washed out of the condenser. The sharp increase in temperature indicates that distillation of the lower-boiling-point component of the mixture is complete.

References

1. Lide, D. R. (Ed.) *Handbook of Chemistry and Physics*; 85th ed. CRC Press: Boca Raton, FL, 2004.

2. Perry, E. S.; Weissberger, A. (Eds.) *Techniques of Organic Chemistry*; 2nd ed.; Wiley-Interscience: New York, 1965, Vol. 4.

Questions

1. Explain why the observed boiling point for the first drops of distillate collected in the simple distillation of a 1:1 molar solution of pentane and hexane illustrated in Figure 11.5 will be above the boiling point of pentane.

2. A mixture contains 80% hexane and 20% pentane. Use the phase diagram in Figure 11.4 to estimate the composition of the vapor over this liquid. This vapor is condensed and the resulting liquid is heated. What is the composition of the vapor above the second liquid?

3. A student carried out a simple distillation on a compound known to boil at 124°C and reported an observed boiling point of 116°–117°C. Gas chromatographic analysis of the product showed that the compound was pure, and a calibration of the thermometer indicated that it was accurate. What procedural error might the student have made in setting up the distillation apparatus?

4. The directions in an experiment specify that the solvent, diethyl ether, be removed from the product by using a simple distillation. Why should the heat source for this distillation be a steam bath, not an electrical heating mantle?

5. The boiling point of a compound is 300°C at atmospheric pressure. Use the nomograph (Figure 11.17) to determine the pressure at which the compound would boil at about 200°C.

6. Azeotropes can be used to assist chemical reactions. Treatment of 1-butanol with acetic acid in the presence of a nonvolatile acid catalyst results in formation of the ester butyl acetate and water. The mixture of 1-butanol/butyl acetate/water forms a ternary azeotrope that boils at 90.7°C. This azeotrope separates into two layers; the upper is largely ester and the lower layer is largely water. The ester forms by an equilibrium reaction that does not especially favor product formation. Describe an apparatus that you could use to take advantage of azeotrope formation to drive the equilibrium toward the products, thus maximizing the yield of ester.

12

SUBLIMATION

Before most solid organic compounds can evaporate, they melt, a process that usually requires a reasonably high temperature. However, some substances, such as iodine, camphor, and 1,4-dichlorobenzene (mothballs), exhibit appreciable vapor pressure below their melting points. You may already have seen iodine crystals evaporate to a purple gas during gentle heating and smelled the characteristic odors of camphor or mothballs. These substances all change directly from the solid phase to the gas phase without forming an intermediate liquid phase by a process called *sublimation.*

The process of sublimation seems somewhat unusual in that, unlike normal phase changes from solid to liquid to gas, no liquid phase forms between the solid and gas phases. The conversion of the solid form of carbon dioxide (also called *dry ice*) directly into CO_2 gas may be the best-known example of sublimation. Carbon dioxide does not have a melting point at atmospheric pressure. More than 5 atm of pressure are necessary before dry ice melts at $-57°C$. The sublimation point for CO_2 at atmospheric pressure is $-78°C$, well below room temperature.

Purification by Sublimation

In the laboratory we can use sublimation as a purification method for an organic compound (1) if it can vaporize without melting, (2) if it is stable enough to vaporize without decomposition, (3) if the vapor can be condensed back to the solid, and (4) if the impurities present do not also sublime. Many organic compounds that do not sublime at atmospheric pressure sublime appreciably at reduced pressure, thus enabling their purification by sublimation. Use of reduced pressure, supplied by a vacuum source, also makes decomposition and melting less likely to occur during the sublimation.

12.1   Assembling the Apparatus for a Sublimation

The apparatus for a sublimation consists of an outer vessel and an inner vessel. The outer vessel holds the sample being purified and is connected to a vacuum source. An inner container, sometimes called a "cold finger," provides a cold surface on which the vaporized compound can recondense as a solid.

Two simple arrangements for sublimation under reduced pressure are shown in Figure 12.1. The inner test tube, which contains cold water or ice and water, serves as a condensation site for the sublimed solid. The outer vessel, a side-arm test tube or a filter flask, holds the substance being purified, and the side arm provides a connection to the vacuum source. The inner and outer vessels are sealed together by a neoprene filter adapter. The distance between the bottom surfaces of the inner tube and the outer tube or filter flask should be 0.5–1.0 cm.

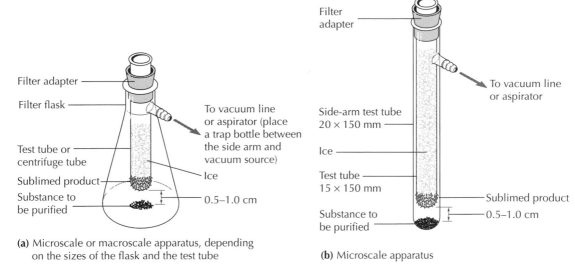

FIGURE 12.1

Two simple apparatuses for sublimation.

If the vapor has to travel a long distance, a higher temperature is needed to keep it in the gas phase, and decomposition of the solid sample may very well occur. If the surfaces are too close, impurities can spatter and contaminate the condensed solid on the surface of the inner tube. Connect the side arm of the test tube or filter flask to a water aspirator or vacuum line, using a safety flask between the aspirator and the sublimation apparatus.

The side-arm test tube apparatus serves well for 10–150 mg of material. The filter flask apparatus can be sized to suit the amount of material being purified. For example, microscale quantities of 10–150 mg can be sublimed in a 25-mL filter flask, whereas 1 g of material would require a 125-mL filter flask with a correspondingly larger test tube for the cold finger. The apparatus shown in Figure 12.2 is a commercially available sublimation apparatus used for gram quantities of material.

FIGURE 12.2

Sublimation apparatus for gram quantities of material.

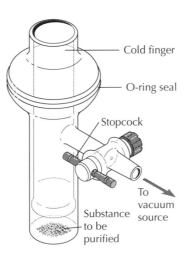

12.2 | Carrying Out a Sublimation

SAFETY PRECAUTION

The lip of the inner test tube must be large enough to prevent it from being pushed through the bottom of the filter adapter by the difference in pressure created by the vacuum. Slippage of the inner test tube could cause both vessels to shatter as the inner test tube hits the outer test tube or flask. Placing a microclamp on the inner test tube above the filter adapter helps to keep the test tube from moving once it is positioned in the filter adapter.

*Ice and water are placed in the inner test tube **after** the vacuum is applied to prevent condensation of moisture from the air on the tube before sublimation takes place.*

Place the sample (10–150 mg) to be sublimed in a 25-mL filter flask or a side-arm test tube. Fit the inner test tube through the filter adapter and adjust the position of the inner tube so that it is 0.5–1.0 cm above the bottom of the flask or side-arm test tube. Turn on the water aspirator or vacuum line. After a good vacuum has been achieved, fill the inner test tube with ice and water, then proceed to heat the sublimation tube gently using a sand bath [see Technique 6.2]. If a filter flask (25 mL for 10–150 mg, 125 mL for 0.25–1.0 g) is used as the outer container, heat it gently on a hot plate or with a sand bath.

During sublimation, you will notice material disappearing from the bottom of the outer vessel and reappearing on the cool outside surface of the inner test tube. If the sample begins to melt, briefly withdraw the heat source from the apparatus. If all the ice melts, remove half the water from the inner test tube with a Pasteur pipet and then add additional ice.

After sublimation is complete, remove the heat source and slowly let air back into the system by gradually removing the rubber tubing from the water aspirator or other vacuum source. Then turn off the water flow in the aspirator or turn off the vacuum source and slowly disconnect the rubber tubing from the side arm of the filter flask or test tube. Carefully remove the inner test tube and scrape the purified solid onto a tared weighing paper. After weighing the sublimed solid, store it in a tightly closed vial.

Questions

1. Which of the following three compounds is the most likely to be amenable to purification by sublimation: polyethylene, menthol, or benzoic acid?

2. A solid compound has a vapor pressure of 65 torr at its melting point of 112°C. Give a procedure for purifying this compound by sublimation.

3. Hexachloroethane has a vapor pressure of 780 torr at its melting point of 186°C. Describe how the solid would behave while carrying out a melting-point determination at atmospheric pressure (760 torr) in a capillary tube open at the top.

TECHNIQUE

13

REFRACTOMETRY

A beam of light traveling from a gas into a liquid undergoes a decrease in its velocity. If the light strikes the horizontal interface between gas and liquid at an angle other than 90°, the beam bends downward as it passes from the gas into the liquid. Application of this phenomenon allows the determination of a physical property known as the *refractive index,* a measure of how much the light is bent, or *refracted,* as it enters the liquid. The refractive index can be determined quite accurately to four decimal places, making this physical property useful for assessing the purity of liquid compounds. The closer the experimental value approaches the value reported in the literature, the purer the sample. Even trace amounts of impurities (including water) change the refractive index, so unless the compound has been extensively purified, the experimentally determined value may not agree with the literature value past the second decimal place.

13.1 Theory of Refractometry

The *refractive index, n,* represents the ratio of the velocity of light in a vacuum (or in air) to the velocity of light in the liquid being studied. The velocities of light in both media are related to the angles that the incident (incoming) and the refracted (outgoing) beams make with a theoretical line drawn 90° to the liquid surface (Figure 13.1):

$$n = \frac{V_{air}}{V_{liq}} = \frac{\sin \theta}{\sin \theta'}$$

where V_{air} is the velocity of light in a vacuum (or air), V_{liq} is the velocity of light in the liquid, θ is the angle of incident light in a vacuum (or air), and θ' is the angle of refracted light in the liquid. The velocity of light in a liquid sample is always less than that of light in air, so refractive index values are numerically greater than 1.

The variables of temperature and the wavelength of the light being refracted influence the refractive index for any substance. The

FIGURE 13.1
Refraction of light in a liquid.

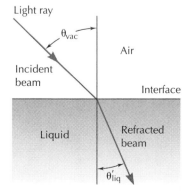

Light ray

θ_{vac}

Air

Incident beam

Interface

Liquid

Refracted beam

θ'_{liq}

temperature of the sample affects its density. The density change, in turn, affects the velocity of the light beam as it passes through the sample. Therefore, the temperature (20°C in the following example) at which the refractive index was determined is always specified by a superscript in the notation of n:

$$n_{\mathrm{D}}^{20} = 1.3910$$

The wavelength of the light used also affects the refractive index, because light of differing wavelengths refracts at different angles. The two bright yellow, closely spaced lines of the sodium spectrum at 589 nm, commonly called the **sodium D line,** usually serves as the standard wavelength for refractive index measurements and is indicated by the D in the subscript of the symbol n. If light of some other wavelength is used, the specific wavelength in nanometers appears in the subscript.

13.2 The Refractometer

The instrument used to measure the refractive index of a compound is called a *refractometer.* Figure 13.2 shows an Abbe refractometer, an example of the type commonly found in undergraduate organic chemistry laboratories. The instrument includes a built-in thermometer for measuring the temperature at the time of the refractive index reading, as well as a system for circulating water at a constant temperature around the sample holder. This type of refractometer uses a white light source instead of a sodium lamp and contains a series of compensating prisms that give a refractive index equal to that obtained with 589-nm light (the D line of sodium).

A few drops of sample are introduced between a pair of hinged prisms (Figure 13.3). The light passes through the sample and is

FIGURE 13.2
Abbe refractometer.
(Courtesy of Leica, Inc.,
Optical Products Division,
P.O. Box 123, Buffalo,
NY 14240.)

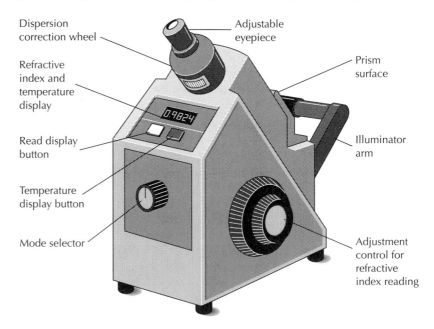

Dispersion correction wheel

Adjustable eyepiece

Refractive index and temperature display

Prism surface

Read display button

Illuminator arm

Temperature display button

Mode selector

Adjustment control for refractive index reading

FIGURE 13.3
Hinged prism of the Abbe refractometer, shown in the open position. (Courtesy of Leica, Inc., Optical Products Division, P.O. Box 123, Buffalo, NY 14240.)

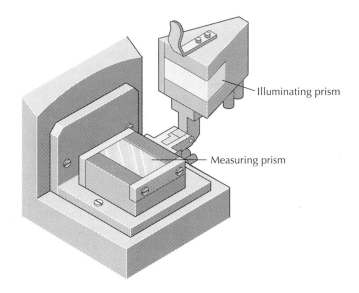

Illuminating prism

Measuring prism

reflected by an adjustable mirror. When the mirror is properly aligned, the light is reflected through the compensating prisms and, finally, through a lens with crosshairs to the eyepiece.

Completely automated refractometers are also commercially available. If an automated refractometer is used in your laboratory, consult your instructor about its operation.

13.3   Steps in Determining a Refractive Index

The following directions apply to the use of a refractometer such as the one shown in Figure 13.2. Consult your instructor about using an automated refractometer.

Do not use acetone to clean prisms because it can dissolve the adhesive holding the prisms.

1. Check the surface of the prisms for residues from previous determinations. If the prisms need cleaning, place a few drops of methanol on the surfaces and blot (do not rub) the surfaces with lens paper. Allow the residual methanol to evaporate completely before placing the sample on the lower prism.
2. With a Pasteur pipet held 1–2 cm above the prism, place 4–5 drops of the sample on the measuring (lower) prism. Do not touch the prism with the tip of the dropper because the highly polished surface scratches very easily, and scratches ruin the instrument. Lower the illuminating (upper) prism carefully so that the liquid spreads evenly between the prisms.
3. Rotate the adjustment control until the dark and light fields are exactly centered on the intersection of the crosshairs in the eyepiece (Figure 13.4). If color (usually red or blue) appears as a horizontal band at the interface of the fields, rotate the chromatic adjustment or dispersion correction wheel until the interface is sharp and uncolored (achromatic). Occasionally the sample evaporates from the prisms, making it impossible to produce a

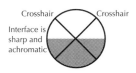

Crosshair   Crosshair

Interface is sharp and achromatic

FIGURE 13.4
The view through the eyepiece when the refractometer is adjusted correctly.

sharp achromatic interface between the light and dark fields. If evaporation occurs, apply more sample to the prism and repeat the adjustment procedure.

4. Press the read display button and record the refractive index in your notebook. Then press the temperature display button and record the temperature.

5. Open the prisms, blot up the sample with lens paper, and follow the cleaning procedure with methanol outlined in step 1.

13.4   Temperature Correction

Values reported in the literature are often determined at a number of different temperatures, although 20°C has become the standard. To compare an experimental refractive index with a value reported at a different temperature, a correction factor must first be calculated. The refractive index for a typical organic compound increases by 4.5×10^{-4} for each $1°$ decrease in temperature. Refractive index values vary inversely with temperature because the density of a liquid almost always decreases as the temperature increases. This decrease in density produces an increase in the velocity of light in the liquid, causing a corresponding decrease in the refractive index at higher temperatures.

To compare an experimental refractive index measured at 25°C to a reported value at 20°C, a temperature correction needs to be calculated:

$$\Delta n = 4.5 \times 10^{-4} \times (T_1 - T_2)$$

where T_1 is the observation temperature in degrees Celsius and T_2 is the temperature reported in the literature in degrees Celsius.

The correction factor, including its sign, is then added to the experimentally determined refractive index. For example, if your experimental refractive index is 1.3888 at 25°C, then you obtain a corrected value at 20°C of 1.3911 by adding the correction factor of 0.0023 to the experimental refractive index.

$$\Delta n = [4.5 \times 10^{-4} \times (25 - 20)] = 0.00225 \text{ (rounds to 0.0023)}$$
$$n^{20} = n^{25} + 0.0023 = 1.388 + 0.0023 = 1.3911$$

The correction needs to be applied before comparing the experimental value to a literature value reported at 20°C.

If an experimental refractive index is determined at a temperature lower than that of the literature value to which it is being compared, the correction has a negative sign and the corrected refractive index is lower than the experimental value.

Questions

1. A compound has a refractive index of 1.3191 at 20.1°C. Calculate its refractive index at 25.0°C.

2. Using ethanol or methanol but not acetone or water to clean the glass surfaces of a refractometer is usually recommended. Why?

OPTICAL ACTIVITY AND ENANTIOMERIC ANALYSIS

Optical activity, the ability of substances to rotate plane-polarized light, played a crucial role in the development of chemistry as the link between the molecular structures that chemists write and the real physical world. A major development in the structural theory of chemistry was the concept of the three-dimensional shape of molecules. When Jacobus van't Hoff and Joseph le Bel noted the asymmetry possible in tetrasubstituted carbon compounds, they claimed that their "chemical structures" were identical to the "physical structures" of the molecules. Not only was the structural theory of the organic chemist useful in explaining the facts of chemistry, it also happened to be "true." Van't Hoff and le Bel could make this claim because their theories of the tetrahedral carbon atom accounted not only for chemical properties but also for the optical activity of substances, a physical property.

14.1

Mixtures of Optical Isomers: Separation/Resolution

A molecule that possesses no internal mirror plane of symmetry and that is not superimposable on its mirror image is said to be chiral, or "handed." Chirality, a molecular property, is normally indicated by the presence of a *stereocenter*—a tetrahedral atom bearing four different substituents. A stereocenter is sometimes called a *chiral* or *asymmetric center.*

Chiral compounds possess the property of *enantiomerism.* Enantiomers are stereoisomers that have nonsuperimposable mirror images. Chiral compounds such as 2-butanol and the amino acid alanine, which contain only one stereocenter, are simple examples of enantiomers.

2-Butanol   Alanine

Enantiomers and Racemic Mixtures

The enantiomers of 2-butanol have identical physical properties, including boiling points, IR spectra, NMR spectra, refractive indices, and TLC R*f* values, except for the direction in which they rotate plane-polarized light. Both enantiomers are optically active—one of them rotates polarized light in a clockwise direction and is called the **(+)-*isomer*.** The other enantiomer rotates polarized light counterclockwise and is called the **(−)-*isomer.*** The rotational power of (+)-2-butanol is exactly the same in the clockwise direction as that of (−)-2-butanol in the counterclockwise direction. Unfortunately, there is no simple theoretical way to predict the direction of the rotation of plane-polarized light on the basis of the configuration at a carbon stereocenter. Thus, it is not apparent which structure of 2-butanol or alanine is the (+)- or the (−)-enantiomer.

Usually, simple compounds obtained from the stockroom are optically inactive, even when their molecules are chiral. For example, you would normally find that a sample of 2-butanol is optically inactive. To understand this apparent paradox, consider the reduction of 2-butanone with sodium borohydride. This reaction can proceed in two ways. Hydride can react with 2-butanone from either the top side or the bottom side of the carbonyl double bond. It undergoes reaction both ways at equal rates, thus giving rise to a 50:50 mixture of the enantiomers of 2-butanol. The reaction produces a product that is optically inactive:

rate a = rate b   Enantiomers formed in
                                 equal amounts

An equal mixture of (+)- and (−)-enantiomers is called a *racemic mixture.* In the separation or *resolution* of a racemic mixture, the enantiomers are transformed into a pair of *diastereomers*—stereoisomers that have different physical and chemical properties. A mixture of two diastereomers is prepared from a racemic mixture by its reaction with an optically active substance. The diastereomers can then be separated by recrystallization, for example, because of the differential solubility of the two diastereomers.

Resolution with Acids or Bases

The simplest reaction for preparing diastereomers from racemic mixtures is that of an acid with a base to form a salt. For resolution or separation of the two enantiomers to occur, the added reagent in the acid/base reaction must be optically active. Neutralization of a racemic amine, for example, by an optically active carboxylic acid is a method for resolving the amine. Neutralization of a carboxylic acid by an optically active amine is a way of resolving the acid. In the neutralization, two salts are produced. The salts are diastereomers and, as such, differ in their solubilities in various solvents. Therefore, they can be separated by fractional crystallization. The less soluble diastereomeric salt is the more easily obtained. The process for resolution of an amine is represented in the following steps:

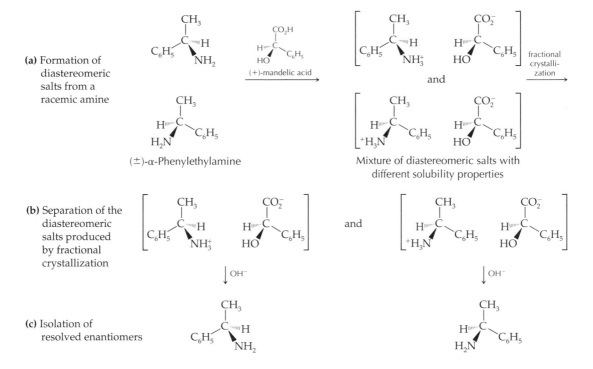

(a) Formation of diastereomeric salts from a racemic amine

(±)-α-Phenylethylamine

(+)-mandelic acid

Mixture of diastereomeric salts with different solubility properties

fractional crystallization →

(b) Separation of the diastereomeric salts produced by fractional crystallization

and

(c) Isolation of resolved enantiomers

If you examine the diastereomeric salts in (a) and (b), you will see that each salt has two stereocenters. When you compare their structures, you will find that the carbon stereocenters bearing the—OH group have the identical configuration in each salt, whereas the stereocenters bearing the—NH$_3^+$ have opposite configurations. Thus the two diastereomeric salts are stereoisomers that are not mirror images.

Optically active acids and bases isolated from plant materials are frequently used for the resolution of racemic mixtures (Table 14.1). The diastereomers necessary for resolution do not need to be salts. For example, diastereomeric esters, formed by reaction of the enantiomers of an alcohol with an optically active carboxylic acid, can also be used.

TABLE 14.1 Optically active acids and bases used for resolutions

Bases	Acids
Brucine	Tartaric acid
Strychnine	Mandelic acid
Quinine	Malic acid
Cinchonine	Camphor-10-sulfonic acid
α-Phenylethylamine	

Enzymatic Resolution

An increasingly useful method for the resolution of racemic mixtures utilizes an enzyme that selectively reacts with one of the enantiomers. Because all enzymes are chiral molecules, the transition states for the reaction of an enzyme with two enantiomers are diastereomeric and the energies of these two transition states differ. Thus one of the chiral enantiomers reacts faster than the other. In many cases an enzyme reacts specifically with only one enantiomer. This specificity provides an excellent method for resolving a racemic mixture if a suitable enzyme is available.

Resolution by Chiral Chromatography

Resolution of a racemic mixture can also be carried out using a chiral chromatographic separation, by either gas chromatography [see Technique 16] or liquid chromatography [see Technique 17]. When a mixture of enantiomers passes through a chiral chromatographic column, each enantiomer has a different attraction for the chiral stationary phase—differences that lead to separation of the enantiomers. Typical stationary phases that produce this effect are proteins or β-cyclodextrins, often immobilized by bonding to silica gel. The less tightly coordinated enantiomer passes through the column more rapidly than the more tightly coordinated enantiomer.

14.2   Polarimetry

The traditional way to measure optical activity is with a polarimeter, a schematic description of which is shown in Figure 14.1. All commercially available polarimeters have the same general features. The analyzer of a simple polarimeter is adjusted manually, whereas an automated instrument has all the components housed in a case with a digital readout of the observed rotation.

How a Polarimeter Works

The light beam approaching the polarizer in Figure 14.1 has wave oscillations in all the planes perpendicular to the direction in which the beam is traveling. When the light beam hits the polarizer, which has ranks and files of molecules arranged in a highly ordered fashion, only the light whose oscillations are in one plane

FIGURE 14.1
Schematic diagram of a polarimeter.

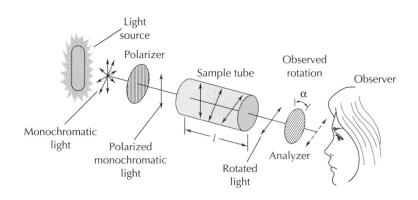

is transmitted through the polarizer. The light that gets through is called *plane-polarized light.* The remaining waves are refracted away or absorbed by the polarizer. In a rough analogy, the light beam hits the polarizer whose molecules are ordered like the slats of wood in a picket fence. Only the light waves whose oscillations are parallel to the slats pass through the polarizer and into the sample tube.

The analyzer is a second polarizer whose ranks and files of molecules must also be lined up for the polarized light waves to be transmitted. If the polarized light has been rotated by an optically active substance in the sample tube, the analyzer must be rotated the same amount to let the light through. The rotation is measured in degrees, indicated by α in Figure 14.1.

Use of Monochromatic Light

Monochromatic light is preferred in polarimetric measurements because the optical activity or rotatory power of chiral compounds depends on the wavelength of the light used. For example, the rotation of 431-nm (blue) light is 2.8 times greater than the rotation of 687-nm (red) light. A common light source is a sodium lamp, which has two very intense emission lines at 589 and 589.6 nm. This closely spaced doublet is called the *sodium D line.* Another common light source is a mercury lamp, using the intense 546.1-nm emission line. The human eye is more sensitive to the mercury emission in the green region than to the sodium line in the yellow region of the visible spectrum.

Reading a Manual Polarimeter

A number of techniques are used to detect the rotation of polarized light with a manual polarimeter. The simplest way is to rotate the analyzer until no light at all comes through the eyepiece. However, this method depends not only on the sensitivity of our eyes but also on our ability to remember quantitatively the amount of brightness we have just seen. In practice, this is difficult to do.

Various optical devices can be used to make the measurement of rotation easier. They depend on a sudden change of contrast when the minimum amount of light is transmitted by the analyzer. Manual polarimeters have a split-field image or two fields divided through the middle (Figure 14.2a). The analyzer is rotated in a clockwise or counterclockwise direction until a point is reached

FIGURE 14.2
Representative images in the light field of a manual polarimeter.

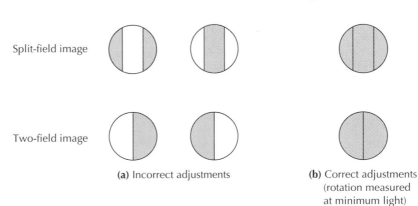

Split-field image

Two-field image

(a) Incorrect adjustments

(b) Correct adjustments (rotation measured at minimum light)

where every field is of equal minimum intensity and the divided fields are no longer visible (Figure 14.2b).

14.3 Using Polarimeter Tubes

Polarimeter tubes are expensive and must be handled carefully. They come in different lengths; 1-dm and 2-dm tubes are the most common. The periscope tube and the tube with an enlarged ring near one end allow removal of any air bubbles from the light path that can be tedious to remedy when using a straight tube. The tubes shown in Figure 14.3 are closed with a glass plate and a rubber washer, both held in place by a one- or two-part screw cap. Be careful not to screw the cap too tightly, because strain in the glass end plate can produce an apparent optical rotation.

Cleaning a Polarimeter Tube

Unless the polarimeter tube is clean and dry, you should first clean the tube with some care. When the tube is clean, rinse it with the solvent you plan to use for the solution of your optically active compound. After the tube has been well drained, rinse it with two or three small portions of your solution to ensure that the concentration of the solution in the polarimeter tube is the same as the concentration of the solution you have prepared. You may want to save these optically active rinses, because your chiral compound can be recovered from them later.

Air Bubbles and Suspended Particles

When you fill a polarimeter tube with a solution, make sure that the tube has no air bubbles trapped in it; bubbles will refract the light coming through. Also make sure that there are no suspended particles in a solution whose rotation you wish to measure, or you may get so little transmitted light that measurement of the rotation will be very difficult. If you have a solution that you suspect is too turbid for polarimetry measurements, filter it by gravity through a small plug of glass wool.

Standardizing the Polarimeter

A polarimeter can be standardized by filling a tube with an optically inactive solvent such as distilled water or with the solvent being used for your sample. Adjust the instrument to the minimum-light position (see Figure 14.2)

If you are using a manual polarimeter, check your ability to use it properly by analysing a 5.00% or 10.00% solution of sucrose in water. Determine the specific rotation [see Technique 14.4] of your sample based on the average of five to seven readings of the optical rotation. Automatic polarimeters normally do not require multiple determinations of the experimental rotation. Consult your instructor about the operation of the polarimeter in your laboratory.

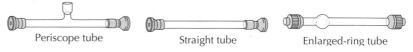

Periscope tube   Straight tube   Enlarged-ring tube

FIGURE 14.3
Different styles of polarimeter tubes.

14.4   Specific Rotation

The magnitude of the optical rotation depends on the concentration of the optically active compound in the solution, the length of the light path through the solution, the wavelength of the light, the nature of the solvent, and the temperature. A typical rotation of common table sugar or sucrose is written in the following manner:

$$[\alpha]_D^{20} = +66.4°(H_2O)$$

The symbol $[\alpha]_\lambda^{T°}$ is called the *specific rotation* and is an inherent property of an optically active compound. T° signifies the temperature of the measurement in degrees Celsius, and λ is the wavelength of light used. In the sucrose example, the sodium D line was used. The specific rotation is calculated from the observed angle of rotation:

$$[\alpha]_\lambda^{T°} = \frac{\alpha}{l \cdot c}$$

where α is the observed angle of rotation, l is the length of the light path through the sample in decimeters, and c is the concentration of the sample $(g \cdot mL^{-1})$.

The cell length is always given in decimeters (dm, 10^{-1} m) in the calculation of specific rotation. When a pure, optically active liquid is used as the sample, its concentration is simply the density of the liquid.

Sometimes a rotation of an optically active substance is given as a *molecular rotation:*

$$[M]_\lambda^{T°} = \frac{M}{100}[\alpha]$$

where M is the molecular weight of the optically active compound.

The value of the specific rotation can change considerably from solvent to solvent. It is even possible for an enantiomer to have a different sign of rotation in two different solvents. Such solvent effects are due to specific solvent/solute interactions. The four most common solvents for polarimetry are water, methanol, ethanol, and chloroform.

The intrinsic specific rotation of a compound is generally considered to be a constant in dilute solutions at a particular temperature and wavelength. However, if you wish to compare the optical activity of a sample with that obtained by other workers, you should use the same concentration in the same solvent. Sucrose makes an excellent reference compound for polarimetry because its specific rotation in water is essentially independent of concentration up to 5–10% solutions.

A change in the specific rotation due to temperature variation may be caused by a number of factors, including changes in molecular association, dipole-dipole interactions, conformation, and solvation. When nonpolar solutes are dissolved in nonpolar solvents, variation in the specific rotation with temperature may not be large. But for some polar compounds, the specific rotation varies markedly with temperature. Near room temperature, the rotation of tartaric acid may vary by more than 10% per degree Celsius.

14.5   Enantiomeric Excess (Optical Purity)

The purity of optically active compounds is reported in terms of enantiomeric excess (or, less accurately, optical purity). *Enantiomeric excess (% ee)* is calculated from the expression

$$\% \text{ ee} = \left(\frac{[\alpha]_{\text{observed}}}{[\alpha]_{\text{pure}}} \right) \times 100\%$$

Thus, if we determine a specific rotation of 6.5° for 2-butanol, we can calculate the enantiomeric excess (% ee) of the sample if we know the specific rotation of pure 2-butanol ($[\alpha] = +13.00°$):

$$\% \text{ ee} = \left(\frac{6.5}{13.00} \right) \times 100\% = 50\%$$

It is instructive to examine the composition of 100 molecules of the mixture with a % ee = 50%. We have an excess of 50 (+)-molecules, which causes the optical activity. The remaining 50 molecules, because they have no net optical activity, are composed of 25 (+)-molecules and 25 (−)-molecules. Thus we have a total of 25 (−)-molecules and 50 + 25 = 75 (+)-molecules.

14.6   Modern Methods of Enantiomeric Analysis

Rather than using polarimetry, it can be useful to convert a mixture of enantiomers to a corresponding mixture of diastereomers and to use high-performance liquid chromatograpy (HPLC) [see Technique 17] or nuclear magnetic resonance (NMR) spectroscopy [see Technique 19] for measuring the composition. These methods can be used to determine how successful a resolution has been or how stereoselective a chemical reaction is. They have the advantage of needing much smaller samples to determine enantiomeric excess than polarimetry usually requires.

Use of Chiral Acid/Base Chemistry for NMR Analysis

If the enantiomers of a chiral carboxylic acid undergo reaction with an optically active amine in an NMR tube, a mixture of diastereomeric salts is produced; these diastereomers can have subtly different NMR spectra. Neutralization of a mixture of enantiomers of a chiral amine by an optically active carboxylic acid can serve the same purpose. The NMR spectra will be fairly complex, and the chiral enantiomers generally need to have a clean singlet for one of its NMR signals so that integration can be used reliably to determine the enantiomeric composition.

Chiral Shift Reagents for NMR Analysis

Chiral lanthanide shift reagents are often used to produce a diastereomeric mixture for NMR analysis. Derivatives of camphor provide shift reagents that are rich in chiral character. $Eu(hfc)_3$, called tris[3-heptafluoropropylhydroxymethylene)-(+)-camphorato] europium (III), is such a compound. This compound undergoes rapid and reversible coordination with Lewis bases, (B:), establishing the following equilibrium:

$$Eu(hfc)_3 + B: \rightleftharpoons B: Eu(hfc)_3$$

The complex B: $Eu(hfc)_3$ brings a paramagnetic ion, Eu^{3+}, into close proximity to the chiral organic base (B:), which induces changes in the ^1H NMR chemical shifts of the chiral base. The chemical shifts are different in each of the two coordinated enantiomers because the formation of the diastereomeric pair causes the protons of the two enantiomers to become nonequivalent.

Eu(hfc)$_3$
(tris[3-heptafluoropropyl-
hydroxymethylene-(+)-
camphorato]europium)

Complex

Identification of the NMR signals of the α-protons and integration of their areas allows determination of the composition of the B: $Eu(hfc)_3$ complex, which equals the enantiomeric composition of the original mixture.

Chiral HPLC

Another modern approach to the determination of enantiomeric excess is the use of chiral high-performance liquid chromatography HPLC. As discussed in Technique 14.1 in the section on chiral chromatography, when a mixture of enantiomers passes through a chiral chromatographic column, different interactions occur between each enantiomer and the chiral stationary phase, which lead to separation of the enantiomers.

References

1. Gordon, A. J.; Ford, R. A. *The Chemist's Companion: A Handbook of Practical Data, Techniques, and References;* Wiley-Interscience: New York, 1972.

2. Morrill, T. C. (Ed.) *Lanthanide Shift Reagents in Stereochemical Analysis;* VCH Publishers: New York, 1986.

Questions

1. A sample of 2-butanol has a specific rotation of $+3.25°$. Determine the % ee and the molecular composition of this sample. The specific rotation of pure (+)-2-butanol is $+13.0°$.

2. A sample of 2-butanol (see question 1) has a specific rotation of $-9.75°$. Determine the % ee and the molecular composition of this sample.

3. An optical rotation study gives a result of $\alpha = +140°$. Suggest a dilution experiment to test whether the result is indeed $+140°$, not $-220°$.

4. The structures of strychnine $(R = H)$ and brucine $(R = CH_3O)$ are examples of alkaloid bases that can be used for resolutions. These molecules are rich sources of chirality (respectively, $[\alpha]_D = -104°$ and $-85°$ in absolute ethanol). Assume that nitrogen inversion is slow and identify the eight stereocenters in each of these two nitrogen heterocyclic compounds.

R = H, strychnine
R = CH$_3$O, brucine

5. Only one of the two nitrogens in strychnine and brucine acts as the basic site for the necessary acid/base reaction for a resolution. Which nitrogen, and why?

PART

2

Chromatography

Few experimental techniques rival chromatography for versatility or usefulness in separating complex mixtures of compounds. Chromatography got its name from the fact that it was originally used to separate mixtures of different-colored substances. Once chemists realized that chromatography could be used to separate colorless substances as well, its development took off.

Three types of chromatography used in organic chemistry are discussed in detail in Part 2:

- Thin-layer chromatography (TLC)
- Gas-liquid chromatography (GC)
- Liquid (column) chromatography (LC), including flash chromatography and high-performance liquid chromatography (HPLC)

Principles of Chromatography

All chromatographic methods depend on the distribution of the substances being separated between the two phases of the chromatographic system, a *mobile phase* and a *stationary phase.* The mobile phase consists of a liquid or gas that carries the sample through the solid or liquid that forms the stationary phase. For example, in both thin-layer and liquid chromatography, a liquid (often coating a finely ground solid) forms the stationary phase. A liquid solvent provides the mobile phase. In gas-liquid chromatography, the mobile phase is a gas.

The compounds in a mixture separate because of differences in their affinities for the stationary phase and their solubility in the mobile phase. A dynamic equilibrium exists between the sample components bonded to the stationary phase and those dissolved in the mobile phase. The compounds being separated move through the mobile phase, interacting with the stationary phase along the way. Separation occurs by two processes: adsorption onto the stationary phase followed by desorption into the mobile phase or by partition between the stationary and mobile phases.

In *adsorption chromatography,* the compounds being separated adsorb onto and desorb from a solid stationary phase many, many times as the solvent passes through the stationary phase. The tighter they adsorb to the stationary phase, the slower they travel through the chromatography column. In *partition chromatography,* the compounds to be separated partition themselves between a stationary liquid phase and a mobile liquid or gas phase. This partitioning occurs in the same way a solute partitions itself between two immiscible solvents used for an extraction [Technique 8]. The greater the attraction a compound has for the stationary liquid phase the slower it travels through the chromatography column.

15

THIN-LAYER CHROMATOGRAPHY

If Technique 15 is your introduction to chromatographic analysis, read the introduction to chromatography on pages 175–176 before you read Technique 15.

Thin-layer chromatography has become a widely used analytical technique. It is simple, inexpensive, fast, efficient, reasonably sensitive, and requires only milligram quantities of material. Thin-layer chromatography is especially useful for determining the number of compounds in a mixture, for possibly establishing whether or not two compounds are identical, and for following the course of a reaction.

15.1   Introduction to Thin-Layer Chromatography (TLC)

In thin-layer chromatography, glass, metal, or plastic plates coated with a thin layer of adsorbent serve as the stationary phase. The mobile phase is a pure solvent or a mixture of solvents; the appropriate composition of the mobile phase depends on the polarities of the compounds in the mixture being separated. Most nonvolatile solid organic compounds can be analyzed by thin-layer chromatography. However, TLC does not work well for many liquid compounds because their volatility can lead to loss of the sample by evaporation from the TLC plate.

Overview of TLC Analysis

To carry out a TLC experiment, a small amount of the mixture being separated is dissolved in a suitable solvent and applied or spotted on the adsorbent near one end of a TLC plate. Then the plate is placed in a closed chamber, with the edge nearest the applied spot immersed in a shallow layer of the mobile phase called the *developing solvent* (Figure 15.1). The solvent rises through the stationary phase by capillary action, a process called *developing the chromatogram.*

As the solvent ascends the plate, the sample is distributed between the mobile phase and the stationary phase. Separation during the development process occurs as a result of the many equilibrations taking place between the mobile and stationary phases and the compounds being separated. **The more tightly a compound binds to the adsorbent, the more slowly it moves on the TLC plate.** The developing solvent moves nonpolar substances up the plate most rapidly. Polar substances travel up the plate slowly or sometimes not at all as the solvent ascends.

The TLC plate is removed from the developing chamber when the *solvent front* (leading edge of the solvent) is about 1 cm from the top of the plate. The position of the solvent front is marked immediately with a pencil line, before the solvent evaporates. The plate is then placed in a hood to dry.

Several methods are available to *visualize* the compounds in the sample. If the TLC plate is impregnated with a fluorescent indicator, the plate can be illuminated by exposure to ultraviolet light.

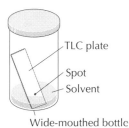

TLC plate

Spot

Solvent

Wide-mouthed bottle

FIGURE 15.1
Developing chamber containing a thin-layer plate.

Alternatively, the compounds can be visualized using a reagent that produces colored spots. The developed and visualized plate is then ready for analysis of the separation.

Determination of the R_f

The analysis of a thin-layer chromatogram consists of determining how far each compound has traveled on the plate relative to the distance the solvent has traveled. Under a constant set of experimental conditions, a given compound always travels a fixed distance relative to the distance traveled by the solvent front (Figure 15.2). This ratio of distances is called the R_f *(ratio to the front)* and is expressed as a decimal fraction:

$$R_f = \frac{\text{distance traveled by compound}}{\text{distance traveled by developing solvent front}}$$

The R_f value for a compound depends on its structure and is a physical characteristic of the compound, just as its melting point is a physical characteristic. Whenever a chromatogram is done, the R_f value should be calculated for each substance and the experimental conditions recorded. The important data that need to be recorded include the following:

- Brand, type of backing, and adsorbent on the TLC plate
- Developing solvent
- Method used to visualize the compounds
- R_f value for each substance

Calculation of an R_f value.   To calculate the R_f value for a given compound, measure the distance that the compound has traveled from where it was originally spotted and the distance that the solvent front has traveled from where the compound was spotted (see Figure 15.2). The measurement is made from the center of a spot. The best data are obtained from chromatograms in which the spots are less than 5 mm in diameter. If a spot shows "tailing," measure from the densest point of the spot. The R_f values for the two substances shown on the developed TLC plate in Figure 15.2 are calculated as follows:

$$\text{Compound 1:} \quad R_f = \frac{26 \text{ mm}}{55 \text{ mm}} = 0.47$$

$$\text{Compound 2:} \quad R_f = \frac{40 \text{ mm}}{55 \text{ mm}} = 0.73$$

Identical R_f values.   When two samples have identical R_f values, you should not conclude that they are the same compound without doing further analysis. There are perhaps 100 R_f values that can be distinguished from one another, whereas there are greater than 10^8 known organic compounds. You could conclude that the samples are different compounds if subsequent (TLC) analyses with different developing solvents reveal different R_f values for each sample. Further analysis by infrared (IR) or nuclear magnetic resonance (NMR) spectroscopy would be needed to provide definitive evidence about whether the compounds are identical or not.

FIGURE 15.2
Measurements for the
R_f value.

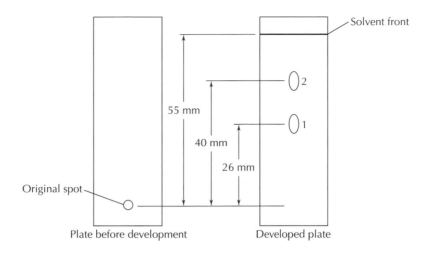

15.2 Plates for Thin-Layer Chromatography

Adsorbents

Three polar solid adsorbents—silica gel ($SiO_2 \cdot xH_2O$), aluminum oxide (Al_2O_3, also called alumina), and cellulose—are commonly used as stationary phases for thin-layer chromatography. A number of intermolecular forces cause organic molecules to bind to these polar stationary phases. Only weak van der Waals forces bind nonpolar compounds to the adsorbent, but polar molecules can also adsorb by dipole-dipole interactions, hydrogen bonding, and coordination to the highly polar metal oxide surfaces. The strength of the interaction varies for different compounds, but one generality can be stated: **the more polar the compound, the more strongly it binds to silica gel or alumina.**

Silica gel and aluminum oxide. Silica gel and alumina are prepared from activated, finely ground powder. Activation usually involves heating the powder to remove adsorbed water. Silica gel is acidic, and it separates acidic and neutral compounds that are not too hydrophilic ("water-loving"). Aluminum oxide is available in acidic, basic, and neutral formulations.

If the plastic seal on a package containing precoated silica gel or alumina sheets has been broken for some time, the TLC plates should be activated before use to remove adsorbed water. This practice enhances the reproducibility of R_f values. Activation is done simply by heating the sheets in a clean oven at the temperature recommended by the manufacturer for 15–30 min.

Cellulose. Cellulose is less polar than silica gel and alumina and is used for the partition chromatography of water-soluble and quite polar organic compounds, such as sugars, amino acids, and nucleic acid derivatives. Cellulose can adsorb up to 20% of its weight in water; the substances being separated partition themselves between the developing solvent and the water molecules that are hydrogen-bonded to the cellulose particles. Paper chromatography is another example of using cellulose as a stationary phase.

Adsorbents for reverse-phase TLC.   The adsorbents used on plates for reverse-phase thin-layer chromatography are based on silica gel modified with alkyl chains, such as C_{18}, C_8, or C_2. The alkyl chains provide a nonpolar surface on the stationary phase that is useful for separating polar compounds. The solvents used in reverse-phase TLC are quite polar, for example, methanol or acetonitrile, often mixed with water. In reverse-phase TLC, the order of movement up the TLC plate is reversed; more polar compounds travel farther up the TLC plate than less polar compounds, which bind more tightly to the nonpolar adsorbent surface.

Backing for TLC Plates

A number of manufacturers sell TLC plates that are precoated with a layer of adsorbent. Plates are available with plastic, glass, or aluminum backing. R_f values for the same adsorbent and solvent may differ if plates of different backings are used.

Plastic backing.   Plastic-backed silica gel plates are usually the least expensive. They can be cut to any desired size with a paper cutter or scissors. The adsorbent surface is of uniform thickness, usually 0.20 mm. Results are quite reproducible, and sharp separation is normal. The plastic backing is generally a solvent-resistant polyester polymer. The adsorbent is bound to the plastic by solvent-resistant polyvinyl alcohol, which binds tightly to both the adsorbent and the plastic. Precoated plastic plates impregnated with a fluorescent indicator are also available; these plates facilitate the visualization of many colorless compounds with a UV lamp [see Technique 15.5].

We suggest using TLC plates of 2.5 × 6.7 cm; 24 plates can be cut from a standard 20 × 20 cm sheet.

Glass and aluminum backing.   TLC plates with a glass or aluminum backing are also available in the standard 20 × 20 cm sheets. Both types can be heated without melting the backing; this property is important if the plate is to be visualized with a reagent that requires heating [see Technique 15.5]. Aluminum sheets can be cut with scissors into convenient sizes for TLC plates. Glass sheets can be cut with a special diamond-tipped tool.

Trimming the edges of plastic- and aluminum-backed plates.   When plastic- or aluminum-backed TLC plates are cut with scissors or a paper cutter, the silica gel coating can become separated from the backing along the cut. A channel forms along the separation where solvent can preferentially migrate and cause uneven flow up the plate. To minimize the effect of any separation of the silica gel coating, trim the sides of the plate at the bottom, as shown in Figure 15.3, before using it for a TLC analysis. Wear gloves to keep skin oils and moisture off the plate; hold the plate only by the edges. Cut down on the coating with the top blade of the scissors (Figure 15.3a). Start the first cut 2–3 mm from the side and taper the cut toward the edge so it ends about 2 cm up the plate (Figure 15.3b). Turn the plate around and begin the second cut along the side of the plate about 2 cm from

FIGURE 15.3
Trimming a plastic- or aluminum-backed TLC plate to minimize separation of the silica gel coating from the backing.

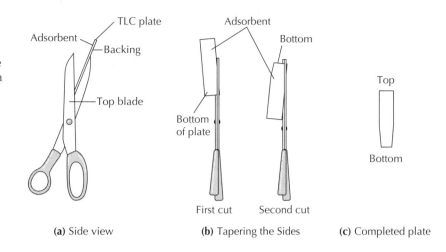

(**a**) Side view   (**b**) Tapering the Sides   (**c**) Completed plate

the bottom, tapering the cut so it removes about 2–3 mm from the width of the plate at the bottom. The resulting plate has slightly tapered sides (Figure 15.3c).

15.3   Sample Application

For TLC analysis, dissolve 10–20 mg of the solid in 1 mL of solvent.

The sample must be dissolved in a volatile organic solvent; a 1–2% solution works best. The solvent needs a high volatility so that it evaporates almost immediately. Acetone or ethyl acetate is commonly used. If you are analyzing a solid, dissolve 10–20 mg of it in 1 mL of the solvent. If you are analyzing a nonvolatile liquid, dissolve about 10 μl of it in 1 mL of the solvent.

Micropipets for Spotting TLC Plates

Commercial micropipets are available in 5- and 10-μL sizes and work well for applying samples onto plastic-backed plates. Glass and aluminum-backed plates require micropipets of a smaller interior diameter. Narrow capillary tubes of 0.7 mm internal diameter are commercially available.

A micropipet can be made easily from an open-ended, thin-walled, melting-point capillary tube. The capillary tube is heated at its midpoint. (A microburner is ideal because only a small flame is required, but a Bunsen burner may also be used.) The softened glass tube is stretched and drawn into a narrower capillary.

SAFETY PRECAUTION

Be sure there are no flammable solvents in the vicinity when you are using a microburner or Bunsen burner.

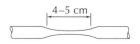

4–5 cm

FIGURE 15.4
Constricted capillary tube.

While heating the tube, rotate it until it is soft on all sides over a length of 1–2 cm. When the tubing is soft, draw out the heated part until a constricted portion 4–5 cm long is formed (Figure 15.4).

After cooling the tube for a minute or so, score it gently at the center with a file and break it into two capillary micropipets. The diameter at the end of a micropipet should be about 0.3 mm. The break must be a clean one, at right angles to the length of the tubing, so that when the tip of the micropipet is touched to the plate, liquid is pulled out by the adsorbent.

Spotting a TLC Plate

Tiny spots of the dilute sample solution are carefully applied with a micropipet near one end of the plate. Keeping the spots small assures the cleanest separation. It is also important not to overload the plate with too much sample; overloading leads to large tailing spots and poor separation.

No type of pen should be used for marking TLC plates because components of the ink separate during development and may obscure the samples.

Before spotting a TLC plate, measure 1 cm from the bottom edge of the plate and lightly mark both edges with a 0.3-cm or shorter pencil mark (Figure 15.5). The imaginary line between these marks indicates the compound's starting point for the R_f calculation.

The micropipet is filled by dipping one end of the capillary tube into the solution to be analyzed. Only $1-5$ μL of the sample solution are needed for most TLC analyses. Hold the micropipet vertically and apply the sample by **touching the micropipet gently and briefly to the plate** on the imaginary line between the two pencil marks (see Figure 15.5). It is important to touch the micropipet to the plate very lightly so that no hole is gouged in the adsorbent and to remove it quickly so that only a very small drop is left on the adsorbent. The spot delivered should be no more than $2-3$ mm in diameter to avoid excessive broadening of the spot during the development. If you need to apply more sample, touch the micropipet to the plate a second time at exactly the same place. Allow one spot to dry before applying the next. The spotting procedure may be repeated numerous times, if necessary.

If you are using 2.5×6.7 cm TLC plates, two spots can be applied to one plate (see Figure 15.5); each spot should be one-third of the distance from the side of the plate. Three spots require a 3.0-cm-wide plate. The spots become larger by diffusion during development, and if they get too close to each other or to the edge of the plate, the chromatograms become difficult to interpret.

FIGURE 15.5
Spotting a thin-layer plate.

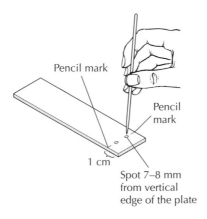

Pencil mark

Pencil mark

1 cm

Spot 7–8 mm from vertical edge of the plate

Testing the amount of sample to spot. You can quickly test for the proper amount of solution to spot on the plate by spotting two different amounts on the same plate; develop the plate as directed in Technique 15.4 and decide which spot gives better results.

Using known standards. If available, an authentic standard should be included on the TLC plate for comparison. If two compounds have the same R_f value, they may be the same compound; if the R_f values differ significantly, they most definitely are not the same compound. If R_f values are quite close, it is best to run the chromatogram again, using a longer TLC plate or a different solvent.

15.4	Development of a TLC Plate

If a developing solvent is not specified for the system you are analyzing, read Technique 15.7 on how to choose a suitable developing solvent before undertaking your TLC analysis.

Preparing the Developing Chamber

To ensure good chromatographic resolution, the developing chamber must be saturated with solvent vapors to prevent the evaporation of solvent from the TLC plate as the solvent rises up the plate. Inserting a piece of filter paper three-quarters of the way around the inside of the developing chamber helps to saturate its atmosphere with solvent vapor by wicking solvent into the upper region of the chamber (Figure 15.6). The paper wick should be a little shorter than a TLC plate so that the plate does not touch the paper. After adding the correct amount of developing solvent, shake the capped TLC chamber briefly to ensure that the paper wick is saturated with solvent.

The solvent depth in the developing chamber must be less than the height of the spots on the TLC plate.

Use enough developing solvent to allow a shallow layer (3–4 mm) to remain on the bottom after the closed chamber has been shaken to wet the filter paper. If the solvent level in the jar is too high, the spots on the plate may be below the solvent level. Under these conditions, the spots leach into the solvent, thereby ruining the chromatogram.

Carrying Out the TLC Development

Uncap the developing chamber and place the TLC plate inside with a pair of tweezers. Recap the chamber, and let the solvent move up the plate. The adsorbent will become visibly moist. **Do not lift or otherwise disturb the chamber while the TLC plate is being developed.**

Do not touch the adsorbent side of the TLC plate with your fingers. Hold the plate by the edge with a pair of tweezers.

The development of a chromatogram usually takes 5–10 min if the chamber is saturated with solvent vapor. When the solvent front is about 1 cm from the top of the plate, remove it from the developing chamber with a pair of tweezers and immediately mark the adsorbent at the solvent front with a pencil. To get accurate R_f values, the final position of the front must be marked before any evaporation occurs. Let the developing solvent evaporate from the plate before visualizing the results.

FIGURE 15.6
Typical developing
chamber.

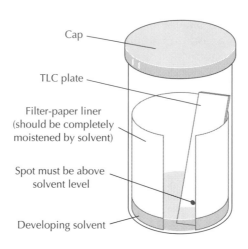

Cap

TLC plate

Filter-paper liner
(should be completely
moistened by solvent)

Spot must be above
solvent level

Developing solvent

S A F E T Y   P R E C A U T I O N

Evaporate the solvent from a developed chromatogram in a fume
hood.

15.5   Visualization Techniques

Chromatographic separations of colored compounds can usually
be seen directly on the TLC plate, but colorless compounds
require indirect methods of visualization. Fluorescence and visualization reagents are commonly used to visualize TLC plates. No
matter which visualization method is used, you should calculate the
R_f value for each compound and record it in your notebook, along
with the experimental conditions of the chromatogram.

Fluorescence   The simplest visualization technique involves the use of adsorbents
that contain a fluorescent indicator. The insoluble inorganic indicator
rarely interferes in any way with the chromatographic results and
makes visualization straightforward. When the output from a short-
wavelength ultraviolet lamp (254 nm) is used to illuminate the adsorbent side of the plate in a darkened room or dark box, the plate
fluoresces visible light. The separated compounds appear as dark
spots on the fluorescent field because the substances forming the spots
usually quench the fluorescence of the adsorbent, as shown in Figure
15.7a. Sometimes substances being analyzed are visible by their own
fluorescence, producing a brightly glowing spot. Outline each spot
with a pencil while the plate is under the UV source to give a perma-
nent record of the analysis and allow the calculation of R_f values.

S A F E T Y   P R E C A U T I O N

Never look directly at an ultraviolet radiation source. Like the sun, it
can cause eye damage.

FIGURE 15.7
Visualization.

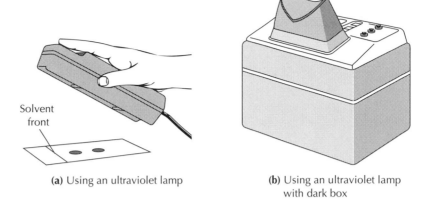

(**a**) Using an ultraviolet lamp

(**b**) Using an ultraviolet lamp
with dark box

*Visualization
Reagents*

Not all substances are visible on fluorescent silica gel, so visualization by one of the following methods should also be tried on any unknown sample.

Dipping or spraying reagents. TLC plates can be dipped briefly in visualizing solutions containing reagents that react to form colored compounds upon heating. Alternatively, the TLC plates can be sprayed with the visualizing solution. Visualization occurs by heating the dipped or sprayed TLC plates with a heat gun or on a hot plate. Two common visualizing solutions are *p*-anisaldehyde and phosphomolybdic acid.*

The spots fade with time, so they should be outlined with a pencil soon after the visualization process.

Iodine visualization. Another way to visualize colorless organic compounds uses their absorption of iodine (I_2) vapor. The TLC plate is put in a bath of iodine vapor prepared by placing 0.5 g of iodine crystals in a tightly capped jar. Colored spots are gradually produced from the reaction of the substances with iodine vapor. The spots are dark brown on a white to tan background. After 10–15 min, the plate can be removed from the jar containing iodine. Sometimes it is necessary to warm the jar slightly on a steam bath or hot plate to increase the amount of iodine vapor in the visualization chamber. The colored spots disappear in a short period of time, so they should be outlined with a pencil immediately after the plate is removed from the iodine bath. The spots will reappear if the plate is again treated with iodine.

Further information on visualization reagents. Consult the references at the end of the chapter for detailed discussions of visualization reagents.

**p-Anisaldehyde visualizing solution: 2 mL of *p*-anisaldehyde in 36 mL of 95% ethanol, 2 mL of concentrated sulfuric acid, and 5 drops of acetic acid. Phosphomolybdic acid visualizing solution: 20% by weight in ethanol.

15.6   Summary of TLC Procedure

1. Obtain a precoated TLC plate of the proper size for the developing chamber.
2. Lightly mark the edges of the origin line with a pencil. Spot the plate with a small amount of a 1–2% solution containing the compounds to be separated.
3. Add a filter-paper wick to the developing jar. Then add a suitable solvent, cap the jar, and shake it briefly to saturate it with solvent vapors.
4. Develop the chromatogram.
5. Mark the solvent front immediately after removing the plate from the developing chamber.
6. Visualize the chromatogram and outline the separated spots.
7. Calculate the R_f value for each compound.

15.7   How to Choose a Developing Solvent
When None Is Specified

When the developing solvent for a TLC analysis is not indicated, how do you select a suitable solvent? In general, you should use a nonpolar developing solvent for nonpolar compounds and a polar developing solvent for polar compounds. Table 15.1 is a good place to start. It shows the relative polarity of common TLC developing solvents and organic compounds by functional-group class.

Finding a Suitable Developing Solvent

Chromatographic behavior is the result of competition between the compounds being separated and the developing solvent for the surface of the adsorbent. Selecting a suitable solvent is often a trial-and-error process, particularly if a mixture of solvents is required to give good separation. If the solvent is adsorbed too well, the adsorbent is deactivated, R_f values are increased, and separation may be incomplete. Therefore, a solvent that causes all the spotted material to move with the solvent front is too polar, whereas one that does not cause any compounds to move from the original spot is not polar enough. An appropriate solvent for a TLC analysis gives R_f values of 0.20–0.70, with ideal values in the range 0.30–0.60.

With a silica gel plate, nonpolar hydrocarbons should be developed with hydrocarbon solvents, but a mixture containing an alcohol and an ester might be developed with a hexane/ethyl acetate mixture. Highly polar solvents are seldom used with silica gel plates.

If you know the compounds in the mixture you want to separate, use Table 15.1 to select solvents to test. If the composition of the mixture is unknown, begin by testing with a nonpolar solvent and then with a medium-polarity solvent or a mixture of hexane

TABLE 15.1	Relative polarities of common TLC solvents and organic compounds	
Common developing solvents	Increasing polarity	Organic compounds by functional-group class
Alkanes, cycloalkanes		Alkanes
Toluene		Alkenes
Dichloromethane		Aromatic hydrocarbons
Diethyl ether		Ethers, halocarbons
Ethyl acetate		Aldehydes, ketones, esters
Acetone		Amines
Ethanol		Alcohols
Methanol		Carboxylic acids
Acetonitrile		
Water		

and ethyl acetate. If a very polar solvent is required to move spots on a particular TLC adsorbent, better results may be obtained by switching to a less active adsorbent and a less polar solvent. Silica gel is less polar than most grades of alumina.

A Rapid Method for Testing Developing Solvents

As a rapid way to determine the best TLC developing solvent among several possibilities, three or four samples can be spotted along the length of the same plate (Figure 15.8). Fill a micropipet with the solvent to be tested and gently touch one of the spots. The solvent will diffuse outward in a circle, and the sample will move out with it. Mixtures of compounds will be partially separated and approximate R_f values can be estimated.

Consider the separation of an alcohol and an ester. Start with a relatively nonpolar solution of 90:10 (v/v) hexane/ethyl acetate. If the R_f values are below 0.2, test a second spot with 70:30 (v/v) hexane/ethyl acetate, then 50:50 (v/v) hexane/ethyl acetate, then pure ethyl acetate. If an ethyl acetate system does not produce R_f values in the satisfactory range, select a more polar solvent system such as a mixture of diethyl ether and acetone and repeat the test with various proportions.

FIGURE 15.8
Rapid method for DNB determining an effective TLC solvent: (a) good development; (b) and (c) poor development.

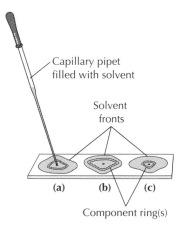

Capillary pipet filled with solvent

Solvent fronts

(a)   (b)   (c)

Component ring(s)

| **15.8** | **Sources of Confusion** |

The R_f Values Are Very Similar

If the R_f values for two compounds are very similar—within ± 0.05—another solvent or mixture of solvents should be tested in order to distinguish between them.

Multiple Overspotting

A question that often arises is how many times to spot a sample on the TLC plate. The answer depends on several factors: the concentration of the spotting solution, the diameter of the capillary spotting tube, how long the capillary tube is in contact with the adsorbent, and the thickness of the adsorbent on the TLC plate. Do a quick trial to determine how many times to spot the sample solution by spotting two different amounts on the same plate and developing the TLC plate. Decide which gives the best results.

No Spots Are Apparent on the Developed Plate

There are several possible reasons why no spots are seen on a developed plate—the origin line may have been submerged, not enough sample was spotted, the UV lamp was set on the wrong wavelength, the wrong side of the plate was irradiated, the dipped plate was not heated long enough, or the compounds being analyzed are volatile.

The solvent level in the developing jar was too high.   Check the solvent level in the developing jar. Was the depth of the solvent high enough to submerge the origin line containing the spots? If so, the spots probably leached into the developing solvent instead of moving up the plate as the solvent ascended.

Not enough sample was spotted.   If the sample solution is too dilute or too little spotting is used, the developed spot might not be visible because there is not enough material to see.

The UV lamp was set at the wrong wavelength.   Most UV lamps have two switches—one for short-wavelength light and one for long-wavelength light. Short-wavelength light is necessary for visualizing TLC plates. Check that you selected the correct switch.

The wrong side of the TLC plate was irradiated by the UV light.   The spots will be visible only if you irradiate the side of the plate containing the TLC adsorbent.

The dipped plate was not heated long enough.   A few minutes of heating are necessary to visualize the spots when p-anisaldehyde or phosphomolybdic acid visualizing solutions are being used.

The compounds being analyzed are volatile.   A liquid sample with a boiling point below 160°C may evaporate from the TLC plate before the plate is visualized. A solid compound that sublimes could also do so before the plate is visualized.

Large, Overlapping, or Tailing Spots

The developed TLC plate may show very large spots, two spots that overlap at the center of the plate, or a spot that shows a long oval tail instead of being circular. Tailing spots, in particular, lead to poor reproducibility of R_f values. These problems are likely to arise because too large a sample of the spotting solution was applied to the TLC plate. Prepare another plate using smaller spots and less overspotting. If the spots are still too large or if they tail, prepare a more dilute spotting solution.

Purity of the Developing Solvent

The purity of the developing solvent is an important factor in the success of a TLC analysis and in obtaining reproducible R_f values. The presence of a soluble impurity can dramatically affect the developing power of the resulting solution compared with that of the pure solvent. For example, the presence of water in acetone changes its developing power appreciably, and therefore the R_f values will differ from values obtained with pure acetone.

Can I Get Quantitative Information from a TLC Analysis?

The size and intensity of the spots can be used as a rough measure of the relative amounts of the substances. These parameters can be misleading, however, especially with fluorescent visualization. Some organic compounds interact much more intensely with ultraviolet radiation than do others, making one spot appear to be more concentrated than another when indeed that may not be the case. Quantitative information is not one of the strengths of thin-layer chromatography.

References

1. Touchstone, J. C. *Practice of Thin Layer Chromatography*; 3rd ed.; Wiley: New York, 1992.
2. Fried, B.; Sherma, J. *Thin-Layer Chromatography: Techniques and Applications*; 4th ed.; Chromatographic Science Series, Vol. 81, Marcel Dekker: New York, 1999.
3. Sherma, J.; Fried, B. (Eds.) *Handbook of Thin-Layer Chromatography*; 3rd ed.; Chromatographic Science Series, Vol. 89, Marcel Dekker: New York, 2003.

Questions

1. When 2-propanol was used as the developing solvent, two substances moved with the solvent front ($R_f = 1$) during TLC analysis on a silica gel plate. Can you conclude that they are identical? If not, what additional experiment(s) would you perform?

2. The R_f value of compound A is 0.34 when the TLC plate is developed in hexane and 0.44 when the plate is developed in diethyl ether. Compound B has an R_f value of 0.42 in hexane and 0.60 in diethyl ether. Which solvent would be better for separating a mixture of compounds A and B by TLC? Explain.

3. A student wishes to analyze a mixture containing an alcohol and a ketone by silica gel TLC. After consulting Table 15.1, suggest a likely developing solvent.

4. If the structures of the two compounds in question 2 are very similar except for the functional group, which would have the larger R_f (a) on silica gel? (b) on alumina?

16   GAS-LIQUID CHROMATOGRAPHY

If Technique 16 is your introduction to chromatographic analysis, read the introduction to chromatography on pages 175–176 before you read Technique 16.

Few techniques have altered the analysis of volatile organic chemicals as much as gas-liquid chromatography (GC). Before GC became widely available just over 50 years ago, organic chemists usually looked for ways to convert liquid compounds into solids in order to analyze them. Gas-liquid chromatography changed all that by providing a quick, easy way for both qualitative and quantitative analysis of volatile organic mixtures. In addition, GC has a truly fantastic ability to separate complex mixtures.

Gas-liquid chromatography does, however, have limitations. It is useful only for the analysis of small amounts of compounds that have vapor pressures high enough to allow them to pass through a GC column, and, like thin-layer chromatography (TLC), gas-liquid chromatography does not identify compounds unless known samples are available. Coupling a gas-liquid chromatograph with a mass spectrometer (GC-MS) combines the superb separation capabilities of GC with the superior identification methods of mass spectrometry [see Technique 20].

Overview of Gas-Liquid Chromatography

In gas-liquid chromatography the stationary phase consists of a nonvolatile liquid, usually a polymer, with a high boiling point. In capillary columns, a thin uniform film of the liquid phase is applied either to the interior wall of a long, narrow capillary tube or to a thin layer of solid support lining the capillary tube. In either case, a clear channel through the center is left for passage of a carrier gas and molecules of the sample (Figure 16.1a). For older packed-column chromatographs, the liquid is coated on a porous, inert solid support that is then packed into a tube (Figure 16.1b).

A flow of inert gas, such as helium or nitrogen, serves as the mobile gas phase. When the mixture being separated is injected into the heated injection port, the components vaporize and are

FIGURE 16.1
Microview of (a) a wall-coated open tubular capillary column and (b) a packed column.

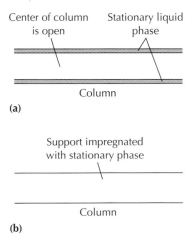

Center of column is open   Stationary liquid phase

Column

(a)

Support impregnated with stationary phase

Column

(b)

FIGURE 16.2
Stages in the separation of a two-component (**A, B**) mixture as it moves through a packed column.

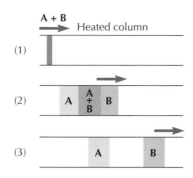

carried by the carrier gas into the column, where separation occurs. The compounds in the mixture partition themselves between the gas phase and the liquid phase in the column, in an equilibrium that depends on the temperature, the rate of gas flow, and the solubility of the components in the liquid phase (Figure 16.2)

Mixtures separate during gas chromatography because their components interact in different ways with the liquid stationary phase. A GC column has thousands of theoretical plates as a result of the huge surface area on which the gas and liquid phases can interact [see Technique 11.4, page 142, for a discussion of theoretical plates]. The partitioning of a substance between the liquid and gas phases depends both on its relative attraction for the liquid phase and on its vapor pressure. The greater a compound's vapor pressure, the greater its tendency to go from the liquid stationary phase into the mobile gas phase. So, in the thousands of liquid-gas equilibria that take place as substances travel through a GC column, a more volatile compound spends more time in the gas phase than does a less volatile compound. In general, lower-boiling-point compounds with higher vapor pressures travel through a GC column faster than higher-boiling compounds.

16.1   Instrumentation for GC

The basic parts of a gas-liquid chromatograph are as follows:

- Source of high-pressure pure *carrier gas*
- *Flow controller*
- Heated *injection port*
- *Column* and *column oven*
- *Detector*
- *Recording device*

These components are shown schematically in Figure 16.3.

A small hypodermic syringe is used to inject the sample through a sealed rubber septum or gasket into the stream of carrier gas in the heated injection port (Figure 16.4). The sample vaporizes immediately and the carrier gas sweeps it into the column—a metal, glass, or fused-silica tube that contains the liquid phase (Figure 16.5). The column is enclosed in an oven whose temperature can be regulated

FIGURE 16.3
Schematic diagram of
a gas-liquid chromato-
graph.

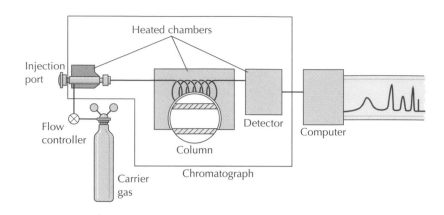

FIGURE 16.4
Injection port during
sample injection.

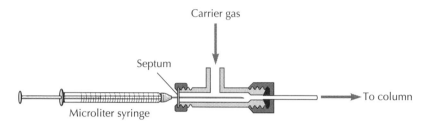

FIGURE 16.5
GC columns.

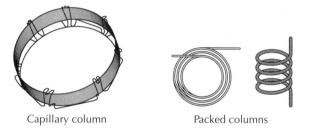

from just above room temperature to greater than 200°C. After the
sample's components are separated by the column, they pass into a
detector, where they produce electronic signals that can be amplified
and recorded.

16.2   Types of Columns and Liquid Stationary Phases

A gas chromatograph can have either capillary or packed
columns. Capillary columns, also called open tubular columns,
have an interior diameter of only 0.2–0.5 mm and a length of
10–100 m. A packed column typically has an interior diameter of
2–4 mm and a length of 2–3 m. Capillary columns usually give
much better separation than do packed columns. The greater
length of capillary columns and the better diffusion of sample
molecules in and out of the liquid phase provide more theoretical
plates whereby equilibration of the sample molecules with the
liquid stationary phase and the gas phase can occur. Capillary

columns not only give better separations, they also do it in a much shorter analysis time.

Types of Columns

Capillary columns.   Several types of capillary columns are available. In a *wall-coated open tubular column (WCOT),* the liquid phase coats the interior surface of the tube, leaving the center open. In a *support-coated open tubular column (SCOT),* the liquid phase coats a thin layer of solid support that is bonded to the capillary wall, again leaving the center of the column open.

Packed columns.   The solid support in packed columns (and SCOT capillary columns) consists of a porous, inert material that has a very high surface area. The most commonly used substance is calcined diatomaceous earth, which contains the crushed skeletons of algae, especially diatoms. Its major component is silica. The efficiency of separation increases with decreasing particle size as a consequence of the expanded surface area available for the liquid coating. With packed columns, however, there is a practical lower limit to the particle size because increased gas pressure is necessary to push the mobile phase through a column packed with smaller particles.

Nature of the Liquid Stationary Phase

The liquid stationary phase interacts with the substances being separated by a number of intermolecular forces: dipolar interactions, van der Waals forces, and hydrogen bonding. These intermolecular forces determine the relative volatility of the adsorbed compounds and play dominant roles in the separation process.

As a general rule, a liquid phase provides the best separation if it is chemically similar to the compounds being separated. Nonpolar liquid coatings are used to separate nonpolar compounds, and polar liquid phases are best for separating polar compounds. In part, this rule is simply a manifestation of the old adage "Like dissolves like." Unless the sample dissolves well in the liquid phase, little or no separation occurs as the substance passes through the column. Table 16.1 lists some commonly used liquid stationary phases for both packed and capillary columns and gives their chemical composition.

Silicones, or polysiloxanes, are polymers with a silicon/oxygen backbone, which can have variation in the R groups attached to the silicon atoms. If all the R groups are methyl, the liquid phase is nonpolar. Substituting benzene rings (phenyl groups) for 5–10% of the methyl groups increases the polarity somewhat. Substitution of other functional groups for the methyl groups of polydimethylsiloxane provides a wide variety of stationary phases suited to almost any application.

Polyethylene glycol, commonly called Carbowax, and diethylene glycol succinate are polymers frequently used as liquid phases for separating polar compounds, which they dissolve in part by being good hydrogen bond acceptors.

Useful Temperature Range of a Liquid Phase

An important characteristic of a liquid phase is its useful temperature range. A stationary phase cannot be used under conditions in which it decomposes or in which its vapor pressure is high enough

TABLE 16.1 Common GC liquid stationary phases

Polarity of column	Chemical composition
Nonpolar	 R = CH$_3$ Polydimethylsiloxane (methyl silicone)
Medium polarity	 R = CH$_3$ or C$_6$H$_5$ Polymethylphenylsiloxane (methylphenyl silicone) Typically, 5–50% of the R groups are phenyl
Polar	—O—CH$_2$—CH$_2$—O—CH$_2$—CH$_2$—O— Polyethylene glycol (Carbowax) Diethylene glycol succinate (DEGS polyester)

that it vaporizes from the column. All liquid stationary phases evaporate, or "bleed," if they are heated to a high enough temperature; this vaporized material then fouls the detector. Therefore, GC columns have specified temperature maxima.

Selecting a Liquid Phase

The proper choice of a liquid stationary phase is often a trial-and-error process. Published experimental procedures usually specify the type of column used for a GC analysis, but eventually you might have to make your own choices. Tables of appropriate liquid phases for specific classes of compounds can be found in the references at the end of the chapter.

16.3 Detectors

Two kinds of detectors are most often used in gas-liquid chromatography: *flame ionization detectors* and *thermal conductivity detectors.* The function of a detector is to "sense" a material and convert the sensing into an electrical signal.

FIGURE 16.6
Chemical reactions in a flame ionization detector.

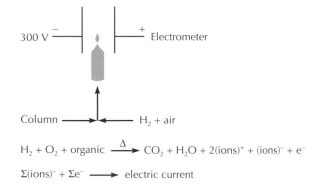

$$H_2 + O_2 + organic \xrightarrow{\Delta} CO_2 + H_2O + 2(ions)^+ + (ions)^- + e^-$$

$$\Sigma(ions)^- + \Sigma e^- \longrightarrow \text{electric current}$$

Flame Ionization Detectors (FIDs)

Flame ionization is a highly sensitive detector system that is commonly used with capillary columns, where the amount of sample reaching the detector is substantially less than that emanating from a packed column. In a flame ionization detector, the organic substances leaving the column are burned in a hydrogen/air flame. The combustion process produces ions that alter the current output of the detector (Figure 16.6). In the chromatograph, the electrical output of the flame is fed to an electrometer, whose response can be recorded.

Thermal Conductivity Detectors (TCDs)

The older thermal conductivity detectors operate on the principle that heat is conducted away from a hot body at a rate that depends on the composition of the gas surrounding it. In other words, heat loss is related to gas composition. The electrical component of a thermal conductivity detector is a hot wire or filament. Most of the heat loss from the hot wire of the detector occurs by conduction through the gas and depends on the rate at which gas molecules can diffuse to and from the metal surface. Helium, the carrier gas most often used with thermal conductivity detectors, has an extremely high thermal conductivity. Larger organic molecules are less efficient heat conductors because they diffuse more slowly. With only carrier gas flowing, a constant heat loss is maintained and there is a constant electrical output. When an organic compound reaches the detector, the gas composition changes and causes the hot filament to heat up and its electrical resistance to increase. The change in electrical resistance creates an imbalance in the electrical circuit that can be recorded.

In practice, the filament of a thermal conductivity detector, a tungsten/rhenium or platinum wire, operates at temperatures from 200°C to over 400°C. An enlarged view of a common thermal conductivity detector is shown in Figure 16.7. Thermal conductivity detectors have the advantages of stability, simplicity, and the option of recovery of the separated materials but the disadvantage of low sensitivity. Because of their low sensitivity, they are unsuitable for use with capillary columns.

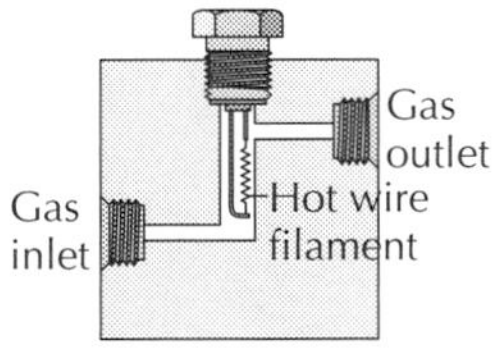

FIGURE 16.7
Thermal conductivity detector.

16.4   Recording the Chromatogram

The recorded response of the detector's electrical signal as the sample passes through it over time is called a *chromatogram*. A typical chromatogram for a mixture of alcohols, which plots the intensity of the detector response against time, is shown in Figure 16.8. The chromatogram shows the changes in the electrical signal as each component of the mixture passes through the detector. You will notice that the later peaks are somewhat broader. This pattern is typical; the longer a compound remains on the column, the broader its peak will be when it passes through the detector.

Most modern gas chromatographs are equipped with computer-based data stations that allow the display and manipulation of the results on the recorder or computer screen in a variety of ways. Not only can the computer show the chromatogram, but it can calculate and display the peak retention times, peak areas, and the percentage composition of the components in the mixture being studied.

Retention Time

Under a definite set of experimental conditions, a compound always travels through a GC column in a fixed amount of time, called the *retention time.* The retention time for a compound, like the *Rf* value in thin-layer chromatography, is an important number, and it is reproducible if the same set of instrumental parameters is maintained from one analysis to another.

Figure 16.9 shows how retention times are determined from a chromatogram. The distance from the time of injection to the time where the peak maximum occurs is the retention time for that compound. If you are not using a modern computer-based data station, you can determine the retention time manually by measuring the distance from the injection to the peak on the chromatogram and dividing it by the recorder chart speed.

The retention time depends on many factors. Of course, the compound's structure is one of them. Beyond that, the kind and amount of stationary liquid phase used in the column, the length of

FIGURE 16.8
GC of a complex mixture of alcohols.

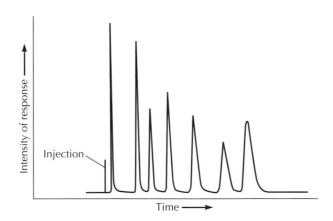

FIGURE 16.9
Measuring retention
times.

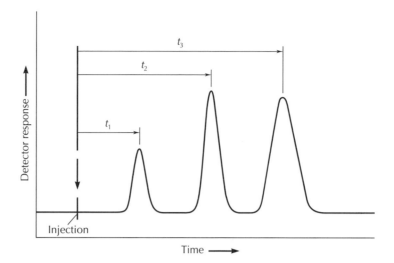

the column, the carrier gas flow rate, the column temperature regimen, the solid support, and the column diameter are most important. To some extent, the sample size can also affect the retention time. Always record these experimental parameters when you note a retention time in your lab notebook.

16.5   Practical GC Operating Procedures

Modern GCs have great analytical power, but they are also complex. You need to learn the functions of many buttons, switches, and dials, and you need to learn the many steps in the procedure for readying the gas chromatograph for an analysis. Your instructor or lab technician will probably have already set a number of the instrumental parameters, but you should always check to ensure that they have been set correctly. Your instructor will show you how to do these operations, because the procedures vary for different instruments.

*Turning On the GC
and Adjusting the
Carrier Gas*

First make sure that the chromatograph and the detector are heated and ready to go and that the carrier gas is on and its pressure is properly set. The necessary pressure depends on the instrument and columns you are using, so check with your instructor before changing the pressure setting. Flow rates for capillary columns generally range from 60 to 70 mL/min; capillary-column chromatographs have built-in flowmeters. With a packed column that is 6 ft long and 1/8 inch in diameter, a flow rate of 20–30 mL/min is common; for a 1/4-inch column of the same length, 60–70 mL/min is usual. A convenient measure of the carrier gas flow rate in a packed-column chromatograph is made at the exit port by using a soap-film (bubble) flowmeter (Figure 16.10).

*Choosing the
Correct GC Column
and Temperatures*

Most modern gas chromatographs have two different columns, only one of which is operational at any time. You can activate the column of your choice with the flick of a switch. Decide whether

FIGURE 16.10
Bubble flowmeter.

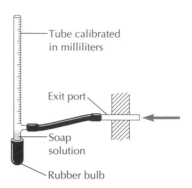

Tube calibrated
in milliliters

Exit port

Soap
solution

Rubber bulb

a polar or nonpolar column is needed to separate the sample being
analyzed and send the signal for that column to the detector. You
also need to see that the GC column oven temperature is set properly and that the detector and the injector port are also at the correct
temperatures. Temperature equilibration can require 20–30 min for
a given set of operating parameters.

The column temperature can be programmed to increase during
an analysis on modern capillary-column GCs. This feature gives the
instrument far greater flexibility compared with the older isothermal
gas chromatographs where a constant column temperature is used.
Having the option of temperature programming allows you to begin
a GC run at 50°C or so and then increase the column temperature at
a selected rate per minute until it reaches a selected maximum temperature. Using temperature programming allows the efficient and
quick separation and analysis of organic mixtures whose components have widely different volatilities.

*Turning on the
Detector*

If you are using a flame ionization detector, the hydrogen and air
tanks must be regulated with the correct flow rates, and the flame
must be lit. It's likely that your instructor will carry out this operation.

Before the sample is injected, the detector circuit must be balanced and the proper sensitivity (attenuation) chosen for the analysis. If you are using a thermal conductivity detector, you must turn
on the inert carrier gas flow 2–3 min before the detector current is
turned on. The thin metal filament of the detector can oxidize and
burn out in the presence of oxygen, much like a tungsten light bulb.

*Sample Size and
Microliter Syringes*

As soon as the instrument is ready and the sample is prepared, you
can inject the sample. Gas-liquid chromatographs take very small
samples, and if too much sample is injected, poor separation will
occur from overloading the column. Injecting the proper amount of
sample may be the most important operation you do in obtaining a
useful gas chromatogram. You will want to consult with your instructor about sample preparation and sample size for the chromatographs
in your laboratory.

For a packed-column GC, 1–3 μL of a volatile mixture are directly
injected through the rubber septum with a special microliter hypodermic syringe. Normally, however, for a capillary-column GC, the

sample must be in a dilute solution. A 2–5% solution in a volatile solvent, such as diethyl ether, works best. Usually 1 drop of a liquid or 20–50 mg of a solid sample diluted with 1 mL of the solvent is sufficient. Then only 0.5–1.0 μL of this dilute sample solution is injected into the GC with a microliter syringe. Even this amount of sample can overload a capillary column, so the injected mixture is split into two highly unequal flows and the smaller one is actually introduced into the column. A split ratio of 1:50 is not uncommon.

For some capillary chromatographs, it may be necessary to pull the plunger back until the entire sample is inside the syringe barrel before inserting the needle. To find out if this is necessary, consult your instructor.

Injection Technique   Proper injection technique is important if you want to get well-formed peaks on the chromatogram. Using both hands, insert the needle all the way into the injection port and immediately push the plunger with a smooth, rapid motion (Figure 16.11). Withdraw the syringe needle without delay after completing the injection. This procedure ensures that the entire sample reaches the column at one time and that there is minimal disturbance of the gas flow. If your GC is equipped with a computer-driven, automatic digital integrator, simply press the start button after withdrawing the syringe needle.

If you are using a noncomputerized packed-column GC, the time of injection can be recorded in several ways. A mark can be made on the recorder base line just after the sample has been injected, but this may be difficult to do reproducibly. If the GC has a thermal conductivity detector, a better way is to include several microliters of air in your syringe. The air is injected at the same time as the sample, and it comes through the column very

FIGURE 16.11
Injecting the sample into the column.

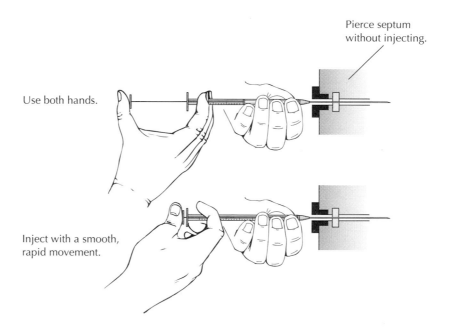

Pierce septum without injecting.

Use both hands.

Inject with a smooth, rapid movement.

quickly as the first tiny peak. Retention times can then be calculated using this air peak as the injection time. It is worth noting that an air peak cannot be used to mark injection time with a flame ionization detector because air does not burn and thus gives no peak.

Completion of a
Chromatographic
Separation

After injection, wait for the peaks to appear on the moving chromatogram. If you are analyzing mixtures with a known number of components, you need wait only until the last component has come through the column before terminating the chromatographic run. If the analysis involves an unknown mixture, it is sometimes difficult to know exactly how long to wait before injecting another sample because components with unexpectedly long retention times may still be present in the column. Determination of the total analysis time for unknown mixtures is a matter of trial and error.

Keeping Microliter
Syringes Clean

A microliter syringe has a tiny bore that can easily become clogged if it is not rinsed after use. If viscous organic liquids or solutions containing acidic residues are allowed to remain in the syringe, you may find that it is almost impossible to move the plunger. For this reason, a small bottle of acetone is often kept beside each GC instrument. One or two fillings of the syringe with acetone will normally suffice to clean it, if done directly after an injection.

However, during a series of analyses it is unnecessary to rinse the syringe with acetone after each injection. This practice may even cause confusion if traces of acetone show up on the chromatogram. For multiple analyses, it is best to rinse the syringe several times with the next sample to be analyzed before filling the syringe with the injection sample.

When you have finished your analyses, thoroughly rinse out the microliter syringe with acetone.

Record Keeping

Attach your GC printouts firmly in your lab notebook, along with a notation of the experimental conditions under which the chromatograms were run. Record the following experimental parameters:

• Injection port temperature
• Column temperature
• Detector temperature
• Carrier gas flow rate
• Injection sample size
• Length of column and identity of its liquid stationary phase

16.6   Sources of Confusion

Modern GCs have great analytical power but they are also complex, and to get good results there are many factors that require careful attention. Using a GC requires thinking and problem-solving skills.

Mastering the operation of a gas chromatograph—with the various adjustments of the column, injector port, and detector temperatures, the carrier gas flow rate, the hydrogen/air fuel mixture, and the sensitivity controls—can seem formidable. Yet it is worth the challenge, because there are few other ways to get quick quantitative data on the composition of organic mixtures.

Instrumental Parameters

A number of the instrumental parameters are likely to be set by your instructor or lab manager, but you should always check to ensure that they have been set correctly. It pays to be careful and systematic in setting up the chromatograph, because if a key factor is overlooked, you have to make the somewhat frustrating decision of how long to wait before you decide to abort a questionable experimental run that is underway. Remember also that compounds from an earlier aborted run may still be in the GC column. They may then come through the detector at unexpected times in the next chromatographic run.

Poor Separation of a Mixture

If the components of your mixture are not well separated, a number of factors can be adjusted. You may have injected too much sample into the column, the column temperature may be too high, or the wrong liquid stationary phase may have been used. Adjust only one parameter at a time until you have achieved a good separation of the mixture.

Trace Amounts of Impurities

If you are using a capillary-column GC, you will probably see many small peaks on your chromatogram that indicate the presence of trace impurities, even if you are analyzing a "pure" compound. There are virtually always tiny amounts of impurities in pure compounds. A GC chromatogram can be a vivid reminder of the immense size of Avogadro's number. Many trillions of molecules pass through the detector of a GC in every chromatographic run. If the detector is sensitive enough, the trace impurities will show up. Usually, you can safely ignore them.

Injection Technique

Developing good injection technique with a microliter syringe is probably the biggest challenge for the GC beginner.

What happens if the plunger is pushed too slowly. If the plunger of the syringe is pushed too slowly, the leading edge of the sample reaches the column before the entire sample has vaporized in the injector port. The components of the sample then move through the column as a series of fronts that overlap with the components that have longer retention times. As a result, the chromatogram shows multiple overlapping peaks and the run must be repeated.

Overlapping and repeating patterns. Overlapping peaks and repeating patterns of peaks can also occur if the sample solution is not drawn into the barrel of the syringe before the injection is made. Otherwise, the solution in the needle can vaporize into the injection port before the rest of the sample is injected. A series of overlapping and repeating peak patterns on the chromatogram signifies that the analysis will have to be done again.

Is the Microliter Syringe Working Properly?

The correct size of the injections and concentration of the sample are crucial to success. You do not want to overload the column with too large a sample. It is also possible to inject virtually no sample because the very narrow bore of the microliter syringe has become plugged. Determining if a microliter syringe is drawing properly can sometimes be difficult to ascertain. The use of packed columns makes it easier to know if the syringe is working properly because a larger sample volume is injected.

16.7   Identification of Components Shown on a Chromatogram

GC analysis can quickly assess the purity of a compound, but as with thin-layer chromatography, a compound cannot be identified by GC unless a known sample is available to use as a standard. Comparison of retention times, peak enhancement, and spectroscopy are among the methods used to identify the components of a mixture.

Comparison of Retention Times

One method of identification compares the retention time of a known compound with the peaks on the chromatogram of the sample mixture. If the operating conditions of the instrument are unchanged, a match of the reference compound's retention time to one of the sample peaks may serve to identify it. This method will not work for a mixture in which the identity of the components is totally unknown, because several compounds could have identical retention times.

Peak Enhancement

When mixtures containing known compounds are being analyzed, *peak enhancement* serves as a method for identifying a peak in the chromatogram. The sample being analyzed is "spiked" with a drop of the known compound and the mixture injected into the chromatograph. If the known that is added is identical to one of the compounds in the mixture, its peak area is enhanced relative to the other peaks on the chromatogram (Figure 16.12).

FIGURE 16.12
Identification by the peak enhancement method.

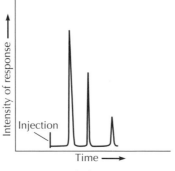

(a) Original chromatogram

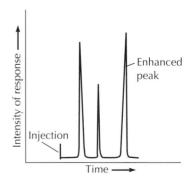

(b) Chromatogram after addition of a known compound identical to a compound in the sample

Spectroscopic
Methods

Positive identification of the compounds in a completely unknown mixture requires the pairing of GC methods with a spectroscopic method such as mass spectrometry (MS), where the mass spectrometer serves as the GC detector. In a GC-MS the two instruments are interfaced so that the separated components pass directly from the chromatograph into the spectrometer.

16.8   Quantitative Analysis

Gas-liquid chromatography is particularly useful for quantitative analysis of the components in volatile mixtures. A comparison of relative peak areas on the chromatogram often gives a good approximation of relative amounts of the compounds.

Determination of
Peak Areas

One great advantage of GC over other chromatographic methods is that approximate quantitative data are almost as easy to obtain as information on the number of components in a mixture. If we assume equal response by the detector to each compound, then the relative amounts of compounds in a mixture are proportional to their peak areas. Most peaks are approximately the shape of either an isosceles or a right triangle, whose areas are simply $A = \frac{1}{2}$ base $\times$ height. Measuring the base of most GC peaks is difficult because abnormalities in their shapes usually occur there. A more accurate estimate of peak area is $A =$ height $\times$ width at half-height (Figure 16.13).

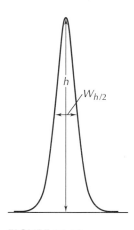

The easiest method for calculating the percentage composition of a mixture is through internal normalization: **The percentage of a compound in a mixture is its peak area divided by the sum of all peak areas.** If you have a two-component mixture

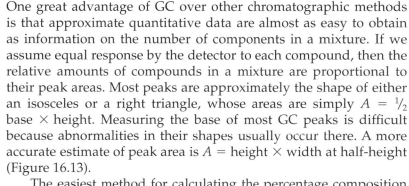

$$\% \text{ compound } 1 = \frac{\text{area}_1}{\text{area}_1 + \text{area}_2} \times 100$$

$$\% \text{ compound } 2 = \frac{\text{area}_2}{\text{area}_1 + \text{area}_2} \times 100$$

FIGURE 16.13
Determining peak area: h, height; $W_{h/2}$, width at half-height.

Electronic digital integrators, common on most modern chromatographs, determine peak areas. Chromatograms produced by these recorders include a table of data that lists both retention times and relative peak areas.

Relative Response
Factors

For accurate quantification of a GC analysis, the response of each component to the detector must be determined from known samples. Each compound has a unique response in a detector, but the detector response varies between classes of compounds. For accurate quantitative interpretation of a chromatogram, analysis of standard mixtures of known concentration must be carried out and a correction factor, called a ***response factor (f),*** must be determined for each compound. The area under a chromatographic peak, A, is proportional to the concentration, C, of the sample producing it; the response factor is the proportionality constant.

$$A = fC \qquad (1)$$

Response factors can be determined as either weight factors or mole factors, depending on the units of concentration used for the standard sample.

In chromatographic analyses, the samples being analyzed usually have more than one component; therefore, the relative response factors of one compound to the other compounds in the sample are usually determined. For a two-component system, the response-factor equation for each component is

$$A_1 = f_1 C_1 \qquad (2)$$
$$A_2 = f_2 C_2 \qquad (3)$$

The relative response factor of compound 1 to compound 2 can be determined by dividing equation 2 by equation 3:

$$\frac{A_1}{A_2} = \frac{f_1}{f_2} \times \frac{C_1}{C_2} \qquad (4)$$

Rearranging equation 4 gives the ratio of response factors, f_1/f_2, the relative response factor of compound 1 to compound 2:

$$\frac{f_1}{f_2} = \frac{A_1}{A_2} \times \frac{C_2}{C_1} \qquad (5)$$

Using data from the chromatogram shown in Figure 16.14 as an example, equation 5 can be used to calculate the molar response factor of compound 1 relative to compound 2; compound 2 is arbitrarily assigned a response factor of 1.00.

$$\frac{f_1}{f_2} = \frac{4.20}{2.18} \times \frac{2.1}{4.5} = \frac{0.90}{1.00}$$

FIGURE 16.14
Chromatogram of a standard mixture containing known concentrations of two compounds.

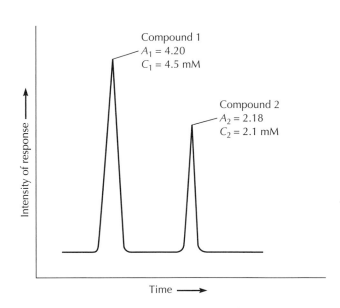

Compound 1
$A_1 = 4.20$
$C_1 = 4.5$ mM

Compound 2
$A_2 = 2.18$
$C_2 = 2.1$ mM

Intensity of response

Time

	TABLE 16.2	Molar percentage composition data for a two-compound mixture uncorrected and corrected for molar response factors, M_f			

Compound	Area (A) (arbitrary units)	Uncorrected % $(A/118.4) \times 100$	M_f	A/M_f	Corrected mol % $(A/(M_f \times 124.0)) \times 100$
Compound 1	50.2	**42.4**	0.90	55.8	**45.08**
Compound 2	68.2	**57.6**	1.00	68.2	**55.08**
Total	118.4	100	—	124.0	100

Therefore, the molar response factor for compound 1 is 0.90 relative to 1.00 for compound 2.

Once relative molar response factors have been determined, the composition of a mixture can be calculated from the areas of the peaks on a chromatogram. Table 16.2 shows how molar response factors (designated M_f) can be used to determine the corrected mole percentage composition of a sample containing compound 1 and compound 2; Table 16.2 also compares these results to the uncorrected composition that was calculated. The differences between the uncorrected and corrected compositions illustrate the necessity of using response-factor corrections for accurate quantitative analysis.

References

1. Miller, J. M. *Chromatography: Concepts and Contrasts*; 2nd ed.; Wiley: Hoboken, NJ, 2005.

2. Skoog, D. A.; Holler, F. J.; Nieman, T. A. *Principles of Instrumental Analysis*; 5th ed.; Saunders: New York, 1998.

3. Ravindranath, B. *Principles and Practice of Chromatography*; Wiley: New York, 1989.

4. Grob, R. L; Barry, E. F. *Modern Practice of Gas Chromatography*; 4th ed.; Wiley: New York, 2004.

Questions

1. Why is a GC separation more efficient than a fractional distillation?

2. What characteristics must a carrier gas have to be useful for GC? What must the characteristics of the column packing be?

3. How do (a) the flow rate of the carrier gas and (b) the column temperature affect the retention time of a compound on a GC column?

4. Describe a method for identifying a compound using GC analysis.

5. Describe a method for identifying a compound purified by and collected from a gas chromatograph.

6. If the resolution of two components in a GC analysis is mediocre but shows some peak separation, what are two adjustments that can be made in the operating parameters to improve the resolution (without changing columns or instruments)?

7. Suggest a suitable liquid stationary phase for the separation of (a) ethanol and water; (b) cyclopentanone (bp 130°C) and 2-hexanone (bp 128°C); (c) phenol (bp 182°C) and pentanoic acid (bp 186°C).

17 LIQUID CHROMATOGRAPHY

Liquid chromatography (LC) and the related methods of flash chromatography and high-performance liquid chromatography (HPLC) complete the triad of chromatographic methods so important in experimental organic chemistry. Liquid chromatography is generally used to separate mixtures of low volatility, whereas gas chromatography works only for volatile mixtures. Unlike thin-layer chromatography (TLC), liquid chromatography can be carried out with a wide range of sample quantities, ranging from a few micrograms for HPLC up to 20 g or more for column chromatography. Like TLC, liquid chromatography can be done under adsorption or partition conditions.

If Technique 17 is your introduction to chromatographic analysis, read the introduction to chromatography on pages 175–176 before you read Technique 17.

Overview of Liquid Chromatography (LC)

In liquid chromatography, also called column chromatography, the stationary phase is a solid *adsorbent* packed into a column. An *elution solvent* serves as the mobile phase and consists of either a pure liquid compound or a solution of liquids. Separation occurs by selective interactions of the compounds in the sample with the adsorbent and the elution solvent. The relative polarities of the adsorbent and the elution solvent determine the order in which compounds in the sample elute from the column. Figure 17.1 illustrates how a mixture of two compounds separates on a chromatographic column. With a polar adsorbent such as silica gel, the compound represented by A would be less polar than

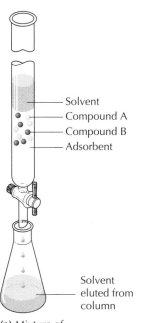

Solvent
Compound A
Compound B
Adsorbent

Solvent eluted from column

(a) Mixture of compounds A and B at top of column

(b) Compounds A and B beginning to separate

(c) Compound A starting to elute from column

FIGURE 17.1
Stages in liquid chromatographic separation of a mixture containing compound A and compound B. Compound A moves faster than does compound B, which is more strongly adsorbed on the stationary phase.

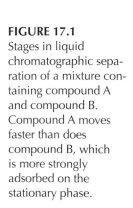

FIGURE 17.1
(continued)

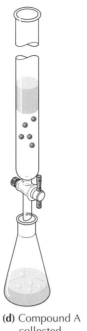

(d) Compound A
collected

(e) Compound B
starting to elute
from column

(f) Compound B
collected

compound B. In reverse-phase chromatography, a relatively non-polar adsorbent would be used, and the compound represented by A would be more polar than compound B.

17.1   Adsorbents

The most common adsorbent for the stationary phase in liquid chromatography is silica gel ($SiO_2 \cdot xH_2O$). Aluminum oxide (alumina, Al_2O_3) is also sometimes used. Most chromatographic separations today use silica gel because it allows the separation of compounds with a wide range of polarities. Silica gel also has the advantage of being less likely than alumina to cause a chemical reaction with the substances being separated. Alumina may be used for separations of compounds of low to medium polarity. Both adsorbents produce a polar stationary phase (aluminum oxide is more polar), and both are generally used with nonpolar to moderately polar elution solvents as the mobile phase. The separation of very polar compounds can be accomplished by reverse-phase chromatography.

Silica Gel

Chromatographic silica gel has 10–20% adsorbed water by weight and acts as the solid support for this water under the conditions of partition chromatography. Compounds separate by partitioning themselves between the elution solvent and the water that is strongly adsorbed on the silica surface. The partition equilibria depend on the relative solubilities of the compounds in the two liquid phases.

The adsorptive properties of silica gel may vary greatly from one manufacturer to another or even within different lots of the same grade from one manufacturer. Therefore, the solvent system previously used for a particular analysis may not work exactly the same way for another separation of the same sample mixture when silica gel from another manufacturer or from a different lot from the same manufacturer is substituted for the stationary phase.

Alumina

Activated alumina, made explicitly for chromatography, is available commercially as a finely ground powder in neutral (pH 7), basic (pH 10), and acidic (pH 4) grades, with a bulk density of about $1 \text{ g} \cdot \text{cm}^{-3}$. Different brands and grades vary enormously in adsorptive properties, mainly because of the amount of water adsorbed on the surface. The strength of the adsorption holding a substance on aluminum oxide depends on the strength of the bonding forces between the substance and the polar surface of the adsorbent. Dipole-dipole and van der Waals forces, as well as stronger hydrogen-bonding and coordination interactions, are important in the adsorption of compounds to aluminum oxide particles.

Adsorbents for Reverse-Phase Chromatography

The separation of very polar compounds may require an adsorbent of lesser polarity than either silica gel or alumina. In this situation, *reverse-phase chromatography,* which uses a stationary phase that is less polar than the mobile phase, may be useful in separating the compounds. Under reverse-phase conditions, elution of the more polar compounds occurs first, with the less polar compounds adsorbed more tightly to the stationary phase. For reverse-phase chromatography, the surface of silica particles is rendered less polar by replacing the hydrogen of terminal Si—OH groups with trialkyl silyl groups such as methyl and long-chain alkyl groups (C_{12}–C_{18}). Reverse-phase columns are commonly used in high-performance liquid chromatography [see Technique 17.9].

$$\sim\text{Si}-\text{O}-\overset{\overset{\displaystyle CH_3}{|}}{\underset{\underset{\displaystyle CH_3}{|}}{\text{Si}}}-(CH_2)_{17}CH_3$$

17.2 Elution Solvents

In liquid chromatography, the *elution solvents* used to dislodge the compounds adsorbed on the column are made increasingly more polar as the separation progresses. Nonpolar compounds bind less tightly than polar compounds on a polar adsorbent, such as silica gel, and dislodge more easily with nonpolar solvents. Therefore, the nonpolar compound in a mixture exits from the column first. The more polar compounds must be eluted, or washed out of the column, with more polar solvents.

Rate of Elution

The polarity of the solvent controls the rate at which materials elute from a silica gel column. If compounds elute too rapidly and poor separation occurs, the elution solvent is too polar. There is no universal series of eluting strengths because this property depends not only on the activity of the adsorbent but also on the compounds being separated.

Selecting an Elution Solvent

Silica gel usually works well as the adsorbent for separating most organic compounds. Thin-layer chromatography on silica gel plates [see Technique 15.7] can be used to determine the best solvent system for separating a mixture by liquid chromatography on silica gel. The separation on a silica gel TLC plate with a particular solvent or combination of solvents reflects the separation that the mixture will undergo with a silica gel column and the same solvent. A solvent that moves the desired compound to an R_f of approximately 0.3 should be a good elution solvent.

The proper choice of elution solvents and the amounts to use are, in part, a trial-and-error process. Polar compounds always require more polar elution solvents than do nonpolar compounds. For example, the separation of 1-decene from 2-chlorodecane requires elution solvents of low polarity, such as alkanes. However, the separation of the alcohol 2-decanol from its oxidation product, 2-decanone, requires more polar solvents, such as a hexane/diethyl ether mixture. Table 17.1 lists common elution solvents and organic compounds by functional-group class in order of increasing polarity.

Purity of Elution Solvents

Elution solvents for column chromatography must be rigorously purified and dried for best results. Small quantities of polar impurities can radically alter the eluting properties of a solvent. For

---

TABLE 17.1 **Relative polarities of common LC solvents and organic compounds on silica gel**

Common elution solvents	Increasing polarity	Organic compounds by functional group class
Alkanes, cycloalkanes		Alkanes
Toluene		Alkenes
Dichloromethane		Aromatic hydrocarbons
Diethyl ether		Ethers, halocarbons
Acetone (anhydrous)		Aldehydes, ketones, esters
Ethyl acetate		Amines
Ethanol (anhydrous)		Alcohols
Methanol		Carboxylic acids

example, the presence of water in a solvent can significantly increase its eluting power. Wet acetone may have an eluting power greater than dry ethanol.

Amount of Adsorbent and Column Dimensions

After deciding which adsorbent to use for a separation, you must decide how much adsorbent to use. The choices naturally depend on the amount of material you wish to separate. In general, for a moderately challenging separation, you should use about 10–20 times as much silica gel or alumina by weight as the material to be separated. More adsorbent should be used for a difficult separation, less for an easy one. If too little adsorbent is used, the column will be overloaded and the separation will be poor. If too much adsorbent is used, the chromatography will take longer, require more elution solvent, and be no more efficient.

A height of 10–20 cm of silica gel often works well, and an 8:1 or 10:1 ratio of the adsorbent height to the inside column diameter is normal. Thus, a 1.5–2.5-cm column diameter is common for liquid chromatography on silica gel. A short, fatter column often produces worse separation, while a tall, thinner column can retain the compounds so tenaciously that the polar solvents required for their elution do not discriminate well between the various compounds on the column.

If you were carrying out a chromatographic separation on a 1.0-gram sample, 15 grams of silica gel would be appropriate. Silica gel has a bulk density of about $0.3 \text{ g} \cdot \text{cm}^{-3}$, so 15 g would occupy a volume of about 50 cm^3 (50 mL); this quantity is called the *column volume*. Aiming for a column height of 15 cm of silica gel, we can calculate the inside diameter of the necessary chromatography column. The column of silica gel is a cylinder with a volume of $\pi r^2 h$. If $h = 15$ cm and $V = 50 \text{ cm}^3$, r is 1.0 cm. Thus, a chromatography column with a 2-cm inside diameter would be appropriate. Common inside diameters for commercial glass columns used in miniscale liquid chromatography are 1.9 cm and 2.5 cm.

Usually one to two column volumes of elution solvent above the adsorbent are used to push the liquid through the silica gel column. Therefore, the chromatographic separation of 1.0 g of material on silica gel would require a glass column 2 cm in diameter and 40 cm long. Either a commercial chromatography column of 2.5-cm diameter and 30-cm length or one of 1.9-cm diameter and 40-cm length would be appropriate for the separation of a 1.0-g sample.

In general, a chromatography column with a stopcock near the bottom is used (Figure 17.2a). However, a glass tube with a

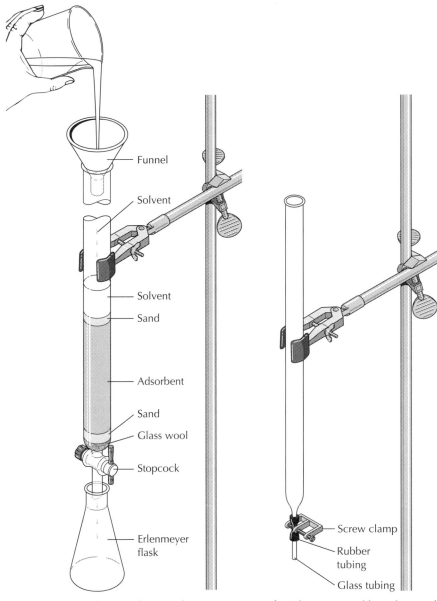

Funnel

Solvent

Solvent

Sand

Adsorbent

Sand

Glass wool

Stopcock

Erlenmeyer flask

Screw clamp

Rubber tubing

Glass tubing

FIGURE 17.2
Chromatographic columns.

(a) Column with stopcock

(b) Column using rubber tubing and screw clamp instead of stopcock

tapered end, fitted with a small piece of flexible tubing and glass tubing, plus a screw clamp, can also work well (Figure 17.2b). The stopcock or clamp is used during the preparation of the column, but after the sample has been added to the top of the column, the results are best when the flow of eluting solvent is not stopped until the separation is completed.

17.4   Preparation of a Chromatography Column

After selecting a chromatography column and weighing the requisite amount of adsorbent, prepare the column (see Figure 17.2a). The packing of a column is just as crucial to the success of the chromatographic separation as is the choice of adsorbent and elution solvents. If the column of adsorbent has cracks or channels or if the top surface is not flat, separation will be poor.

The column must never be allowed to dry out once it is prepared, so the solvent level should never be allowed to fall below the top of the sand above the adsorbent. If the adsorbent becomes dry, it may pull away from the walls of the column and form channels. Once you begin a chromatographic separation, finish it without interruption.

17.4a   Preparation of a Miniscale Column

Clamp the chromatography column in an upright position on a ring stand or vertical support rod and fill it approximately one-half full either with the first developing solvent you plan to use or with a less polar solvent. Add a small piece of glass wool as a plug, and push it to the bottom of the column with a glass rod, making sure all the air bubbles are out of the glass wool. Cover the plug with 3–4 mm of white sand. The glass wool plug and sand serve as a level support base to keep the adsorbent in the column and prevent it from clogging the stopcock or tip. The adsorbent can be added to the column by either the dry method or the slurry method.

Dry Adsorbent Method

Place a powder funnel in the top of the column, and with the stopcock closed, pour the adsorbent slowly into the solvent-filled column. Take care that the adsorbent falls uniformly to the bottom. Do not add the adsorbent too quickly or clumping may occur. The adsorbent column should be firm, but if it is packed too tightly, the flow of elution solvents becomes too slow.

The top of the adsorbent must always be horizontal. Gentle tapping on the side of the column as the adsorbent falls through the solvent prevents the formation of bubbles in the adsorbent. If large bubbles or channels develop in the column, the adsorbent should be discarded, and the column should be repacked. Any irregularities in the adsorbent column may cause poor separation, because part of the advancing sample will move faster than the rest. The time consumed in repacking will be much less than the time wasted trying to make a poor column function efficiently.

After all the adsorbent has been added, carefully pour 3–4 mm of white sand on top to protect the adsorbent from mechanical disturbances when solvents are poured into the column. Leave a small amount of solvent above the sand and close the stopcock.

Slurry Method

If you use a liquid more polar than an alkane in packing the column, you may need to prepare a slurry of the adsorbent and solvent in an Erlenmeyer flask by **slowly** adding the requisite amount of adsorbent to an excess of solvent. The use of a slurry prevents the formation of clumps or gas bubbles in the column, which can form from the heat produced by the interaction between polar solvents and the surface of the adsorbent.

Place a powder funnel in the top of the column and half fill the column with the same solvent used to prepare the slurry. Partially open the stopcock so that the solvent drains slowly into an Erlenmeyer flask. Swirl the flask containing the slurry and pour a portion of it into the column. Tap the side of the column constantly while the slurry is settling. Swirl the slurry thoroughly before each portion is added to the column. Add more solvent as needed so that the solvent level never falls below the level of the adsorbent at any time during the packing procedure. The solvent drained from the column can be reused for this purpose. Once all the adsorbent is in the column, return the collected solvent to the column once or twice to firmly pack the adsorbent.

After all the adsorbent has settled, carefully pour 3–4 mm of white sand on top. The layer of sand protects the adsorbent from mechanical disturbances when new solvents are poured into the column during the separation process. Be sure that there is a small amount of solvent above the sand and close the stopcock.

17.4b Preparation of a Williamson Microscale Column

The Williamson microscale chromatography apparatus is similar to the miniscale apparatus, except that it consists of several pieces fitted together. **Before you start to prepare the column, collect all the reagents and equipment you will need for the entire procedure.** Assemble the plastic funnel, glass column, Buchner microfunnel with a polyethylene frit, and the plastic stopcock as shown in Figure 17.3. With the stopcock closed, fill the column with hexane (or other nonpolar solvent) nearly to the top. Weigh about 2 g of silica gel into a 10-mL Erlenmeyer flask and add 7–8 mL of hexane to the flask; swirl the flask to thoroughly wet the adsorbent. Swirl the flask again to suspend the adsorbent and immediately pour the entire contents of the flask into the funnel at the top of the column. Place an Erlenmeyer flask under the column and open the stopcock to collect the solvent as it drains. Use a few milliliters of solvent to rinse the remaining adsorbent from the flask and add the slurry to the funnel. Tap the side of the column gently to help pack the adsorbent. Close the stopcock when the solvent level is a few millimeters above the top of the adsorbent.

FIGURE 17.3
Williamson microscale
column.

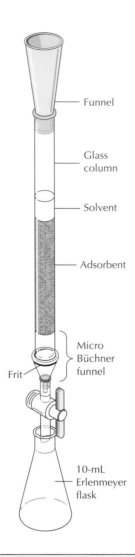

Funnel

Glass
column

Solvent

Adsorbent

Micro
Büchner
funnel

Frit

10-mL
Erlenmeyer
flask

17.4c Preparation and Elution of a Microscale Column

A column suitable for separating 50–100 mg of a mixture can be prepared in a large-volume Pasteur pipet.* Regular-size Pasteur pipets (5 ¾ in) can be used for separating 10–30 mg of sample. **Assemble all equipment and reagents for the entire chromatographic procedure before you begin to prepare the column.** The entire procedure of preparing the column and collecting the fractions must be done without interruption.

Obtain about 50 mL of the elution solvent in an Erlenmeyer flask fitted with a cork. Dissolve the mixture being separated in a small test tube by adding approximately 1 mL of the elution solvent or another solvent that is less polar than the elution solvent. Cork

---

*Available from Fisher-Scientific, catalog item 13678-8; the pipets have a capacity of 4 mL.

the test tube. Label a series of 10 test tubes (13 × 100 mm) for fraction collection. Pour 5 mL of elution solvent into one test tube and mark the liquid level on the outside of the tube. Place a corresponding mark on the outside of the other 9 test tubes.

Pack a small plug of glass wool into the stem of the Pasteur pipet, using a wooden applicator stick or a thin stirring rod (Figure 17.4, step 1). Clamp the pipet in a vertical position and place a 25-mL Erlenmeyer flask underneath it to collect the drained solvent. Weigh approximately 3 g of silica gel adsorbent in a tared 50-mL beaker; add enough elution solvent to make a thin slurry. Transfer the adsorbent slurry to the column using a 5-in Pasteur pipet (Figure 17.4, step 2). Continue adding slurry until the column is two-thirds full of adsorbent. Fill the column four to five times with the elution solvent to pack the adsorbent well. This eluted solvent can be reused.

Allow the solvent level to almost reach the top of the adsorbent and place the test tube labeled "Fraction 1" under the column. Draw the sample mixture into a 9-in Pasteur pipet, hold the pipet tip just above the surface of the adsorbent, and add the sample 1 drop at a time to the center of the column. When the entire sample

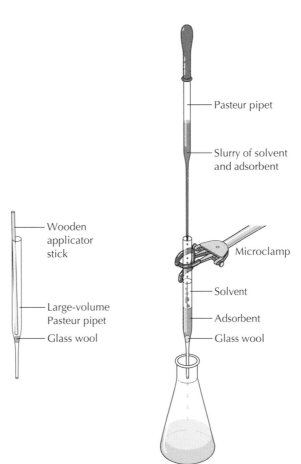

FIGURE 17.4

Setting up a microscale column.

1. Pack a glass wool plug in a large-volume Pasteur pipet.

2. Add a slurry of solvent and adsorbent.

is on the column, use a Pasteur pipet to add elution solvent by gently running it down the interior wall of the pipet. Maintain a column of solvent above the silica gel while you collect fractions of approximately 5 mL in labeled test tubes.

| 17.5 | **Sample Application and Elution Techniques** |

Application of Sample

Liquid samples can be applied directly on the column, but a mixture of solids must be added to the column, either dissolved in a solvent or preadsorbed onto a small amount of silica gel. Before a liquid or solution sample is applied to a miniscale or a Williamson microscale column, the solvent used in packing the column should be allowed to drain from the column until its level is just at the top of the upper sand layer and the stopcock is closed.

Preparation of a sample solution. For a solid sample, the solvent used in packing the column or another solvent of similar polarity is preferred for dissolving the sample mixture. If the sample's components do not dissolve in the first elution solvent, a small amount of a more polar solvent can be used to prepare the sample solution. However, the sample solution should be as concentrated as possible, preferably less than 5 mL in volume for a miniscale column and 1 mL for a microscale column. If the sample volume is too large, the compounds will have begun to move down the column while the sample is still entering at the top, which results in poor separation.

Application of a liquid sample or sample solution onto a column. Draw the sample into a 9-in Pasteur pipet, hold the pipet with the tip just above the level of the sand, and add the sample **one drop at a time to the center of the column.** Reopen the stopcock and allow the upper level of the sample solution to just reach the top of the sand; then close the stopcock. Fill the column with elution solvent carefully so that the upper layer of the column is not disturbed. The use of a chromatography funnel with a closed bottom and small holes in the stem wall provides a gentle flow of solvent down the wall of the tube that does not disturb the sand and adsorbent (Figure 17.5). Finally, proceed to elute the compounds from the column. Follow the same procedure when changing elution solvents.

FIGURE 17.5
Chromatography funnels.

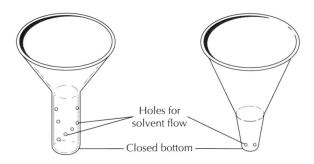

Holes for solvent flow

Closed bottom

Sample adsorbed on silica gel. An alternative to preparing a sample solution is to preadsorb the sample onto a small amount of silica gel, remove the solvent, and carefully pour the dry mixture onto the top of the column. For a miniscale sample, add 1–2 g of silica gel to a solution of the sample, remove the solvent using a rotary evaporator [see Technique 8.10], and carefully add the dry powder to the top of the column. For a microscale sample, add 300 mg of silica gel to a solution of the sample and, in a hood, evaporate the solvent by warming the sample container in a hot-water bath while stirring the mixture to prevent bumping.

Elution of the Compounds in the Sample

Fresh solvent needs to be added to the top of the column continuously during the elution process. Do not allow the level of solvent to drop below the top of the adsorbent column or the top surface of the adsorbent to be disturbed by the addition of solvents; use the type of funnel shown in Figure 17.5. Elution of the compounds in the sample is done by using a series of increasingly polar elution solvents. The less polar compounds elute first with the less polar solvents. Polar compounds usually come out of a column only after a switch to a more polar solvent. As the elution proceeds, the compounds in the mixture separate into a series of bands in the column (Figure 17.6). With colorless compounds, the bands are invisible; with colored compounds, the bands are seen.

Changing elution solvents during a separation. A mixture of two solvents is commonly used for elution. Addition of small amounts of a polar solvent to a less polar one increases the eluting power in a gentle fashion. For example, the development of the column can begin with hexane, and if nothing elutes from the column with this

FIGURE 17.6
Chromatography column during elution.

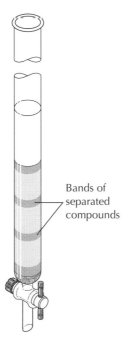

Bands of separated compounds

solvent, a 2–5% solution of diethyl ether in hexane can be used next, followed by increasing concentrations of ether, then pure diethyl ether for the most polar compounds.

If the change of solvent is made too abruptly, enough heat may be generated from adsorbent/solvent bonding to cause cracking or channeling of the adsorbent column. In some cases, a low-boiling-point elution solvent may actually boil on the column. The bubbles that form will degrade the efficiency of the column.

Flow rate of elution solvent. A greater solvent height above the adsorbent layer provides a faster flow rate through the column. An optimum flow rate is about 2–3 mL $\cdot$ min^{-1}. If the flow is too slow, poor separation may result from diffusion of the bands. A reservoir at the top of a column can be used to maintain a proper height of elution solvent above the adsorbent so that an adequate flow rate is maintained. A separatory funnel makes a good reservoir. It can be filled with the necessary amount of solvent and clamped directly above the column. The stopcock of the separatory funnel can be adjusted so that elution solvent flows into the column as fast as it flows out at the bottom.

Size of elution solvent fractions. The size of the elution solvent fractions collected at the bottom of the column depends on the particular experiment. Common fraction sizes range from 10 to 50 mL for miniscale columns and from 2 to 10 mL for microscale columns. If the separated compounds are colored, it is a simple matter to tell when the different fractions should be collected. However, column chromatography is not limited to colored materials.

With an efficient adsorbent column, each compound in the mixture being separated is eluted separately. After one compound has come through the column, there is a time lag before the next one appears. Hence, there are times when only solvent drips out of the column. To ascertain when you should collect a new fraction of eluent, either note the presence of crystals forming on the tip of the column as the solvent evaporates or collect a few drops of liquid on a watch glass and evaporate the solvent in a hood. Any relatively nonvolatile compounds that are being separated will remain on the watch glass.

Removing the adsorbent from the column.   When you are finished eluting the sample from the column, allow any remaining solvent to drain from the column. The chromatography tube can then be emptied by opening the stopcock, inverting the column over a beaker, and using gentle air pressure at the tip to push out the adsorbent.

Recovery of Separated Compounds

Ascertain the purity of each fraction by GC or TLC analysis and combine the fractions containing each pure component. Recover the compounds by evaporation of the solvent. Evaporation methods include using a rotary evaporator [see Technique 8.10] or blowing off the solvent with a stream of nitrogen or air in a hood.

17.6	**Summary of Column Chromatography Procedure**

1. Prepare a properly packed column of adsorbent.
2. Carefully add the sample mixture to the column as a small volume of solution or liquid, or as a solid adsorbed on silica gel.
3. Elute the adsorbed compounds with progressively more polar solvents.
4. Collect the eluted compounds in fractions from the bottom of the column.
5. Evaporate the solvents to recover the separated compounds.

17.7	**Flash Chromatography**

Gravity liquid chromatography, described in Techniques 17.4–17.5, can be extremely time consuming, and it has been largely replaced by flash chromatography in research laboratories. In *flash chromatography,* pressure is used to push the elution solvent through the adsorbent column. The flash technique is not only much faster but is also more efficient because the silica gel adsorbent has a smaller particle size, 38–63 μm (230–400 mesh), compared to 63–210 μm (70–230 mesh) for gravity columns. The total time to prepare and elute a column can be less than 30 min. The smaller particle size of the stationary phase requires pressures up to 20 pounds per square inch (psi), thus necessitating a chromatography column that does not leak and a source of nitrogen gas or compressed air. Although it is desirable to have an R_f difference of ≥ 0.35 for the compounds being separated, it is possible to separate compounds with an R_f difference of ~0.15.

Gas pressure controls the flow rate of the elution solvent through the column. One type of apparatus consists of a glass column topped by a variable bleed device (Figure 17.7). The bleed device has at its top a Teflon needle valve that controls the pressure applied to the top of the solvent in the column. Table 17.2 provides column and solvent dimensions for preparation of a flash silica gel column of 12–15 cm in height. Either the available flash column sets the range of sample sizes that can be accommodated or the size of the sample to be separated indicates the column size needed. Table 17.2 also shows that a smaller column diameter requires that the collected fraction sizes be correspondingly smaller. In addition, the smaller the difference in R_f values, the smaller the size of the sample that can be placed on the column. Elution fractions must be analyzed by TLC or GC.

Before running a flash column, the TLC characteristics of the sample's components should be determined. Ideally a solvent system that provides an R_f difference of ≥ 0.35 should be used. Systems that have been found useful include petroleum ether (30°–60°C) mixed with one of the following: diethyl ether, ethyl acetate, or

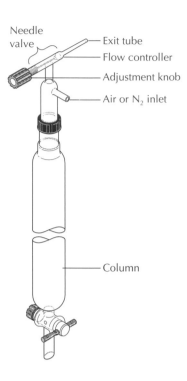

FIGURE 17.7
Apparatus for flash
chromatography.

Needle valve — Exit tube
— Flow controller
— Adjustment knob
— Air or N$_2$ inlet
— Column

acetone. As in gravity liquid chromatography, the composition of the elution solvent can be changed during the course of elution.

Preparation of the Column

The preparation of a flash chromatography column is done very much like the preparation of a gravity column. Begin by placing a glass wool plug at the bottom of the flash column (a long glass tube may be used to insert the plug) and covering it with a thin layer (3–4 mm) of 50–100-mesh sand. With the stopcock open, add, with tapping, 12–15 cm of 230–400-mesh silica gel to the solvent-filled column.* Alternatively, the adsorbent can be packed by the slurry method [see Technique 17.4a]. The necessary calculation for finding

TABLE 17.2   Column dimensions and solvent volumes for flash chromatography

		Typical sample size, mg		
Column diameter, mm	Volume of eluent, mL	$\Delta R_f \geq 0.2$	$\Delta R_f \geq 0.1{-}0.2$	Recommended fraction size, mL
10	100	100	40	5
20	200	400	160	10
30	400	900	360	20
40	600	1600	600	30
50	1000	2500	1000	50

Source: Still, W. C.; Kahn, M.; Mitra, A. J. *Org. Chem.* **1978,** *43,* 2923–2925.

*Aldrich and other suppliers indicate whether the silica gel is suitable for flash chromatography.

the mass of silica gel to use for this column height is discussed in Technique 17.3. Add a second layer of sand (3–4 mm) at the top of the silica gel and level it with gentle tapping.

Fill the column with the elution solvent. Use of a chromatography funnel, which has a closed bottom and small holes in the stem wall, provides a gentle flow of solvent down the wall of the column that does not disturb the packing of the sand and adsorbent (see Figure 17.5). Insert the flow controller, and with the needle valve open, gently turn on the flow of pressurized gas. Control the pressure by placing your finger (wear gloves) over the end of the exit tube, and manipulate the pressure so that the column is packed tightly. When the solvent has just reached the level of the sand, close the stopcock, and remove the flow controller.

Application of the Sample

Prepare a concentrated solution of your sample (25% or more) dissolved in the elution solvent. If the sample is not very soluble in the elution solvent, use a small amount of a more polar solvent. Draw the sample solution into a 9-in Pasteur pipet, hold the pipet with the tip just above the level of the sand, and add the sample **1 drop at a time to the center of the column.**

Elution of the Column

After the sample is on the column, fill it with the first elution solvent, using a chromatography funnel. Reinsert the flow controller and adjust the needle valve to reach an equilibrium pressure that causes the level of solvent to drop at a rate of 5 cm · min^{-1}. **Never let the column run dry**—the solvent should never go below the level of the top sand layer. Collect the proper fraction volumes of eluent solution (see Table 17.2) until all the solvent you planned to use has passed through the column of adsorbent or until fraction monitoring indicates that the desired components have been eluted.

Recovery of Separated Compounds

The purity of each fraction can be ascertained by gas chromatographic or thin-layer chromatographic analysis. Each of the fractions containing the same pure component should be combined before the compounds are recovered by evaporation of the solvent. Evaporation methods include using a rotary evaporator [see Technique 8.10] or blowing off the solvent in the hood with a stream of nitrogen or air.

17.8 Sources of Confusion

Polarity of Elution Solvent

If the elution solvent is too polar, the sample mixture will elute too quickly and poor separation will result. If the solvent is not polar enough, the sample will elute too slowly and the bands of compounds will broaden by diffusion, again resulting in poor separation along with a waste of time and solvent. An elution solvent that produces an R_f of about 0.3 for the desired compound on silica gel TLC is best if the separation of the other components is adequate.

Packing the Column Unevenly

For a chromatography column to work successfully in separating a mixture, the adsorbent must be packed uniformly without air bubbles, gaps, or surface irregularities. If the packing is not satisfactory, the sample mixture will not separate well.

Nonhorizontal bands. Nonhorizontal bands result if the adsorbent surface at the top of the column is not flat and horizontal, if the column is not clamped in a perfectly vertical position, or if the sample is not evenly applied to the column (Figure 17.8a). If nonhorizontal bands are present, poor separation can result because the lower part of one band can coelute with the upper part of the next band.

Channeling. If a depression or other irregularity is present at the top of the adsorbent surface, if cracks occur in the adsorbent, or if an air bubble is trapped in the column, part of the advancing front of a band will move ahead of the rest of the band, a process called *channeling* (Figures 17.8b and 17.8c). If the fronts of two bands are close together, they may elute together, rendering the chromatographic separation ineffective.

Applying the Sample Improperly

Achieving a good separation with a chromatography column also depends on how the sample mixture is prepared and applied to the column.

Adding the sample to the column. It is essential not to disturb the surface of the adsorbent column while the sample is applied. Slowly add a liquid sample or a concentrated sample solution with a 9-in Pasteur pipet 1 drop at a time to the center of the column. Hold the tip of the pipet just above the adsorbent surface. Alternatively, a sample solution can be mixed with a small amount of silica gel, the solvent evaporated, and the sample-coated silica gel poured onto the top of the column.

FIGURE 17.8
Problems that occur as a result of a poorly packed column.

(a) Nonhorizontal bands   **(b)** Channeling caused by irregular surface   **(c)** Channeling caused by air bubble

Overloading the column.   If the amount of sample is too large for the amount of adsorbent used in packing the column, the column will be overloaded and incomplete separation of the mixture's components will occur. Calculate the correct amount of adsorbent to use with the information in Technique 17.2 for gravity chromatography or the information in Table 17.2 for flash chromatography.

Too much solvent in the sample solution.   Prepare the sample in a minimal amount of solvent. If too much solvent is used to dissolve the sample, the excess will behave as an elution solvent and start to carry the mixture's components down the column. Separation will be incomplete because the entire sample was not on the column before its components started to move down the column.

The Column Becomes Dry

If the level of solvent falls below the top of the column, the adsorbent can become dry and pull away from the column wall. The channels that form compromise the effectiveness of the column. Be sure that the adsorbent is covered with solvent throughout the chromatographic procedure. Have all solvents at hand before starting the elution so that the separation can be completed without interruption.

Changing the Solvent Polarity Too Quickly

The polarity of the elution solvent often needs to be increased as the elution proceeds. However, the increase in polarity must be made gradually. If the polarity change is made too rapidly, enough heat may be generated from adsorbent/solvent bonding to cause gas bubbles that lead to channeling or even open cracks in the adsorbent column. The first change in polarity should add only 2–5% of the more polar solvent to the original elution solvent.

Diffuse or Broad Bands

If the elution solvent flows through the column at too slow a rate or if it is not polar enough to displace the desired compounds at a reasonable rate, poor separation may result from diffusion of the bands at a faster rate than the substance moves down the column. The optimum flow rate is about $2–3 \text{ mL} \cdot \text{min}^{-1}$ for a gravity column and $15–20 \text{ mL} \cdot \text{min}^{-1}$ for a flash column.

17.9   High-Performance Liquid Chromatography

High-performance liquid chromatography (HPLC) allows separations and analyses to be completed quickly and with superior separation and sensitivity compared with other liquid chromatography techniques. However, like gas chromatography (GC), HPLC generally utilizes small samples and is most often used for the analysis of mixtures rather than for preparative purposes. Unlike GC, HPLC can be used equally well with volatile and non-volatile mixtures.

The stationary phase in an HPLC column has a particle size of only 3–10 μm. The enhanced separation and sensitivity of the column come from the increased surface area provided by these very small particles. However, particles of this small size pack very tightly, a condition that severely restricts the flow of solvent through the column. Pressures of 1000–6000 psi are required to force the solvent through the column at a rate of $1-2$ mL $\cdot$ min^{-1}.

Instrumentation

The instrumentation for high-performance liquid chromatography consists of a column, an automatic sample injection system, a solvent reservoir, a pump, a detector, and a recorder or computer. Figure 17.9 is a block diagram of a typical high-performance liquid chromatograph. The length of the column can range from 5 to 30 cm with an inner diameter of 1–5 mm for analytical HPLC of 0.01–1.0-mg samples. Preparative HPLC columns, which have a larger diameter, can be used to separate samples of 100 or more mg.

The solvent is stored in a reservoir and passes through a filtration system before being pumped through the injector port, the column, and the detector. The solvents used for HPLC must be of very high purity because impurities degrade the column by irreversible adsorption on the particles of the solid phase. More sophisticated instruments have several solvent reservoirs and a gradient elution system that allows the composition of solvent mixtures to be changed during the course of a separation.

The detectors used for HPLC have a high sensitivity, usually in the microgram-to-nanogram range. The two most common types are ultraviolet (UV) and refractometer detectors. A UV detector is relatively inexpensive and can be used with gradient elution. It detects any organic compound that absorbs in the UV region. The limitations of a UV detector preclude its use with solvents that themselves absorb in the UV region or with samples that do not have a suitable chromophore for UV absorption. Refractometer detectors measure changes in the refractive index of the eluent as a sample's components move off the column and through the detector. A refractometer detector cannot be used with gradient elution because the base line would change as the solvent composition changes.

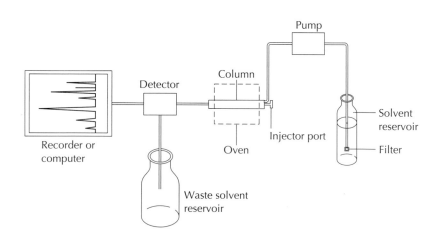

FIGURE 17.9
Schematic representation of a typical high-performance liquid chromatograph.

Columns and Solvents for Reverse-Phase HPLC

HPLC columns usually have a liquid covalently bonded to microporous silica (SiO_2) particles as the stationary phase.

$$\text{Silica particle} -O-\underset{\underset{\displaystyle OCH_2CH_3}{|}}{\overset{\overset{\displaystyle OCH_2CH_3}{|}}{Si}}-R$$

In the commonly used ***reverse-phase columns*** the R groups are long-chain alkyl groups such as $(CH_2)_{17}CH_3$. Reverse-phase columns are used to separate moderately polar to polar compounds. The more polar compounds elute first because the solvent is more polar than the hydrophobic stationary phase. Two useful polar elution solvents for reverse-phase chromatography that do not interfere with UV detection are methanol and acetonitrile, often mixed with water.

Sample Preparation

The ideal solvent for sample preparation is the same solvent as that used for the mobile liquid phase. Methanol is probably the most common solvent. Approximately 10–20 μL of a very dilute solution (0.0001–0.01 M) are normally used for the injection sample. Therefore, the amount of sample necessary for analytical HPLC columns is usually less than 1 mg. Prepare a sample solution of 1 mg or less of the sample in approximately 5 mL of solvent. The sample solution must be filtered through a micropore filter of about 0.5-μm pore size to remove any solid impurities that could clog the HPLC column. The filtration is done by drawing about 1 mL of the sample solution into a syringe and injecting it through the micropore filter into a small vial. After filtration, the vial is usually capped with a rubber septum. Consult your instructor about specific operating procedures for the HPLC instrument in your laboratory.

References

1. Miller, J. M. *Chromatography: Concepts and Contrasts*; 2nd ed.; Wiley: New York, 2005.
2. Still, W. C.; Kahn, M.; Mitra, A. "Rapid Chromatographic Technique for Preparative Separations with Moderate Resolution"; *J. Org. Chem.* **1978**, *43*, 2923–2925.
3. Harris, D. C. *Quantitative Chemical Analysis*; W. H. Freeman and Company: New York, 2003, Chapter 25, "High-Performance Liquid Chromatography."
4. Snyder, L. R.; Kirkland, J. J.; Glajch, J. L. *Practical HPLC Method Development*; 2nd ed.; Wiley: New York, 1997.

Questions

1. Once the adsorbent is packed in the column, it is important that the level of the elution solvent not drop below the top of the adsorbent. Why?

2. What precautions must be taken when you introduce the mixture of compounds to be separated onto the adsorbent column?

3. What effect will the following factors have on a chromatographic separation: (a) too strong an adsorbent; (b) collection of large elution fractions; (c) solvent level below the top of column adsorbent?

4. Arrange the following compounds in order of decreasing ease of elution from a column of silica gel: (a) 2-octanol; (b) 1,3-dichlorobenzene; (c) *tert*-butylcyclohexane; (d) benzoic acid.

5. Compare the advantages and limitations of liquid chromatography with those of thin-layer and gas-liquid chromatography.

PART

3

Spectroscopic Methods

Part 3 introduces the three spectroscopic methods most frequently used by organic chemists to determine the structures of organic compounds:

- Infrared (IR) spectroscopy
- Nuclear magnetic resonance (NMR) spectroscopy
- Mass spectrometry (MS)

Throughout the study of organic chemistry, you are asked to think in terms of structure, because structure determines the properties of molecules. The experienced organic chemist can anticipate many of the physical and chemical properties of various compounds by a glance at their structures.

Fifty and more years ago, structure was determined largely by time-consuming chemical methods. Modern organic spectroscopic methods have produced a revolution in the determination of the structures of complex organic molecules. What used to take months can now often be done in a day. For organic molecules having molecular weights of 300 or less, the job can often be done within an hour or so. The new techniques are based in large part on the absorption of radiation from various portions of the electromagnetic spectrum. In effect, spectroscopic techniques provide "snapshots" of a molecule.

Infrared spectroscopy is the oldest of the three spectroscopic methods, and in some ways the newer NMR spectroscopy and mass spectrometry have outshone it. However, IR spectroscopy provides quick and valuable information on functional groups present in a molecule (for example, $C=O$, OH, and CO_2H).

^1H and ^{13}C NMR spectroscopy reveal many details of an organic compound's structure, particularly the interrelated connectivity of its hydrogen and carbon atoms. Chemical shifts, spin-spin coupling patterns, and integration can be invaluable in structure determination. Today, NMR is arguably the most powerful spectroscopic method in organic chemistry. It is no surprise that it is the major focus of this spectroscopic methods section.

Mass spectrometry, which uses highly energetic electrons to ionize molecules, allows chemists to determine the molecular weight of a compound, and the fragmentation pattern of the ionized molecule provides data that can assist in the identification of the compound.

Integrating the data obtained from the three different spectroscopic methods discussed in Part 3 is important in the characterization of an organic compound. One spectral method may reveal features about a compound that may not be clear from another method, or one spectral method may confirm the existence of a structural unit suggested by another method. Appendix C contains a number of integrated spectroscopic problems with data obtained by several spectroscopic techniques.

18

INFRARED SPECTROSCOPY

Infrared (IR) spectroscopy is one of the most useful spectroscopic techniques in organic chemistry. It is a rapid and effective method for identifying the presence or absence of simple functional groups. When infrared energy is passed through a sample of an organic compound, absorption bands are observed. The positions of these IR absorption bands have been correlated with types of chemical bonds, which can provide key information about the nature of functional groups in the sample.

The *mid infrared,* extending from 4000 to 600 cm^{-1}, is the region of most interest to organic chemists. This is the region in which absorptions from typical organic compounds appear. When coupled with other spectroscopic techniques, such as nuclear magnetic resonance [see Technique 19], infrared spectroscopy allows organic chemists to systematically and confidently determine the molecular structures of organic compounds.

18.1   IR Spectra

An IR spectrum has energy, measured as frequency or wavelength, plotted along the horizontal axis and intensity of the absorption plotted along the vertical axis. There are several different formats for plotting the data depending on the scales used for the axes. Figure 18.1 shows examples of IR spectra of cyclopentanone recorded on two different IR spectrometers.

The horizontal scale in Figure 18.1a is linear in wavelength of the infrared radiation, which is the default axis used by older IR spectrometers. Many of the original libraries of infrared spectra were plotted using this format. The horizontal scale in Figure 18.1b is linear in wavenumbers, the standard frequency scale for infrared radiation used by most modern IR spectrometers. Microcomputers incorporated into modern IR spectrometers can quickly interchange data between the two formats. The shapes of the absorption bands appear quite different in Figures 18.1a and 18.1b, but their actual positions in the spectrum are the same.

In the two IR spectra of cyclopentanone, the major absorption band appears at 5.72 micrometers in Figures 18.1a and at 1747 cm^{-1} in Figure 18.1b. These IR bands are characteristic of the carbonyl group (C=O), one of the major functional groups in organic chemistry.

18.2   Molecular Vibrations

The atoms making up a molecule are in constant motion. The movements of the atoms relative to each other can be described as vibrations, and infrared spectroscopy has been called *vibrational spectroscopy.* The photons of IR radiation absorbed by an organic

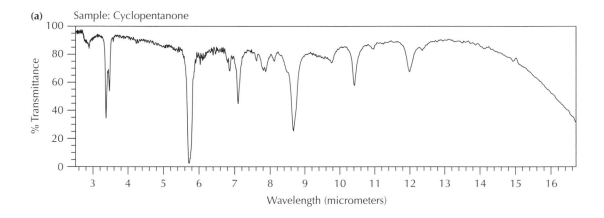

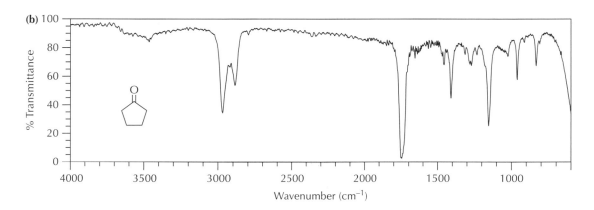

FIGURE 18.1
Infrared spectra of cyclopentanone recorded with (a) the horizontal (energy) scale linear in micrometers (wavelength) and (b) the horizontal scale linear in wavenumbers (frequency).

molecule have just the right amount of energy to stretch or bend its covalent bonds. The energy of infrared radiation is on the order of 8–40 kJ/mole (2–10 kcal/mole). This amount is not enough energy to break a covalent bond, but it is enough to increase the amplitude of bond vibrations. When infrared radiation is absorbed, the sample becomes warm as its molecules increase their kinetic energy. This is how infrared heat lamps work.

An absorption band appears in an infrared spectrum at a frequency where a molecular vibration occurs in the molecule. Energy levels of molecular vibrations are quantized, which means that only infrared energy with the same frequencies as the molecular vibrations can be absorbed. The energy levels available to a molecular vibration are expressed as

$$E = h\nu_0(\nu + \tfrac{1}{2}) \text{ for } \nu = 0, 1, 2, 3 \ldots$$

where h = Planck's constant and ν_0 = the zero-point vibrational level of the bond. The energy (ΔE) of the absorbed radiation that will promote a vibration of frequency (v) from one energy level to the next energy level is

$$\Delta E = hv$$

The frequency (v) and wavelength (λ) of light are related by

$$v = c/\lambda$$

where c = the speed of light. Substituting this relationship into the equation for the absorbed radiation yields

$$\Delta E = hc(1/\lambda)$$

The quantity ($1/\lambda$) is called the *wavenumber* ($\bar{v}$) and is usually expressed in units of reciprocal centimeters (cm^{-1}). A wavenumber defines the number of wave crests per unit length. It is proportional to the frequency as well as to the energy of an IR absorption.

$$\Delta E = hc\bar{v}$$

An IR absorption band is often called a peak, and its maximum is defined as the position of maximum absorption in wavenumber units (cm^{-1}). Frequency in units of wavenumbers, cm^{-1}, and wavelengths in units of micrometers, μm (10^{-6} meters, called microns in the older literature), can be interconverted by the following relationship:

$$\bar{v} = 10,000/\lambda$$

Fundamental Molecular Vibrations

There are two kinds of fundamental molecular vibrations: stretching and bending. In a *stretching vibration,* the distance between two atoms increases and decreases in a rhythmic manner, but the atoms remain aligned along the bond axis. Figure 18.2 shows a *symmetric stretching vibration* in which the atoms stretch in and out simultaneously. In a *bending vibration,* the positions of atoms change relative to the bond axis, as shown in Figure 18.3. A nonlinear molecule made up of n atoms has $3n-6$ possible fundamental stretching and bending vibrations.

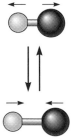

FIGURE 18.2
Fundamental stretching vibrational mode of a diatomic molecule.

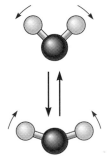

FIGURE 18.3
Fundamental bending vibrational mode of a triatomic molecule.

FIGURE 18.4
The three fundamental
vibrational modes
of water.

Symmetric stretching   Asymmetric stretching   Scissoring

EXERCISE

Water (H—O—H) is a nonlinear molecule consisting of three atoms.
(a) How many fundamental vibrations does it have? (b) Describe them.
Answer: (a) Water has three fundamental vibrations ($3n - 6 = 3$). Two of the
vibrations are stretching vibrations and one is a bending vibration. (b) The
vibrations are shown in Figure 18.4. The first is a symmetric stretching vibration. The second stretching vibration is an ***asymmetric stretching vibration*** in
which one hydrogen atom moves out as the other hydrogen atom moves in.
The bending vibration involves a kind of scissoring motion in which the
H—O—H bond angle changes back and forth.

For molecules containing many atoms, there are numerous fundamental vibrations. The stretching and bending vibrations of a
methylene (CH_2) group are shown in Figure 18.5.

Complexity of IR
Spectra

Organic compounds, which contain 10 or 20 atoms or more, can
manifest substantial numbers of IR peaks, and the spectra of
organic compounds can be complex. The total number of observed
absorption bands is generally different from the total number of
possible fundamental vibrations. Some fundamental vibrations are
not IR active and do not absorb energy. However, additional
absorption bands that occur as a result of overtone vibrations, combination vibrations, and coupling of vibrations more than make up
for the decrease.

Overtone bands are observed when fundamental vibrations
produce intense absorption bands. Overtone frequencies are multiples

FIGURE 18.5
Vibrational modes of
the methylene group
(CH_2).

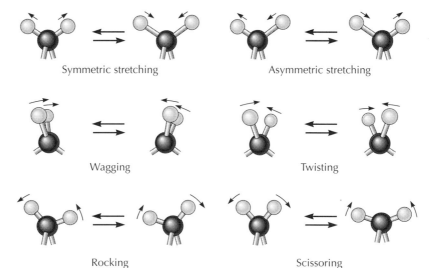

Symmetric stretching   Asymmetric stretching

Wagging   Twisting

Rocking   Scissoring

of the fundamental frequency, and they result from the change of more than one vibrational energy level.

Combination bands appear at frequencies that correspond to sums and differences of two or more fundamental vibrational frequencies. The intensities of the overtone and combination absorption bands are usually less than the intensities of fundamental vibrations.

A coupling interaction called *Fermi resonance* can occur in compounds where an absorption band due to an overtone or combination band is close to the frequency of a fundamental vibration. Their interaction causes the intensity of the fundamental vibration to decrease and the intensity of the overtone or combination band to increase.

Fortunately, many of the peaks in an IR spectrum can usually be ignored. The large number of fundamental vibrations, their overtones, and combinations of vibrations make it far too difficult to understand quantitatively entire IR spectra of most organic compounds, but as you will see, they can easily yield a great deal of qualitative information about functional groups. Moreover, the complexity of an IR spectrum imparts a unique pattern for each compound, allowing an IR spectrum to be used as a "fingerprint" for identification.

Correlation of Peaks with Specific Bond Vibrations

The absorptions corresponding to specific molecular vibrations appear in definite regions of the IR spectrum, regardless of the particular compound. For example, the stretching region of O—H bonds in all alcohols appears at nearly the same frequency. In the same way, the C=O vibrations of all carbonyl compounds appear within a narrow frequency range.

What determines the frequency and intensity of IR peaks? Following are the most important factors:

- Type of vibration, stretching or bending
- Strength of the bond connecting the atoms, particularly the bond order
- Masses of the atoms attached by the covalent bonds
- Electronegativity difference between the two atoms or groups of atoms in a bond

Type of vibration. In general, the stretching of covalent bonds takes more energy than bending vibrations. So stretching vibrations in the infrared appear at higher frequencies.

Type of vibration	Frequency (cm^{-1})
C—H stretching	3000–2800
—CH_2— bending	1470–1430

Bond order. Bond order is simply the amount of bonding between two atoms. For example, the bond order between carbon atoms increases from one to two to three for ethane (CH_3—CH_3), ethene (ethylene, CH_2=CH_2), and ethyne (acetylene, HC≡CH), respectively. In general, the higher the bond order, the greater the energy required to stretch the bond. Higher bond order produces a higher-frequency IR absorption.

Bond order	Type of bond	Stretching frequency (cm^{-1})
1	C—C, C—O, C—N	1300–800
2	C=C, C=O, C=N	1900–1500
3	C≡C, C≡N	2300–2000

Atomic mass. The frequency of the IR absorption also relates to the atomic masses of the vibrating atoms. Covalent bonds to hydrogen occur at high frequencies compared to bonds between heavier atoms—a light weight on a spring tends to oscillate faster than a heavy weight.

Type of bond	Stretching frequency (cm^{-1})
O—H	3650–2500
N—H	3500–3150
C—H	3300–2850

Electronegativity differences and peak intensities. Bond polarity does not significantly affect the position of IR absorption, but it greatly influences the intensity of IR peaks. If a vibration (stretching or bending) induces a significant change in the dipole moment, an intense IR band will result. Thus, when bonds are between atoms having different electronegativities, such as C—O, C=O, and O—H, the IR stretching vibrations are very intense. A symmetric molecule such as ethylene, on the other hand, does not show an absorption band for the C=C stretching vibration.

The intensity (peak size) of an IR absorption can be reported in terms of either transmittance (T) or absorbance (A). ***Transmittance*** is the ratio of the amount of infrared radiation transmitted by the sample to the intensity of the incident beam. Percent transmittance is $T \times 100$. In practice, peak intensities are reported in a more qualitative fashion.

A properly prepared sample produces an IR spectrum in which the most intense peak nearly fills the vertical height of the chart. Peaks of that magnitude are termed *strong (s);* smaller peaks are called either *medium (m)* or *weak (w).* Peaks can also be described as *broad (br)* or sharp. It is important that the most intense peak in an IR spectrum be above 0% transmittance (5–10% is good) so that its peak maximum can be measured accurately.

18.3 IR Instrumentation

There are two major classes of instruments used to measure IR absorption: dispersive and Fourier transform (FT) spectrometers. ***Dispersive spectrometers*** were developed first and for a long time were the standard infrared instruments. The advent of computers allowed the development of ***Fourier transform infrared (FTIR) spectrometers*** in the 1960s. In recent years, instruments incorporating powerful and relatively inexpensive microcomputers have allowed most laboratories to convert to FTIR instruments.

Dispersive Spectrometers

In a dispersive IR spectrometer, the source of radiation, often a heated filament, provides a beam of IR radiation that is split into two beams. The beams are directed by mirrors through both sample and reference cells. The sample and reference beams are alternately selected for measurement by means of a special rotating sector mirror. The selected beam components are recombined into a single beam and focused onto a diffraction grating, which separates the beam into a continuous band of infrared frequencies. A slit allows only a narrow range of these frequencies to reach the detector. By continuously changing the angle of the diffraction grating, the entire infrared spectrum is scanned, and the instrument records the intensity of the radiation as a function of frequency.

Fourier Transform Spectrometers

Unlike the older dispersive instruments, *FTIR spectrometers* gather data at all IR wavelengths at the same time. A simplified diagram of an FTIR spectrometer is shown in Figure 18.6. Infrared radiation from a heated source is directed to a *beam splitter,* a thin film of the element germanium sandwiched between two highly polished plates of potassium bromide. The beam splitter separates the radiation into two beams. One beam is reflected off the beam splitter and directed to a fixed mirror. The other beam is transmitted through the beam splitter and directed to a moving mirror, which is controlled by a laser. The mirrors reflect their respective beams of infrared energy back to the beam splitter, where the beams recombine. The two beams traveled different distances to the mirrors, so their frequencies are now out of phase. The constructive and destructive combination of the out-of-phase frequencies produces an *interferogram.* The beam splitter and mirror assembly is known as a *Michelson interferometer.*

The interferogram is an array of signal intensities that reveals the difference in the two optical paths. Information about every infrared frequency is contained in the interferogram. The beam of infrared energy, encoded as an interferogram, is directed through a sample to the detector. On interacting with the sample, specific frequencies of infrared energy are absorbed through excitation of molecular vibrations. Fourier transform mathematics is then used to sort out the frequencies of infrared energies encoded in the modified interferogram. The result is an infrared spectrum plotted as an array of intensities versus frequencies measured in cm^{-1}.

In actual practice, two scans are required: a scan of the empty sample compartment referred to as the *background* scan and a scan

FIGURE 18.6
Diagram of a single-beam FTIR spectrometer. The interior of the instrument is isolated from the ambient environment by purging with dry nitrogen or dry, carbon-dioxide-free air.

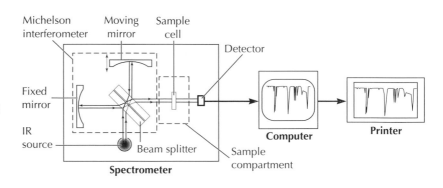

FIGURE 18.7
The collection and processing of data required for the creation of an infrared spectrum with a single-beam FTIR spectrometer.

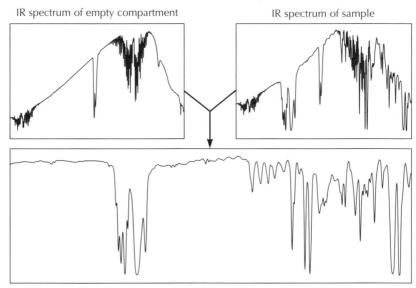

IR spectrum of empty compartment   IR spectrum of sample

IR spectrum of sample corrected for background signals

with the sample in the beam of infrared energy. The background scan contains signals due to water vapor and gaseous carbon dioxide in the atmosphere, the emission profile of the source, and film coatings of the optics, among other things. The background spectrum is subtracted from the sample spectrum to produce a spectrum displaying only absorptions due to the sample. The steps involved in creating a spectrum from the data are outlined in Figure 18.7.

Although it is more complicated than dispersive IR spectroscopy, there are numerous advantages to the FTIR method. Results of multiple scans can be combined to average out random noise, and excellent spectra can be obtained rapidly from very small samples. FTIR spectrometers have few mirror surfaces and because more energy gets to the detector, they are much more sensitive. Also, the resolution of the spectrum from an FTIR spectrometer is much higher. FTIR data are digitized; the quality of a spectrum can often be improved by baseline correction or the subtraction of peaks resulting from impurities.

18.4   Operating an FTIR Spectrometer

The most difficult step in taking the IR spectrum of a sample is usually the preparation of the sample. However, an FTIR spectrometer is a modern instrument with many capabilities. It needs to be used with respect.

1. Prepare a sample. Methods for preparing samples for IR spectroscopy are described in Technique 18.5.
2. Briefly open the sample compartment and confirm that there is nothing in the sample beam. Close the compartment.
3. Run a background scan. The data are collected, processed, and stored in the instrument's computer memory. The instrument indicates when this operation is completed.

4. Briefly open the sample compartment and place your sample in the sample beam. Close the compartment.

5. Run a sample scan. The data are collected and processed. The background scan is automatically subtracted from the sample scan. The result, an infrared spectrum of the sample, is displayed on the monitor.

If the sample was prepared as a solution, it is necessary to compensate for solvent absorptions by carrying out steps 6 and 7. **For all other samples, skip to step 8.**

6. Repeat steps 1–5 using pure solvent as the sample or call up a spectrum of the solvent from the IR instrument's memory.

7. Use the instrument's software to subtract the solvent signals from the spectrum of the sample. It may require some experimentation to determine the appropriate weighting for the solvent signals.

8. Use the instrument's software to mark the frequency of each major peak in the region of 4000–1500 cm^{-1}. Having the exact frequencies (wavenumbers) of these peaks on the printed spectrum can be helpful in analyzing the spectrum.

9. Format the spectrum and print out a copy for analysis and for inclusion in your laboratory notebook.

| 18.5 | **Techniques of Sample Preparation** |

IR spectra can be obtained from liquid, solid, or gas samples. Solid and liquid compounds are often prepared as thin films or as solutions that allow infrared radiation to pass through them. Various additional methods for preparing IR samples of solids and liquids are also described in this section. Gas samples require a special gas cell for sampling. Gas samples are encountered infrequently in organic chemistry and are not included in the discussion.

IR Sample Cells

The windows of the sample cells must be transparent to IR radiation in the mid-infrared region. Because glass absorbs IR radiation, it cannot be used to make IR sample cells. Most cells are made from alkali halides, in particular polished sodium chloride disks that, for the most part, are transparent in the mid-infrared region.

It is important to be aware that alkali halide sample cells are very susceptible to water damage and that care must be taken to ensure that all samples are completely dry. Water etches and clouds the surface of cells and disks, rendering them useless. Also, touching the polished surfaces of the disks with fingers leaves indelible fingerprints from skin moisture and oils. **The sodium chloride disks should be handled only by the edges.** The disks are much softer than glass and they break easily if dropped even a short distance. When preparing an IR sample, avoid touching the polished surface of sodium chloride disks with a glass pipet because the pipet can nick and scratch the surface. The only way to remove nicks, scratches, and fingerprints is to repolish the disks.

Thin Films for Liquid Compounds

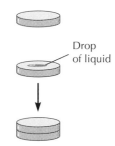

Drop of liquid

(a) Preparing sample

(b) Disc holder with sample

FIGURE 18.8
Preparation of thin film sample for IR spectroscopy.

Solution IR for Liquid and Solid Compounds

A thin film pressed between NaCl disks is the most convenient method for preparing a liquid for IR analysis (Figure 18.8). A drop of *neat* sample (liquid with no added solvent) is placed on one disk; the other disk is placed on top of the drop. The disks are gently rotated and then gently squeezed together to form a film approximately 0.01 mm in thickness. The sandwich is placed in a holder that is subsequently positioned in the sample compartment of the IR spectrometer. When the sample has a low viscosity, the holder shown in Figure 18.9 is a better choice because it keeps the sample film tightly in contact with the salt plates.

Steps in Preparing and Using a Thin Film

SAFETY PRECAUTION

Wear gloves and handle all solvents only in a hood.

1. Clean the disks with a **dry solvent**—acetone or dichloromethane.
2. Place a folded tissue on the lab bench. Place one disk on top of the tissue pad.
3. Using a Pasteur pipet, place 1 drop of the liquid sample on the center of the disk. Be careful not to touch the surface of the disk with the pipet.
4. Place the second disk on top of the first and gently rotate it; then gently press the disks together.
5. Obtain the IR spectrum.
6. Clean the disks with a **dry** solvent—acetone or dichloromethane. Store the disks in a desiccator to protect them from moisture.

Liquid as well as solid compounds can be dissolved in a solvent and analyzed in a solution cell such as the one shown in Figure 18.10. In a solution cell, a spacer between the salt plates provides a fixed path

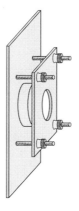

FIGURE 18.9
IR sample plate holder for low-viscosity liquids and solutions.

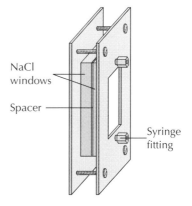

NaCl windows

Spacer

Syringe fitting

FIGURE 18.10
Sealed solution cell for IR spectroscopy.

length, typically 0.1 mm. Solutions are injected into the cells using a syringe (without the needle). A spectrum of the solvent and a spectrum of the solution are obtained separately, and the spectrum of the solvent is then subtracted from that of the solution by software provided with the IR instrument. Subtraction of the solvent peaks produces the spectrum of the compound itself. However, in frequency regions where the solvent has strong absorptions, sample peaks or their absence will not be reliable; these frequency regions should be ignored when interpreting an IR spectrum.

Unfortunately, no single organic solvent is transparent over the entire IR region. It is important to use a solvent that has a minimum number of IR absorptions in the regions of interest. Chloroform has few intense IR absorptions and is often used as an IR solvent. However, it does have intense absorptions due to the C—Cl stretching vibrations as well as absorptions due to C—H stretching and bending. Such a small amount of IR radiation passes through $CHCl_3$ in these regions that any signals due to the sample are obscured. The regions of strong absorbance for solvents frequently used in preparing IR samples are given in Table 18.1.

Steps in Preparing and Using an IR Sample Solution

SAFETY PRECAUTION

Wear gloves and handle all solvents only in a hood.

1. Prepare a 10% solution of your compound in an appropriate solvent.
2. Draw the solvent used to prepare the sample solution into the barrel of a glass syringe (no needle).
3. Rinse the solution cells with the solvent used to prepare the solution.
4. Draw the sample solution into the syringe. Fit the tip of the syringe into the bottom solution-cell fitting and completely fill the cell with the solution. There should be no air bubbles left in the cell when you insert a stopper into each fitting.
5. Obtain the IR spectrum.
6. Clean the cell in a hood by flushing it with solvent. Use a gentle stream of clean, dry air or nitrogen to blow the residual solvent out of the cell. Store the dry cell in a desiccator to protect it from moisture.

Cast Films for Solid Compounds

A thin film of solid can be prepared by placing a drop of a concentrated solution of the compound in the center of a clean sodium chloride plate. The best solvent to use for this solution is one that has a high vapor pressure at room temperature and does not dissolve NaCl. Diethyl ether, dichloromethane, and ethyl acetate work well; methanol, ethanol, and water must be avoided. For best results the salt plate must have a smooth, polished surface because scratched and pitted plates lead to uneven distribution of the sample.

TABLE 18.1	Absorption regions of common solvents and mulling compounds
Carrier	Absorption region (in cm^{-1})
Solvents	
Chloroform[a]	3125–2940
	1250–1190
	835–625
Dichloromethane[a]	3300–2850
	2450–2300
	1550–1150
	<930
Mulling compounds	
Fluorolube	1300–1080
	1000–920
	910–870
	<670
Nujol	3000–2800
	1490–1450
	1410–1360
	750–720
Potassium bromide	Transparent

a. Toxicity hazard.

Steps in Preparing and Using a Cast-Film IR Sample

SAFETY PRECAUTION

Wear gloves and handle all solvents only in a hood.

1. In a small test tube, prepare 0.3–0.5 mL of a 10–20% sample solution in a volatile organic solvent. Cork the test tube.
2. Clean an NaCl disk with a **dry** solvent—acetone or dichloromethane.
3. Place a folded tissue on the lab bench. Place the clean disk on top of this tissue pad. Make sure the disk is level.
4. Using a Pasteur pipet, place 1 drop of the sample solution at the center of the disk. Be careful not to touch the surface of the disk with the pipet.
5. Allow the solvent to completely evaporate. It may be necessary to repeat steps 4 and 5 up to four or five times to build up a film of the compound thick enough to produce an acceptable IR spectrum.
6. Place the NaCl disk in a sample holder like that shown in Figures 18.8 and 18.9.
7. Obtain the IR spectrum.
8. Clean the disk with a **dry** solvent—acetone or dichloromethane. Store the disk in a desiccator to protect it from moisture.
9. If your sample compound is especially valuable, you can evaporate the solvent from the solution remaining in the small test tube and recover the compound.

KBr Pellets for Solid Compounds

Potassium bromide (KBr) does not absorb mid-region IR radiation. Thus, a solid compound can be prepared for IR spectroscopy by mixing the sample with anhydrous KBr powder and pressing the mixture into a thin, transparent disk. Potassium bromide disks are excellent for IR analysis, but their preparation is challenging and requires great care.

The solid sample must be ground exceedingly fine because large particles scatter IR radiation—exhibited on the spectrum as a dramatically sloping base line. The sample is ground with a polished mortar and pestle made of agate or some other nonporous material or by vibrating the mixture in a small ball mill, similar to the mill used by dentists to mix amalgam fillings.

Care must be taken to maintain anhydrous conditions. The smallest trace of water in the disk can disrupt homogeneous sample preparation and can also produce spurious O—H peaks at 3450 cm^{-1} and 1640 cm^{-1}. The mixture is pressed into transparent disks with a special die. In the research laboratory, the KBr/compound mixture is subjected to 14,000–16,000 psi in a high-pressure disk press. A convenient alternative to a high-pressure press is the minipress shown in Figure 18.11.

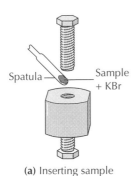

Spatula — Sample + KBr

(a) Inserting sample

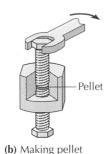

—Pellet

(b) Making pellet

FIGURE 18.11
Preparation of a KBr pellet with a minipress.

Steps in Preparing and Using a KBr Pellet

1. Using a small, nonporous mortar and pestle, grind a small quantity of the solid compound (0.5–2.0 mg) until it is an exceedingly fine powder. Use a small, flat spatula to scrape the ground solid from the surface of the mortar and grind it thoroughly with 100 mg of completely dry potassium bromide.
2. Thread one bolt halfway into the minipress die.
3. Add the sample/KBr mixture to the minipress die. Tap the side of the minipress to encourage all the solid mixture to fall to the bottom of the die. Try to cover the bottom of the die with a thin, even coating of the mixture. Too much material can produce poor-quality pellets, which are thick and opaque.
4. Thread the second bolt into the minipress die by hand as far as it will go.
5. Secure the minipress die in a vise or similar device.
6. Apply pressure to the sample using a wrench to tighten the second bolt.
7. Remove the bolts.
8. Place the minipress die containing the KBr pellet into a sample holder like the one used for the thin film sample, shown in Figure 18.8.
9. Obtain the IR spectrum.
10. Clean the minipress die and bolts and store them in a container to protect them from moisture.
11. Clean the equipment used for grinding the sample.

Mulls for Solid Compounds

A *mull* is not a true solution but is a fine dispersion of a solid organic compound in a viscous liquid. The most common liquids used for IR mulls are Nujol (a brand of mineral oil, which is a mixture of long-chain alkanes) and Fluorolube (a mixture of completely fluorinated alkanes). The fluorinated mulling substances are often used for more polar compounds. As is the case with solution infrared spectroscopy, Nujol and Fluorolube display IR peaks that may obscure peaks due to the dispersed compound (see Table 18.1). The preparation of a good mull requires care and practice.

Steps in Preparing and Using a Mull

SAFETY PRECAUTION

Wear gloves and handle all solvents only in a hood.

1. Using a small agate or nonporous ceramic mortar and pestle, grind 10–15 mg of the solid until the sample is exceedingly fine and has a caked, glassy appearance. Use a small flat spatula to scrape the ground solid from the surface of the mortar.
2. Add 1 drop of mulling liquid to the ground solid in the mortar. **Be careful!** Err on the side of adding too little mulling liquid because it is impossible to remove it if you add too much. Grind the mixture to make a uniform paste with the consistency of toothpaste; it should not be grainy but must not be runny.
3. Transfer the paste to an NaCl disk with a small flat spatula, as in Figure 18.8. Press the disks together gently and place them in a sample holder.
4. Obtain the IR spectrum.
5. Clean the disks with a **dry** solvent—acetone or dichloromethane. Store the disks in a desiccator to protect them from moisture.
6. Clean the equipment you used for grinding the sample.

Sample Cards for Solid Compounds

A relatively new innovation in IR spectroscopy is the use of a disposable sampling card. The sample is applied to an inert, microporous matrix in the middle of the card, but first the sample card is scanned in the IR spectrometer and its IR spectrum saved in the instrument's memory. Liquids are applied neat (without solvent). Solids are applied in solution and the solvent is allowed to evaporate. The card is placed in the sample beam and scanned. The spectrum of the blank sample card is then subtracted from that of the card with the compound applied by software provided with the IR instrument. This subtraction produces the spectrum of the compound itself.

Polyethylene or polytetrafluoroethylene is usually used for the solid support matrix on the card. Polyethylene has strong absorptions in the regions $2918-2849 \text{ cm}^{-1}$, $1480-1430 \text{ cm}^{-1}$, and $740-700 \text{ cm}^{-1}$. Polytetrafluoroethylene has strong absorptions in the regions $1270-1100 \text{ cm}^{-1}$ and $660-460 \text{ cm}^{-1}$. As is the case with the solvent in solution IR and the mulling agent in mulls, the infrared peaks of the sample card matrix may obscure peaks due to your sample.

A technique analogous to sample cards using Teflon tape as a solid support is described in Ref. 5.

Attenuated Total Reflectance (ATR) for Liquid and Solid Compounds

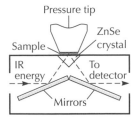

FIGURE 18.12
Cross section of single-reflection attenuated total reflectance (ATR) accessory.

FTIR instruments are extremely sensitive, and techniques have been developed that make it unnecessary to prepare KBr disks, solutions, or Nujol mulls or to use sample cards. One technique is *attenuated total reflectance (ATR)*, which operates by bouncing a beam of infrared radiation directly off the surface of a liquid or a powdered solid. However, ATR requires specialized sampling accessories. A single-reflection ATR accessory is shown in Figure 18.12.

The sample is placed in close contact with a high-refractive-index crystal such as zinc selenide. Close contact is achieved by screwing down a pressure tip onto the sample. After entering the zinc selenide crystal, the beam of infrared energy penetrates a small distance (a few micrometers) into the sample before being reflected. The reflected energy is attenuated (becomes less intense) in regions of the IR spectrum where the sample absorbs.

An IR spectrum from an ATR accessory is similar to a transmission IR spectrum, but there are some differences. The frequencies of the absorptions are the same, but the relative intensities of the peaks differ. A comparison of the transmission spectrum and the ATR spectrum of polystyrene is shown in Figure 18.13. The differences occur because lower frequency infrared energy penetrates further into the sample than higher frequency IR energy does. Because the lower frequencies interact with more sample, their absorbance bands are more intense.

Software is available that can correct for the different intensities at different wavelengths. Use of this software produces IR spectra that more closely resemble transmission spectra, which makes it easier to compare the ATR and transmission spectra. For high-quality IR spectra, a multiple-reflection ATR accessory with a long crystal

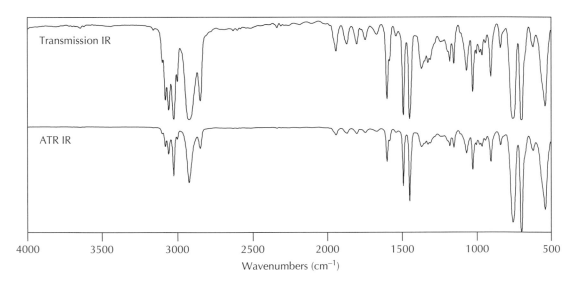

FIGURE 18.13
Comparison of (top) a transmission spectrum of polystyrene with (bottom) an ATR spectrum of polystyrene.

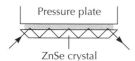

FIGURE 18.14
Multiple-reflection ATR crystal and sample.

and a trough that can be filled with sample is often used. A multiple-reflection crystal is shown in Figure 18.14.

Steps in Obtaining an ATR IR Spectrum

1. Place a small amount of solid sample on top of the ZnSe crystal in the ATR accessory. Use a wooden stick or other nonabrasive tool for this operation because a metal spatula can easily scratch the surface of the crystal.
2. Lower the pressure tip so that it is in contact with the solid. (**Note:** To avoid contamination of the tip, a small piece of paper can be placed between the tip and the sample.)
3. Apply approximately 10 pounds of pressure to the sample. The mechanism and appropriate pressure vary for different accessories, so find out from your instructor the procedure for your ATR accessory.
4. Obtain the IR spectrum.
5. Raise the pressure tip from the sample. Wipe the sample from the crystal and the pressure tip with a tissue. Then wipe the crystal and pressure tip with a methanol-moistened tissue.

18.6 Interpreting IR Spectra

Confirming the identity of a compound is one of the most important uses of IR spectroscopy. Because of the numerous and interactive vibrations of a typical organic molecule, no two compounds are known to have identical IR spectra. The unique pattern of each compound allows an IR spectrum to be used as a "fingerprint" for identification. Databases of IR spectra of known compounds can be searched for a match with the spectrum of an unknown compound—an identification method frequently used in forensic and quality control laboratories. Comparing an IR spectrum obtained in the laboratory to the spectra available in a compendium of IR spectra can be useful (Ref. 1); however, it can also be time consuming.

Often there is no sample spectrum available for comparison, and it is necessary to interpret the IR spectrum. The most basic interpretation consists of an inventory of the functional groups in the molecule. Systematic examination of the IR spectrum and identification of absorption bands due to fundamental stretching vibrations are used to construct the functional group inventory. Combining IR data with structural information from other techniques, such as nuclear magnetic resonance spectroscopy, usually allows unequivocal assignment of a molecular structure to an organic compound.

Regions of the IR Spectrum

As shown in Figure 18.15, an IR spectrum can be broken down into three regions:

- Functional group region
- Fingerprint region
- Aromatic region

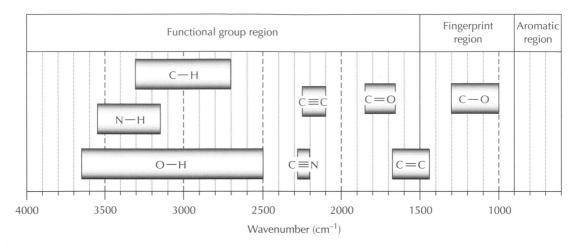

FIGURE 18.15
Approximate regions of chemical bond stretches in an IR spectrum.

Functional group region. The functional group region (4000–1500 cm^{-1}) provides unambiguous, reasonably strong peaks for most major functional groups. Figure 18.15 shows the approximate regions in which peaks appear as a result of important bond-stretching vibrations. Structurally similar compounds that contain the same functional groups are virtually identical in this region. For example, the absorption bands for the carbonyl groups of 2-butanone and 3-hexanone both appear at the same frequency, 1715 cm^{-1}. Since many functional groups, such as C=C, C≡C, C≡N, N—H, O—H, and C=O bonds, show IR bands in the 4000–1500 cm^{-1} region, **both the presence and absence of peaks in this region are significant.** The absence of appropriate IR bands in the functional group region argues against the presence of a functional group, except in the rare cases when a stretching vibration has no associated dipole change.

Fingerprint region. The fingerprint region (1500–900 cm^{-1}) is normally complex because of the many bending vibrations and combination bands that appear in this region. Before the development of NMR spectroscopy, when IR spectroscopy was the major structural probe available to organic chemists, much effort went into analyzing and assigning characteristic vibrations in this region. NMR spectroscopy now provides detailed structural information more directly and reliably. Except for a few intense absorptions, such as C—O stretching vibrations, IR peaks in the fingerprint region are now primarily used for fingerprint pattern matching.

Aromatic region. The aromatic region (900–600 cm^{-1}) provides information about the substitution pattern of benzenes and other aromatic compounds, although it is generally easier to determine these aromatic substitution patterns by NMR spectroscopy.

EXERCISE

All organic compounds have IR absorptions because of C—H and C—C stretching and bending vibrations. For each of the following compounds, identify the additional bond-stretching vibrations that should be observed. Using Figure 18.15 as a guide, identify regions of the IR spectrum where you would expect to see characteristic absorptions for each compound.

(a) 2-propanol   (c) phenylethyne   (e) 4-methylphenylamine
(b) propanoic acid   (d) 1-hexene   (f) benzonitrile

Answer

(a) 2-propanol (O—H, 3650–2500 cm^{-1}; C—O, 1300–1000 cm^{-1})
(b) propanoic acid (O—H, 3650–2500 cm^{-1}; C=O, 1850–1650 cm^{-1})
(c) phenylethyne (C≡C, 2250–2100 cm^{-1}; C=C, 1680–1440 cm^{-1})
(d) 1-hexene (C=C, 1680–1440 cm^{-1})
(e) 4-methylphenylamine (N—H, 3550–3150 cm^{-1}; C=C, 1680–1440 cm^{-1})
(f) benzonitrile (C≡N, 2280–2200 cm^{-1}; C=C, 1680–1440 cm^{-1})

FOLLOW-UP ASSIGNMENT

Using Figure 18.15 as a guide, identify regions of the IR spectrum in which you would expect to see characteristic functional group absorptions for each of the following compounds: (a) cyclopentanone, (b) methyl acetate, (c) methoxybenzene, (d) acetamide, (e) 1-aminohexane.

Because the intensities of IR absorptions can vary a good deal, the use of an absorption frequency table, such as Table 18.2, has limitations, particularly for peaks listed as *m (medium)* and *w (weak)* intensity. As in the analysis of other experimental data, you must think about the significance of your conclusions, rather than assuming that an algorithm will lead to the correct answer every time.

Where to Begin?

An efficient approach to interpreting an IR spectrum usually starts with a survey of the 4000–1500 cm^{-1} functional group region and the creation of an inventory of bond types present in the molecule. This inventory allows you to get a good idea of which functional groups are in the compound and which functional groups are not.

The functional group region can be subdivided into narrower frequency regions that are characteristic of specific bond types. We will describe each important region and give examples of spectra illustrating the fundamental stretching bonds. Besides correlating a stretching vibration with a frequency (wavenumber), it is important to consider the general appearance of the signal. Is it sharp? Is it broad? Is it weak? Is it strong?

O—H and N—H Stretch of Alcohols and Amines (3650–3200 cm^{-1})

Alcohols and phenols show strong IR bands due to oxygen-hydrogen bond stretching and amines show medium intensity IR bands due to nitrogen-hydrogen bond stretching. The appearance of absorptions in this region is highly varied, which can actually add to their usefulness.

TABLE 18.2   Characteristic infrared absorption peaks of functional groups

Vibration	Position (cm^{-1})	Intensity[a]	
Alkanes			
C—H stretch	2990–2850	m to s	
C—H bend	1480–1430 and 1395–1340	m to w	
Alkenes			
=C—H stretch	3100–3000	m	
C=C stretch	1680–1620 (sat.)[b] 1650–1600 (conj.)[b]	w to m	
=C—H bend	995–685	s	See Table 18.3 for detail.
Alkynes			
≡C—H stretch	3310–3200	s	
C≡C stretch	2250–2100	m to w	
Aromatic compounds			
C—H stretch	3100–3000	m to w	
C=C stretch	1625–1440	m to w	
C—H bend	900–680	s	See Table 18.3 for detail.
Alcohols			
O—H stretch	3650–3550	m	Free
	3550–3200	br, s	Hydrogen bonded
C—O stretch	1300–1000	s	
Amines			
N—H stretch	3550–3250	br, m	1° (two bands) 2° (one band)
Nitriles			
C≡N stretch	2280–2200	s	
Aldehydes			
C—H stretch	2900–2800 and 2800–2700	w	H—C=O Fermi doublet
C=O stretch	1740–1720 (sat.) 1715–1680 (conj.)	s	
Ketones			
C=O stretch	1750–1705 (sat.) 1700–1665 (conj.)	s	
Esters			
C=O stretch	1765–1735 (sat.) 1730–1715 (conj.)	s	
C—O stretch	1300–1000	s	
Carboxylic acids			
O—H stretch	3200–2500	br, m to w	
C=O stretch	1725–1700 (sat.) 1715–1680 (conj.)	s	
C—O stretch	1300–1000	s	

TABLE 18.2 *(continued)*

Vibration	Position (cm^{-1})	Intensity[a]	
Amides			
N—H stretch	3500–3150	m	1° (two bands), 2° (one band)
C=O stretch	1700–1630	s	
Anhydrides			
C=O stretch	1850–1800 and 1790–1740	s	
C—O stretch	1300–1000	s	
Acid chlorides			
C=O stretch	1815–1770	s	
Nitro compounds			
NO$_2$ stretch	1570–1490 and 1390–1300	s	

a. s = strong, m = medium, w = weak, br = broad
b. sat. = saturated; conj. = conjugated

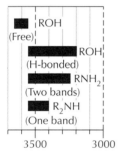

Alcohols. If an alcohol is prepared for IR analysis in any form other than a dilute solution, the hydroxyl group hydrogen bonds with neighboring molecules and the signal caused by the O—H stretch appears as a broad band between 3550 and 3200 cm^{-1}. In dilute solution, the hydroxyl groups are free from intermolecular hydrogen bonding and the signal that results from the O—H stretch is much sharper and shifted to a higher frequency, approximately 3600 cm^{-1}. The IR spectrum of a thin film of 2-propanol, shown in Figure 18.16, exhibits a broad, strong O—H stretching absorption at 3365 cm^{-1}.

Amines. The medium intensity N—H stretching vibrations of primary and secondary amines also appear in this region. In dilute solutions, the number of signals depends on the number of hydrogen

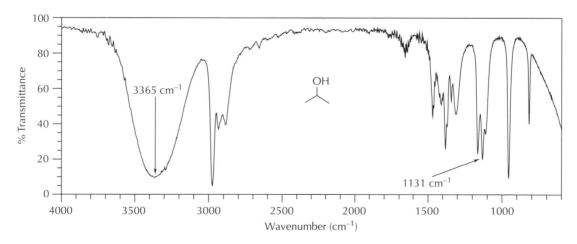

FIGURE 18.16
IR spectrum of 2-propanol (thin film).

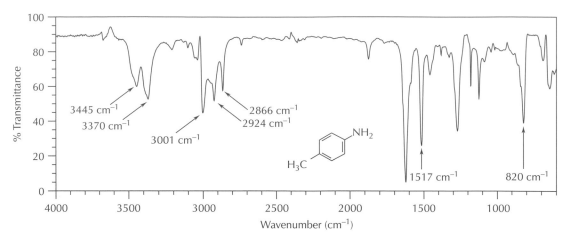

FIGURE 18.17
IR spectrum of 4-methylphenylamine ($CHCl_3$ solution).

atoms attached to the nitrogen atom. Primary amines show two peaks and secondary amines show only one. The IR spectrum of 4-methylphenylamine is shown in Figure 18.17. Because it is a primary amine, there are two absorptions (at 3445 and 3370 cm^{-1}) from symmetric and asymmetric H—N—H stretching vibrations. Amines are also capable of hydrogen bonding, so the position and shape of the absorption may vary. The higher the concentration of the amine and the better it can hydrogen bond, the broader the absorption will be. Hydrogen bonding shifts N—H stretching absorptions to lower frequencies. Alkyl amines are stronger bases than aromatic amines and tend to form stronger hydrogen bonds.

O—H Stretch of Carboxylic Acids (3200–2500 cm^{-1})

As a result of extensive intermolecular hydrogen bonding, carboxylic acids generally show an unusually broad O—H stretching absorption, with the band often tailing from about 3200 cm^{-1} all the way down to 2500 cm^{-1}. The intensity of this band is medium to weak. The spectrum of propanoic acid shown in Figure 18.18 illustrates this behavior. In this spectrum the O—H stretching band is so broad that the sharper C—H stretch at approximately 3000 cm^{-1} is superimposed on top of it. This superimposition is not uncommon with the O—H and C—H stretching vibrations of carboxylic acids. The structure of the intermolecular hydrogen-bonded propanoic acid dimer is

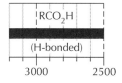

$$CH_3CH_2C \begin{matrix} O \text{------} H\text{—}O \\ \\ O\text{—}H \text{------} O \end{matrix} CCH_2CH_3$$

C—H Stretch (3310–2850 cm^{-1})

Because most organic compounds contain hydrogen atoms, you can expect to find C—H stretching signals in most IR spectra. The position of the C—H stretch depends on the hybridization of the carbon atom to which the hydrogen is bound.

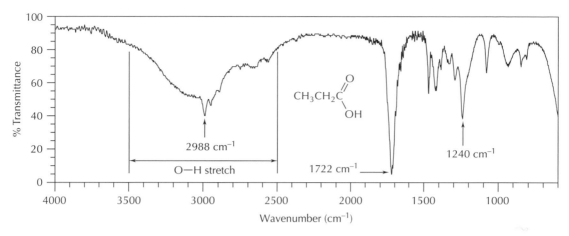

FIGURE 18.18
IR spectrum of propanoic acid (thin film).

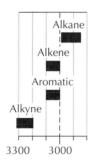

sp **Hybridization.** If the carbon atom is *sp* hybridized, the absorption appears near 3300 cm^{-1}. A good example is found in the spectrum of phenylacetylene shown in Figure 18.19. The C—H stretch of the acetylene appears at 3277 cm^{-1}. This band could be confused with a signal resulting from an O—H or N—H stretch were it not for its shape. The C—H band is much sharper than the typical hydrogen-bonded O—H or N—H stretch found in this region.

sp^2 **Hybridization.** Peaks that occur when hydrogen atoms are attached to *sp^2*-hybridized carbon atoms of alkenes and aromatic compounds appear in the region 3100–3000 cm^{-1}. In the spectrum of phenylacetylene (see Figure 18.19), the aromatic hydrogen stretching vibrations appear from 3066 to 3006 cm^{-1}. In the spectrum of 1-hexene shown in Figure 18.20, the vinyl hydrogen stretch appears at 3084 cm^{-1}.

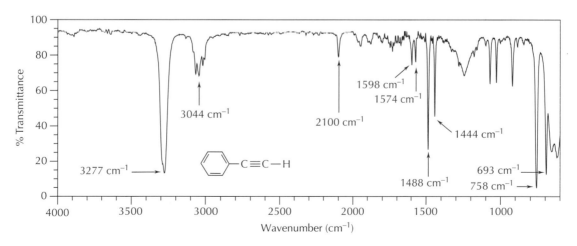

FIGURE 18.19
IR spectrum of phenylacetylene (thin film).

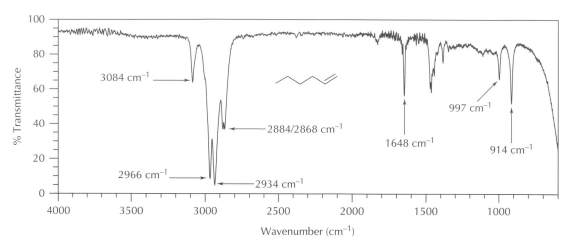

FIGURE 18.20
IR spectrum of 1-hexene (thin film).

sp³ Hybridization. Hydrogen atoms attached to sp^3 hybridized carbon atoms exhibit absorption bands in the 2990–2850 cm^{-1} region. There are usually several alkyl C—H stretching vibration bands in an IR spectrum. In the spectrum of 1-hexene, there are four distinct peaks from 2966 to 2868 cm^{-1} because of C—H stretching vibrations of hydrogen atoms attached to sp^3 carbon atoms.

C≡C and C≡N Stretch (2280– 2100 cm⁻¹)

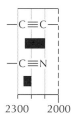

Only the triple bonds of nitriles and alkynes have absorptions in this region. If it is a strong absorption, it is likely to be that of a nitrile, such as the signal at 2230 cm^{-1} in the spectrum of benzonitrile (Figure 18.21). The difference in electronegativity between carbon and nitrogen leads to a highly polarized C≡N bond and thus a strong absorption. Alkynes have weak- to medium-intensity absorption bands in this region. The C≡C bond stretch in phenylacetylene is the small peak at 2100 cm^{-1} (see Figure 18.19).

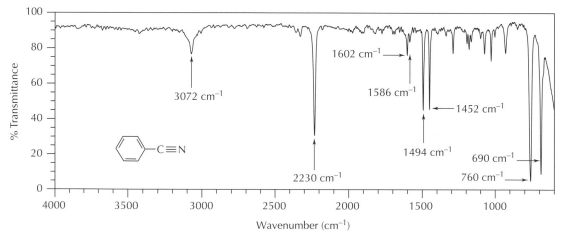

FIGURE 18.21
IR spectrum of benzonitrile (thin film).

C=O stretch (1850–1630 cm⁻¹)

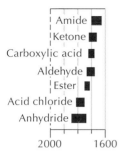

C=O stretch (1850–1630 cm⁻¹)

The carbonyl group is one of the most important functional groups in organic compounds. If there is a $C=O$ present in the molecule, there will be a strong, sharp absorption band in the 1850–1630 cm⁻¹ region. Good examples of $C=O$ stretching are the strong band at 1747 cm⁻¹ in the spectrum of cyclopentanone (see Figure 18.1) and the strong band at 1722 cm⁻¹ in the spectrum of propanoic acid (see Figure 18.18). If there is no strong band in the 1850–1630 cm⁻¹ region, there is no $C=O$ in the molecule. The exact position of the signal within this region, however, depends on what type of functional group contains the $C=O$ group.

Functional group	Example	C=O stretch
Amides	Acetamide	1681 cm⁻¹
Ketones	Acetone	1715 cm⁻¹
Carboxylic acids	Propanoic acid	1722 cm⁻¹
Aldehydes	Acetaldehyde	1727 cm⁻¹
Esters	Methyl acetate	1745 cm⁻¹
Acid chlorides	Acetyl chloride	1806 cm⁻¹
Acid anhydrides	Propanoic anhydride	1827 and 1766 cm⁻¹

Notice that acid anhydrides are characterized by the presence of two peaks in the $C=O$ stretching region. These arise from symmetric and asymmetric $C=O$ stretching vibrations.

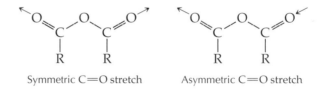

Symmetric C=O stretch   Asymmetric C=O stretch

Effect of ring strain and conjugation. Factors such as ring strain and conjugation cause deviations from the position of the $C=O$ band for saturated acyclic compounds. Ring strain causes the position of the absorption band to move to a higher frequency, indicating that the strength of the bond has increased. Compare the absorption bands of acetone (1715 cm⁻¹) and methyl acetate (1745 cm⁻¹) to the carbonyl absorption bands of the cyclic compounds listed here.

Ketone	C=O stretch	Ester	C=O stretch
Cyclopropanone	1818 cm⁻¹	—	—
Cyclobutanone	1783 cm⁻¹	Propanolactone	1840 cm⁻¹
Cyclopentanone	1747 cm⁻¹	4-Butanolactone	1770 cm⁻¹
Cyclohexanone	1716 cm⁻¹	5-Pentanolactone	1730 cm⁻¹
Cycloheptanone	1702 cm⁻¹	6-Hexanolactone	1732 cm⁻¹

Conjugation with a double bond or with an aromatic ring decreases the bond order of the $C=O$ slightly and causes the position of absorption to move to a lower frequency by 20–30 cm⁻¹. Compare the position of the $C=O$ absorption band of 4-methylpentan-2-one (Figure 18.22) to that of the conjugated compound 4-methyl-3-penten-2-one (Figure 18.23). The absorption band is shifted from 1719 cm⁻¹ to 1695 cm⁻¹. A similar shift is observed

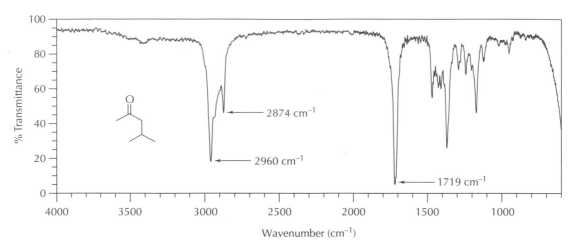

FIGURE 18.22
IR spectrum of 4-methylpentan-2-one (thin film).

when the C=O group is conjugated with a benzene ring. The C=O absorption band in acetophenone appears at 1690 cm^{-1}, whereas the C=O absorption band in acetone appears at 1715 cm^{-1}.

Corroborating IR peaks. Because the C=O bond is present in many functional groups and the position of the stretching vibration is affected by many variables, it is difficult to differentiate between carbonyl-containing functional groups by the C=O stretching frequency alone. It is usually necessary to identify other absorption bands in the IR spectrum to ascertain the identity of the functional group exhibiting the C=O stretch.

Primary and secondary amides exhibit N—H stretching absorption bands in the 3500–3150 cm^{-1} region, two bands for primary amides and one band for secondary amides.

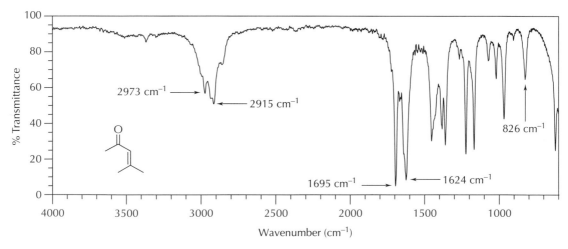

FIGURE 18.23
IR spectrum of 4-methyl-3-penten-2-one (thin film).

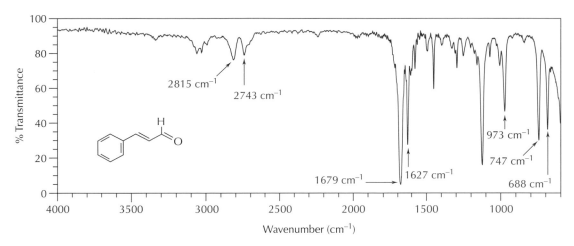

FIGURE 18.24
IR spectrum of cinnamaldehyde (thin film).

Carboxylic acids exhibit an extremely broad absorption band between 3200 and 2500 cm^{-1} because of hydrogen-bonded O—H stretching vibrations (see Figure 18.18).

Aldehydes exhibit two weak but very distinct absorption bands in the C—H stretching regions (2900–2800 cm^{-1} and 2800–2700 cm^{-1}). The two bands are an example of a Fermi doublet as a result of the interaction of the fundamental stretching vibration of the aldehyde C—H bond with an overtone band. The characteristic aldehyde C—H bands at 2815 and 2743 cm^{-1} are evident in the spectrum of cinnamaldehyde shown in Figure 18.24.

Esters exhibit a very strong band in the C—O stretching region (1300–1000 cm^{-1}). In the spectrum of methyl acetate shown in Figure 18.25, the C—O stretching vibration appears at 1246 cm^{-1}.

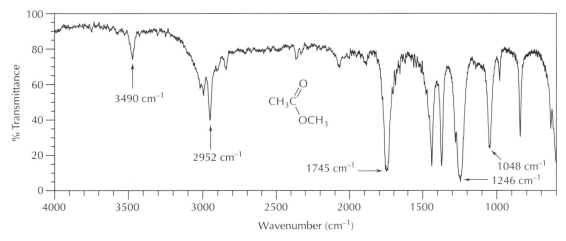

FIGURE 18.25
IR spectrum of methyl acetate (thin film).

C=C Stretch (1680–1440 cm⁻¹)

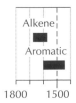

| Alkene |
| Aromatic |

1800   1500

Absorptions in the $1680-1440$ cm^{-1} region occur because of C=C bonds in alkenes as well as C=C bonds in aromatic compounds. Their intensities vary from weak to medium. A typical absorption of this type is the band at 1648 cm^{-1} in the spectrum of 1-hexene (see Figure 18.20). The position of the band and its intensity are affected by conjugation. The position of the C=C stretching absorption in 4-methyl-3-penten-2-one (see Figure 18.23) appears at 1624 cm^{-1}, and its intensity is significantly stronger than the intensity of the band in 1-hexene.

Aromatic compounds have four bands in this region near 1600, 1580, 1500, and 1450 cm^{-1}. The first two bands are generally weak and the second two are generally moderate in intensity. The band at 1450 cm^{-1} can be obscured by CH$_2$ bending vibrations if an alkyl group is present. These bands are evident in the spectra of phenylacetylene (see Figure 18.19) and benzonitrile (see Figure 18.21).

C—O Stretch (1300–1000 cm⁻¹)

Because C—O bonds are highly polarized, their absorption bands are generally very strong. However, the assignment can sometimes be ambiguous because the peaks occur in the fingerprint region $(1500-900$ cm$^{-1})$, which is cluttered with many absorption bands due to bending vibrations, overtone bands, and combination bands. Esters, ethers, and alcohols show bands in this region. In the spectrum of 2-propanol (see Figure 18.16), the signal at 1131 cm^{-1} is attributed to the C—O stretching vibration. Esters often exhibit two C—O stretching vibrations, one for the C—O bond to the carbonyl carbon and one for the C—O bond to the carbon of the alcohol group. In the spectrum of methyl acetate (see Figure 18.25), the bands appear at 1246 and 1048 cm^{-1}. Strong absorptions within this region have been correlated with the degree and type of substitution of an alcohol.

Type of alcohol	C—O stretch
RCH$_2$—OH Primary	$1075-1000$ cm^{-1}
R \\ HC—OH / R′ Secondary	$1130-1000$ cm^{-1}
R″ | R—C—OH | R′ Tertiary	$1210-1100$ cm^{-1}
⟨benzene ring⟩—OH Phenol	$1260-1180$ cm^{-1}

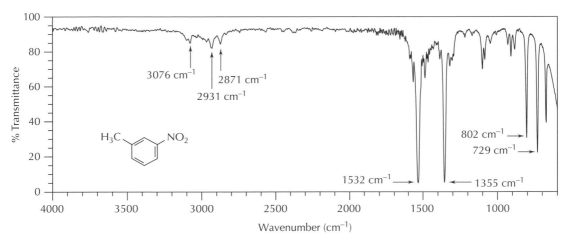

FIGURE 18.26
IR spectrum of 3-nitrotoluene (thin film).

NO₂ Stretches (1570–1490 cm⁻¹ and 1390–1300 cm⁻¹)

Aromatic nitro groups have two very distinctive absorptions due to symmetric and asymmetric O—N—O stretches. The bands are usually the most intense in the spectrum. In the spectrum of 3-nitrotoluene, shown in Figure 18.26, the signals appear at 1532 and 1355 cm⁻¹.

Symmetric stretch of nitro group

Asymmetric stretch of nitro group

Useful Diagnostic Peaks (1000–600 cm⁻¹)

An infrared spectrum can be highly cluttered with peaks, and not every one can be easily or directly correlated to a specific vibration. However, there are some absorption bands, in addition to the fundamental IR stretching vibrations, that can provide structural information. The number and arrangement of substituents on the C=C bond can often be determined from the presence of strong signals below 1000 cm⁻¹, which occur because of C—H bending vibrations. Table 18.3 summarizes these diagnostic peaks in the region 1000–600 cm⁻¹.

Absorptions at 997 and 914 cm⁻¹ in the spectrum of 1-hexene (see Figure 18.20) are characteristic of a monosubstituted alkene. In the spectrum of cinnamaldehyde (see Figure 18.24), the *trans*-disubstituted C=C bond is indicated by the absorption at 973 cm⁻¹. The trisubstituted alkene in 4-methyl-3-penten-2-one is indicated by the absorption appearing at 826 cm⁻¹ in its IR spectrum (see Figure 18.23). Absorptions at 760 and 690 cm⁻¹ in the spectrum of benzonitrile (see Figure 18.21) are characteristic of a monosubstituted benzene ring. The signal at 820 cm⁻¹ in the spectrum of 4-methylphenylamine (see Figure 18.17) is characteristic of a 1,4-disubstituted aromatic compound.

T A B L E 1 8 . 3	**Out-of-plane C—H bending vibrations of alkenes and aromatic compounds**

Structure	Position (cm⁻¹)
$\underset{H}{\overset{R}{\diagdown}}C{=}CH_2$	997–985 and 915–905
$\underset{H}{\overset{R}{\diagdown}}C{=}C\underset{R}{\overset{H}{\diagup}}$	980–960
$\underset{H}{\overset{R}{\diagdown}}C{=}C\underset{H}{\overset{R}{\diagup}}$	730–665
$\underset{R}{\overset{R}{\diagdown}}C{=}CH_2$	895–885
$\underset{R}{\overset{R}{\diagdown}}C{=}C\underset{H}{\overset{R}{\diagup}}$	840–790
⬡—R	770–730 and 720–680
⬡ (R, R ortho)	770–735
R—⬡—R (meta)	810–750 and 725–680
R—⬡—R (para)	860–800

18.7   A Procedure for Interpreting an IR Spectrum

IR spectroscopy is an important tool for determining what functional groups are in a molecule. For most organic compounds, this information alone is not sufficient to unequivocally determine the structure. However, the inventory of functional groups coupled with other data, particularly NMR and mass spectra, usually leads to a definitive elucidation of a compound's structure.

After you have interpreted numerous IR spectra, the need for a structured approach in compiling an inventory of functional groups will not be very great. But in the beginning, a general method that provides a structured and logical approach may be helpful in learning to interpret an IR spectrum.

Step 1. Check 1850–1630 cm^{-1} Region

A strong signal in this region indicates the presence of a carbonyl group. If there are no strong signals, no C=O group is present, and you should proceed to step 2. If a C=O group is present, use signals in other regions of the IR spectrum to identify the specific type of carbonyl functional group.

- Two strong bands centered near 1800 cm^{-1} indicate an acid anhydride group.
- The presence of two weak absorption bands in the region 2900–2700 cm^{-1} indicates an aldehyde group.
- The presence of an extremely broad band extending from 3200 to 2500 cm^{-1} indicates a carboxylic acid group.
- The presence of strong absorption bands in the region 1300–1000 cm^{-1} indicates an ester group.
- The presence of one or two medium-intensity bands in the 3500–3150 cm^{-1} region indicates an amide group.

In the absence of any of the preceding conditions, a single strong C=O stretching absorption near 1700 cm^{-1} is probably the result of a ketone. A single strong absorption near 1800 cm^{-1} is probably the result of an acid chloride.

Step 2. Check 3650–3200 cm^{-1} Region

The presence of a strong, broad signal indicates the hydroxyl group of an alcohol. There should be an accompanying strong band due to C—O stretching in the region 1300–1000 cm^{-1}. The presence of medium-intensity bands indicates an amine group. Primary amines have two bands and secondary amines have one.

Step 3. Check C—H Stretching Region at 3310–2850 cm^{-1}

A strong, sharp band near 3300 cm^{-1} indicates a terminal alkyne group. There should be an accompanying weak- to medium-intensity band due to C≡C stretching near 2200 cm^{-1}. Bands in the region 3100–3000 cm^{-1} are a result of C—H stretching in alkenes or aromatic compounds. Corroborating bands can narrow the choices.

- A medium-intensity band near 1650 cm^{-1} indicates the presence of a C=C bond.
- Several weak- to medium-intensity bands in the region 1620–1450 cm^{-1} may suggest the presence of an aromatic ring.
- If the presence of an alkene or an aromatic ring is indicated, the region 1000–600 cm^{-1} may determine the substitution pattern (see Table 18.3).

Signals in the region 2990–2850 cm^{-1} are caused by C—H stretching in alkyl groups.

Step 4. Check
2280–2100 cm⁻¹
Region

The presence of a strong signal near 2250 cm^{-1} indicates a nitrile group. The presence of a medium- to weak-intensity band near 2170 cm^{-1} indicates a C≡C group.

Step 5. Check
1300–1000 cm⁻¹
Region

If there are one or two strong absorptions in this region and no signals in the O—H or C=O stretching regions, an ether group may be present.

Step 6. Prepare
an Inventory of
Functional Groups

Assemble a list of the functional groups indicated by the IR spectrum. If NMR data or a molecular formula are available, coordinate them with the results from IR spectroscopy. Fit the pieces together into likely chemical structures that are consistent with the data and with the rules of chemical bonding.

18.8	A Case Study

In this section you will see how the information derived from an IR spectrum of an organic compound can help you to determine its molecular structure. The molecular formula of the compound is $C_9H_{10}O$ and its IR spectrum, prepared as a thin film, is shown in Figure 18.27.

Begin by surveying the 4000–1500 cm^{-1} functional group region. The general approach presented in Technique 18.7 can then lead you to the creation of an inventory of bond types present in the molecule. This inventory allows you to get a good idea of which functional groups are present in the compound and which functional groups are not.

The absence of a strong signal in the region 1850–1630 cm^{-1} indicates that there are no C=O groups present. Prominent in the region 3650–3200 cm^{-1} is the intense, broad band at 3327 cm^{-1}. Its intensity and position indicate the presence of a hydroxyl group. There are also strong bands in the 1300–1000 cm^{-1} region, consistent

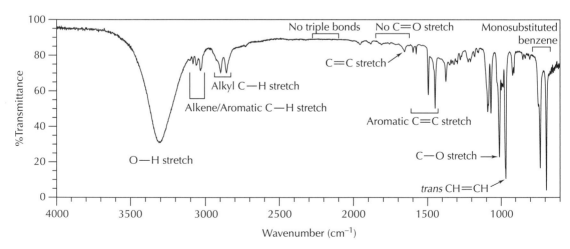

FIGURE 18.27
IR spectrum of $C_9H_{10}O$ (ATR-corrected).

with the C—O stretching vibration of an alcohol, although the cluttered nature of this area makes a definitive assignment of the signals difficult.

In the C—H stretching region, 3310–2850 cm^{-1}, there are signals from 3100 to 3000 cm^{-1} superimposed in part on the broad and intense O—H stretching signal. The signals in the region 3100–3000 cm^{-1} signify C—H stretching in alkenes or aromatic compounds. The presence of a C=C bond is confirmed by a weak intensity band at 1668 cm^{-1}. The signals at 1599, 1578, 1494, and 1449 cm^{-1} are consistent with the presence of an aromatic ring. There are also two signals of medium intensity at about 2900 cm^{-1}, which signify C—H stretching vibrations of at least one alkyl group.

An absence of any signals in the region 2280–2100 cm^{-1} rules out the presence of any functional groups with triple bonds.

Because the presence of both a C=C bond and an aromatic ring are indicated, a check of the region 1000–600 cm^{-1} is warranted. The signal at 967 cm^{-1} indicates that the double bond is *trans*-disubstituted. The signals at 740 and 692 cm^{-1} indicate that the aromatic ring is monosubstituted (see Table 18.3).

In summary, our inventory of functional groups consists of a monosubstituted benzene ring (C$_6$H$_5$—), a *trans*-disubstituted double bond (—CH=CH—), a hydroxyl group (—OH), and at least one sp^3 carbon atom. The molecular formula of the compound is C$_9$H$_{10}$O. If the alkyl carbon atom is part of a methylene group, we have accounted for all of the necessary atoms. There are two ways to put these pieces together:

The structure on the right can be eliminated because it is the unstable enol isomer of an aldehyde. The compound that produced the IR spectrum shown in Figure 18.27 is the structure on the left, (E)-3-phenyl-2-propen-1-ol, commonly called cinnamyl alcohol.

This case study was carefully chosen to show the power of infrared spectroscopy. In most cases it would be difficult if not impossible to reach a definitive structure for a compound given only a molecular formula and an IR spectrum unless one successfully searched a database for a match with the spectrum of the compound. However, even if a complete structure doesn't result from the assembly of an inventory of functional groups, the knowledge of which functional groups are present can be a great help in understanding the compound and its properties.

18.9   Sources of Confusion

The three major sources of confusion in infrared spectroscopy arise from faulty sample preparation, incorrect use of the FTIR spectrometer, and the inherent complexity of IR spectra.

Problems with Sample Preparation

Careful sample preparation is essential to producing a useful IR spectrum.

Water. If the sample is not scrupulously dry, the suspended or dissolved water will result in bands in the O—H stretching region near 3500 cm^{-1} and in the O—H bending region near 1650 cm^{-1}. In addition to producing a spectrum with misleading absorptions, the water will also etch the sodium chloride disks used to contain the sample. Etched disks absorb and scatter infrared radiation and future spectra will have less resolved IR signals.

Intense signals. If the sample is too concentrated or too thick in the case of thin films, the large bands will "bottom out" at 0% transmittance, producing wide absorption bands from which it is impossible to determine an exact absorption frequency. Small signals will appear to have larger significance than they deserve, often leading to erroneous assignments of the IR peaks. The remedy is to prepare a less concentrated solution or a thinner film.

Broad indistinct signals. If you are working with a thin film, it is likely that the sample has evaporated or migrated away from the sampling region of the infrared beam. With mulls and KBr pellets, the solid sample probably has not been ground finely enough.

Sloping base line. A sloping base line, as shown in the spectrum of fluorenone in Figure 18.28, is a problem with Nujol mulls that is difficult to avoid. Often a severely sloping base line is the result of a poorly ground solid, but even with careful grinding, some samples still produce spectra with sloping base lines. With the availability of digitized data on an FTIR spectrometer, the base line can be adjusted with the instrument's software.

Missing peaks. At times, you may encounter spectra that seem internally inconsistent. For example, you may be working with the

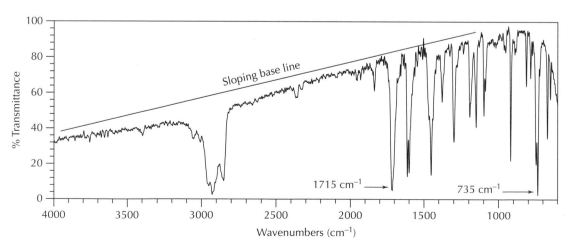

FIGURE 18.28
IR spectrum of fluorenone (Nujol mull).

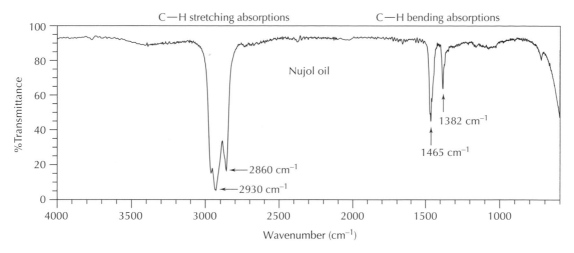

FIGURE 18.29
IR spectrum of Nujol (thin film).

Nujol mull spectrum of a compound that you strongly suspect contains one or more functional groups, and yet there are no signals in its IR spectrum indicating their presence. A common mistake made when preparing Nujol mulls is the addition of too much mineral oil, leading to a spectrum that is virtually indistinguishable from the spectrum of Nujol itself, shown in Figure 18.29.

Problems Using the Spectrometer

Although FTIR spectrometers are not especially difficult to use, two confusing situations are encountered from time to time.

No spectrum. What if the IR spectrum you obtain consists of a flat line at 100% transmittance? The easy explanation is that you forgot to put the sample into the IR beam. But what if you did put the sample into the beam? In all likelihood, you placed the sample in the IR beam before you ran a background scan and then left it in the beam and ran a sample scan. In that case, the background scan and the sample scan are the same. The result of subtracting the background scan from the sample scan is equivalent to 100% transmittance over the entire wavelength range.

Unexpected peaks near 2350 cm^{-1}. You may see a pair of signals near 2350 cm^{-1} on your IR spectrum, which may be either up or down in direction. These signals are absorption bands of carbon dioxide. If the sample compartment is left open for long periods, the ambient atmosphere, which contains CO_2, infiltrates the compartment. If the signals are down, the sample compartment was left open before running the sample scan. If the signals are up (greater than 100% transmittance), the sample compartment was left open before running the background scan. The remedy for this problem is to keep the sample compartment closed except when installing or removing a sample and to allow enough time for the closed sample compartment to be purged with purified air or nitrogen before obtaining the IR spectrum.

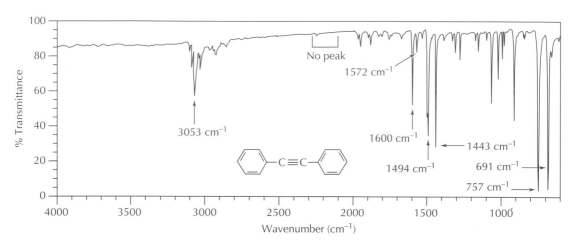

FIGURE 18.30
IR spectrum of diphenylethyne (KBr disk).

Inherent Complexity of IR Spectra

The number of observed absorption bands is generally different from the total number of possible fundamental molecular vibrations. Some vibrations are not IR active and do not absorb energy. Some absorption bands result from overtone vibrations, combination vibrations, and the coupling of vibrations.

Missing peaks. When the IR spectrum of a symmetric or nearly symmetric compound is taken, an expected absorption peak may be missing from the spectrum. For example, the spectrum of diphenylethyne, shown in Figure 18.30, displays no characteristic C≡C stretch near 2200 cm^{-1}. The absence of the C≡C absorption is the result of symmetry; the C≡C bond does not have a dipole bond moment and its stretching vibration is not IR active.

Extra peaks. Extra peaks in unexpected positions can lead to confusion. In most cases, the extraneous signals are overtones of very strong peaks in the spectrum. A good example is seen in the spectrum of methyl acetate (see Figure 18.25). The signal at 3490 cm^{-1} is in the region where O—H stretching absorptions appear, but the peak is clearly not an OH stretch because of its weak intensity and the shape of the absorption. It is an overtone of the intense C=O peak at 1745 cm^{-1}.

References

1. *The Aldrich Library of FT-IR Spectra*; 2nd ed.; Aldrich Chemical Company: Milwaukee, WI, 1992; 3 volumes.
2. Silverstein, R. M.; Webster, F. X.; Kiemle, D. J. *Spectrometric Identification of Organic Compounds*; 7th ed.; Wiley: New York, 2005.
3. Crews, P.; Rodríguez, J.; Jaspars, M. *Organic Structure Analysis*; Oxford University Press: Oxford, 1998.
4. Colthup, N. B.; Daly, L. H.; Wiberly, S. E. *Introduction to Infrared and Raman Spectroscopy*, 3rd. ed.; Academic: Boston, 1990.
5. Oberg, K. A.; Palleros, D. R. *J. Chem. Educ.* **1995,** 72, 857-859.

Questions

1. In each of the sets that follow, match the proper compound with the appropriate set of IR bands and give the rationale for your assignment.

   a. dodecane, 1-decene, 1-hexyne, 1,2-dimethylbenzene

   3311(s), 2961(s), 2119(m) cm^{-1}

   3020(s), 2940(s), 1606(s), 1495(s), 741(s) cm^{-1}

   3049(w), 2951(m), 1642(m) cm^{-1}

   2924(s), 1467(m) cm^{-1}

   b. phenol, benzyl alcohol, methoxybenzene

   3060(m), 2835(m), 1498(s), 1247(s), 1040(s) cm^{-1}

   3370(s), 3045(m), 1595(s), 1224(s) cm^{-1}

   3330(br, s), 3030(m), 2950(m), 1454(m), 1223(s) cm^{-1}

   c. 2-pentanone, acetophenone, 2-phenylpropanal, heptanoic acid, 2-methylpropanamide, phenyl acetate, 1-aminooctane

   3070(m), 2978(m), 2825(s), 2720(m), 1724(s) cm^{-1}

   3372(m), 3290(m), 2925 cm^{-1}

   3070(w), 1765(s), 1215(s), 1193(s) cm^{-1}

   3300—2500(br, s), 2950(m), 1711(s) cm^{-1}

   3060(m), 2985(w), 1690(s) cm^{-1}

   3352(s), 3170(s), 2960(m), 1640(s) cm^{-1}

   2964(s), 1717(s) cm^{-1}

2. Treatment of cyclohexanone with sodium borohydride results in a product that can be isolated using distillation. The IR spectrum of this product is shown in Figure 18.31. Identify the product and assign the major IR bands.

3. In an attempt to prepare diphenylacetylene, 1,2-dibromo-1,2-diphenylethane is refluxed with potassium hydroxide. A hydrocarbon with the chemical formula $C_{14}H_{10}$ is isolated. The infrared spectrum exhibits signals at 3100–3000 cm^{-1} but no signals in the region 2300–2100 cm^{-1}. Is this spectrum consistent with a compound containing a carbon-carbon triple bond? Explain the absence of a signal in the 2300–2100 cm^{-1} region.

4. Review the following experimental data and comment on how the sample preparation techniques affected the appearance of the high-wavenumber end of the IR spectrum.

   a. The IR spectrum of a thin film of 1-pentanol reveals a single broad and strong band centered at 3300 cm^{-1}, whereas the spectrum of a dilute chloroform solution of the same compound shows both the same broad band and a sharp spike at 3650 cm^{-1}.

   b. IR analysis of 2-hydroxyacetophenone shows a broad band centered at 3080 cm^{-1} for a wide range of concentrations in chloroform solution.

   c. When simple alcohols are analyzed in mulling compounds (Nujol and Fluorolube), the IR spectrum shows essentially the same broad, strong band centered at approximately 3300 cm^{-1}. However, dilute solutions of the alcohols show, in addition to the same 3300 cm^{-1} band, a new, sharp peak at approximately 3650 cm^{-1}.

5. When benzene is treated with chloroethane in the presence of aluminum chloride, the product is expected to be ethylbenzene (bp 136°C). During the isolation of this product by distillation, some liquid of bp 80°C was obtained. Identify this product, using its boiling point and the IR spectrum in Figure 18.32.

6. In principle, matched solution cells should remove IR bands due to the solvent and reveal peaks due to solute (sample) in the region where solvents absorb (see Table 18.1). In practice, although solvent band removal is easily achieved, it is often difficult to discern the peaks due to the sample. Explain.

7. Consider the IR spectra shown in Figures 18.33 through 18.40 and match them to the following compounds: biphenyl, 4-isopropyl-1-methylbenzene, 1-butanol, phenol, 4-methylbenzaldehyde, ethyl propionate, benzophenone, acetanilide. (**Note:** Some of the samples were prepared as thin films and some were prepared as Nujol mulls.)

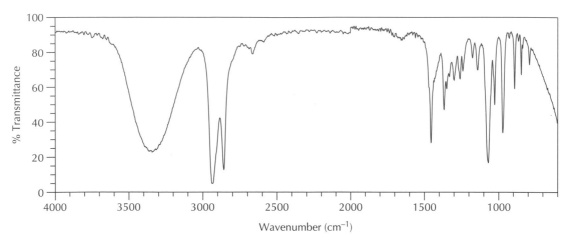

FIGURE 18.31
IR spectrum for question 2 (thin film).

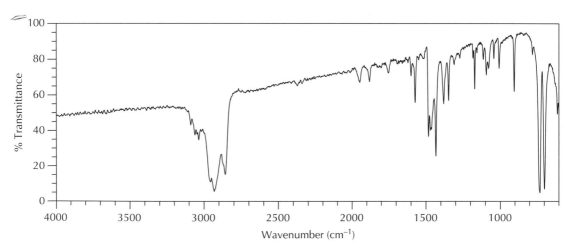

FIGURE 18.32
IR spectrum for question 5 (thin film).

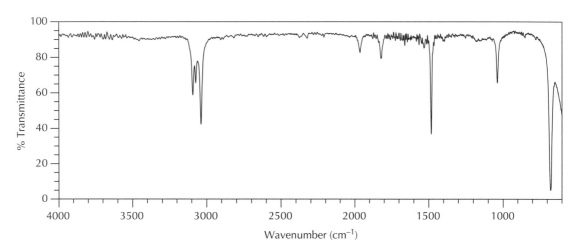

FIGURE 18.33
IR spectrum for question 7 (Nujol mull).

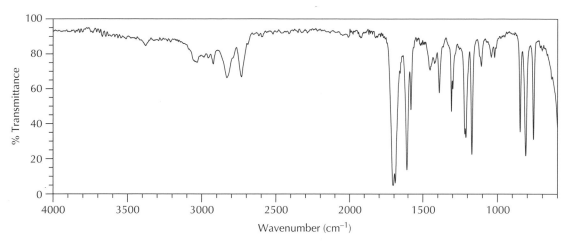

FIGURE 18.34
IR spectrum for question 7 (thin film).

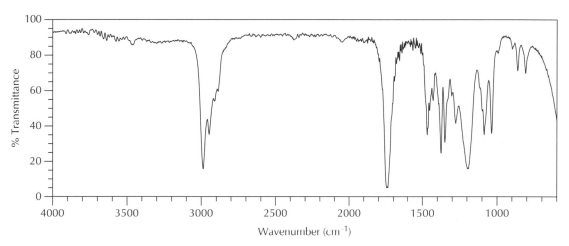

FIGURE 18.35
IR spectrum for question 7 (thin film).

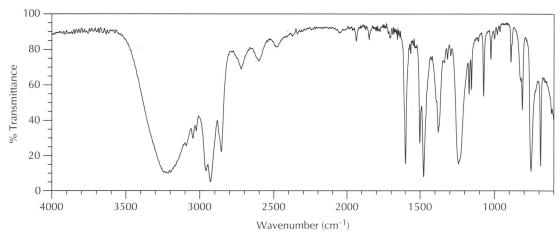

FIGURE 18.36
IR spectrum for question 7 (Nujol mull).

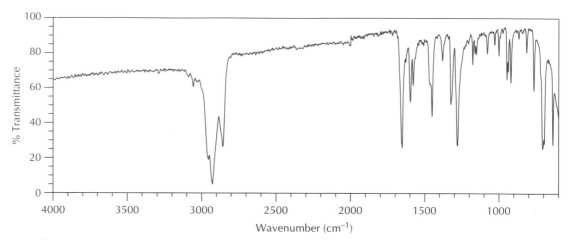

FIGURE 18.37
IR spectrum for question 7 (Nujol mull).

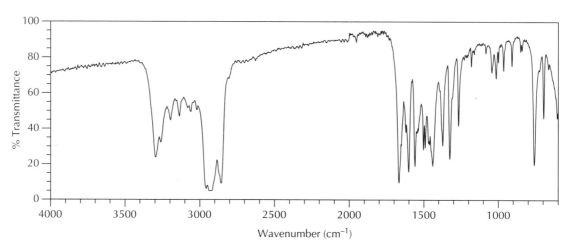

FIGURE 18.38
IR spectrum for question 7 (Nujol mull).

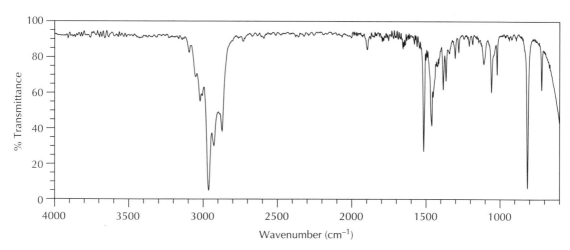

FIGURE 18.39
IR spectrum for question 7 (thin film).

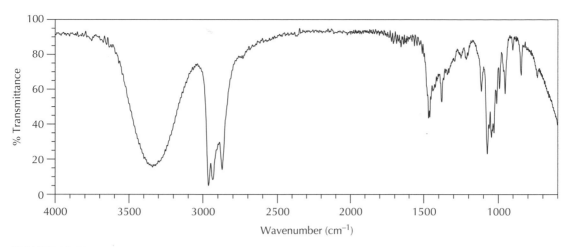

FIGURE 18.40
IR spectrum for question 7 (thin film).

19

NUCLEAR MAGNETIC RESONANCE SPECTROSCOPY

Nuclear magnetic resonance (NMR) spectroscopy is one of the most important modern instrumental techniques used in the determination of molecular structure. For the past 50 years, nuclear magnetic resonance has been in the forefront of the spectroscopic techniques that have completely revolutionized organic structure determination. Like other spectroscopic techniques, NMR depends on quantized energy changes that are induced in molecules when they interact with electromagnetic radiation. The energy needed for NMR is in the radio frequency range of the electromagnetic spectrum and is much lower energy than that needed by other spectroscopic techniques.

Nuclear Spin

The theoretical foundation for nuclear magnetic resonance arises from the **spin, I,** of an atomic nucleus. The value of I is related to the atomic number and the mass number and may be 0, $\frac{1}{2}$, 1, $\frac{3}{2}$, 2, and so forth. Any isotope whose nucleus has a nonzero magnetic moment ($I > 0$) is in theory detectable by NMR spectroscopy. Readily observable nuclei include ^1H, ^2H, ^{13}C, ^{15}N, ^{19}F, and ^{31}P. The most important nuclei for organic structure determination are ^1H and ^{13}C, both of which have spin of $\frac{1}{2}$. ^1H NMR is the major focus of this chapter, followed by a shorter discussion of ^{13}C NMR.

Any nucleus with both an even atomic number and an even mass number has a nuclear spin of 0. Because ^{12}C and ^{16}O have nuclear spins of 0, they do not produce NMR signals and do not interfere with or complicate the signals from ^{1}H and ^{13}C. In addition, ^{12}C is the major isotope of carbon and is present in almost 99% natural abundance. Therefore, the small amount of NMR active ^{13}C does not complicate ^{1}H NMR spectra to any great extent.

Nuclear Energy Levels

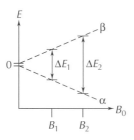

FIGURE 19.1
Influence of an external magnetic field on spin state energy levels.

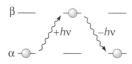

FIGURE 19.2
Excitation of a nucleus from low spin state to high spin state and emission of energy on relaxation of the nucleus.

Magnetic Resonance

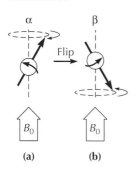

FIGURE 19.3
Nuclear magnetic dipole (a) aligned with an external magnetic field (α) and (b) opposed to an external magnetic field (β).

There are $(2I + 1)$ energy levels allowed for a nucleus with spin of I. Because ^{1}H and ^{13}C have spins of $\frac{1}{2}$, there are two possible energy levels for these nuclei ($2I + 1 = 2$). In the absence of an external magnetic field, the two levels are *degenerate*—they have the same energy. However, in the presence of an applied magnetic field, the energy levels move apart. The separation of degenerate nuclear spin energy levels by an external applied magnetic field is illustrated in Figure 19.1. One energy level, designated α, decreases in energy and the other level, designated β, increases in energy. The difference in energy between the levels, ΔE, is directly related to the strength of the externally applied magnetic field, B_0.

In spectroscopy, the usual convention for expressing energy changes is frequency (ν), as described by Planck's law:

$$\Delta E = h\nu$$

The change in energy of an NMR transition is extremely small by chemical standards—only about 10^{-6} kJ·mol^{-1}, which corresponds to energy in the radio frequency region. With a magnetic field strength of 1.41 tesla, the resonance frequency for ^{1}H nuclei is 60 MHz. If the magnetic field strength is 7.05 tesla, the resonance frequency is 300 MHz.

As shown in Figure 19.2, the absorption of energy can cause excitation of a nucleus to a higher quantized energy level. When a nucleus in the higher energy spin state drops to the lower energy spin state, a process called *relaxation,* it gives up a quantum of energy. The emitted energy, in the radio frequency region, produces an NMR signal.

We can think of any nucleus with a spin number greater than 0 as a spinning, charged body. The principles of physics tell us that a magnetic field is associated with this moving charge. When placed in a magnetic field, a spinning nucleus precesses about an axis aligned in the direction of the magnetic field. The precession of a child's top about a vertical axis as it spins can be used as a mechanical model for the process. The magnetic dipole of the spinning nucleus shown in Figure 19.3a is aligned with the external magnetic field, B_0, whereas the magnetic dipole of the spinning nucleus shown in Figure 19.3b is opposed to the external magnetic field. To flip the magnetic dipole from the aligned position to the opposed position requires a quantized addition of energy to the system. Absorption of energy can occur only if the system is in resonance.

For *resonance* to occur, the applied frequency (ν) must be precisely tuned to the rotational frequency of the precessing nucleus. Then the nucleus can absorb a quantum of energy and flip from the lower energy spin state (α) to the higher energy spin state (β). The energy difference between the two spin states is very small and the number of nuclei in each spin state is nearly equal, but in the large magnetic field of a modern NMR spectrometer there are a few more nuclei, approximately 0.001%, in the lower energy spin state than in the higher energy spin state. Because the spin states are not equally populated, a nuclear magnetic resonance effect can be observed.

If all the ^1H nuclei in a molecule had the same resonance frequency, ^1H NMR spectroscopy would be of little use to organic chemists. In an NMR spectrometer, energies of ^1H nuclei in an organic compound are slightly different because of their different structural environments, and a typical ^1H NMR spectrum is an array of many different frequencies. The same is true for ^{13}C NMR spectra.

NMR Instrumentation and Sample Preparation

19.1   NMR Instrumentation

The first NMR spectrometers were continuous wave (cw) instruments. The sample was irradiated with radio frequency energy as the applied magnetic field was varied. When a match between the radio frequency energy and the energy difference between the two spin states of the nucleus (hv in Figure 19.2) occurred, a signal was detected. The energy required reflected the environment of the nucleus. A radio frequency receiver was used to monitor the energy changes.

Fourier Transform NMR

More recent instruments use a technique known as ***pulsed Fourier transform NMR (FT NMR).*** In this technique, a broad pulse of electromagnetic radiation excites all the ^1H or ^{13}C nuclei simultaneously, resulting in a continuously decreasing oscillation caused by the decay of excited nuclei back to their stable energy distribution. The oscillating, or decaying, sine curve is called a *free-induction decay (FID).* The FID, often referred to as a time domain signal, is converted to a set of frequencies, or a "normal" spectrum, by the mathematical treatment of a Fourier transform. The relatively simple FID of a compound with only a single frequency is shown in Figure 19.4a. In Figure 19.4b, you can see that the FID from a compound with two frequencies is more complex. Constructive combinations of the two frequencies produce enhanced signals and destructive combinations give little or no signal. The Fourier transform of the FID in Figure 19.4b produces two signals.

Most organic compounds are much more complex and the FID is made up of the contributions from hundreds of frequencies. A computer program using Fourier transform mathematics is required to convert the FID to the "normal" spectrum. When the sample size is small, the acquisition of the signal from more than one pulse (or "scan") is necessary to obtain NMR signals with the desired signal-to-noise ratio. Modern FT NMR spectrometers "lock" on a

FIGURE 19.4
FID and Fourier
transform of FID of
(a) one signal and
(b) two signals.

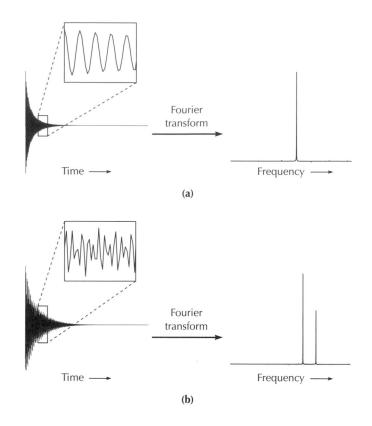

signal from deuterium in the NMR solvent, assuring that multiple acquisition scans are synchronized. The NMR computer programs the multiple pulses and collects the data from them. The components of an FT NMR spectrometer are illustrated in Figure 19.5.

FIGURE 19.5
Block diagram of
a basic FT NMR
spectrometer.

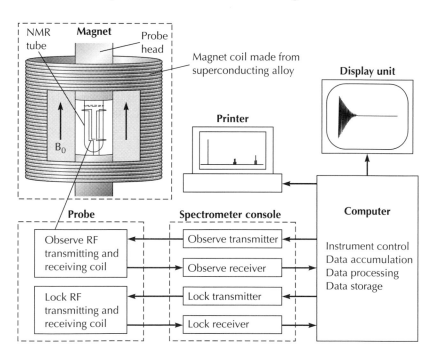

NMR Spectrometers

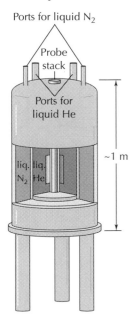

FIGURE 19.6
Typical superconducting electromagnet for a 200- to 400-MHz NMR spectrometer, showing the vacuum-jacketed Dewar vessels for the liquid nitrogen and liquid helium coolants.

There are many different models of NMR spectrometers, and it is common practice to refer to them by their nominal operating frequency. Modern NMR instruments operate at substantially higher frequencies than the 60-MHz instruments that were once the standard. Research instruments routinely operate at 300–500 MHz, and many laboratories have instruments with operating frequencies of 600 and even 800 MHz. The high magnetic fields necessary for these instruments can be achieved only by using superconducting electromagnets. Because the materials used to build the magnets are superconducting only at very low temperatures, these magnets are maintained in double-jacketed Dewar vessels cooled by liquid helium and liquid nitrogen (Figure 19.6).

There are numerous benefits in using higher frequency and field strength. High-field instruments have greater sensitivity because of a greater difference in spin state populations, which translates into stronger sample signals relative to background noise. The higher field strength also means larger energy differences between different nuclei and thus greater signal separation. The advantage of greater signal separation is evident by comparing the spectra of ethyl propanoate, $C_5H_{10}O_2$, shown in Figure 19.7. The NMR spectrum in Figure 19.7a was obtained on a 60-MHz continuous wave (cw) NMR spectrometer. The spectrum in Figure 19.7b was obtained with a 200-MHz FT NMR spectrometer. In the 60-MHz spectrum, a group of signals is centered at 1.15, which in the 200-MHz spectrum separates into two groups of signals, one centered at 1.15 and one centered at 1.26.

19.2 Preparing Samples for NMR Analysis

Almost all NMR analysis is done using dilute solutions, whether the sample is a solid or a liquid. Concentrated solutions, undiluted liquids, and solids usually exhibit broad peaks that are not easy to interpret. Spectra with sharp, well-differentiated signals are obtained only with dissolved samples. It is important to use the required minimum amount of sample for an NMR spectrum but not much more.

NMR Solvents

The choice of solvent for an NMR sample is important. Most of the material in an NMR tube is solvent, so ideally we want a chemically inert solvent that does not absorb energy in the magnetic field. Thus, for ^1H NMR we want a solvent with no protons. Most of the solvents used for preparation of NMR samples are deuterated forms of common solvents, such as chloroform ($CHCl_3$), acetone, and water. Although deuterium does have a magnetic moment, its signal is well removed from the region where protons absorb.

Deuterated chloroform ($CDCl_3$) is the most commonly used NMR solvent because it dissolves a wide range of organic compounds

(a)

(b)

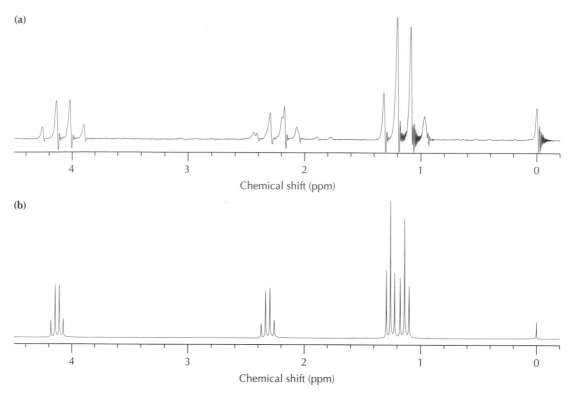

FIGURE 19.7
^{1}H NMR spectra of ethyl propanoate at (a) 60 MHz and (b) 200 MHz.

and is not prohibitively expensive. Deuterated acetone is another commonly used solvent, but it is quite a bit more expensive. Table 19.1 lists several standard NMR solvents.

Deuterated solvents are never 100% deuterated. For example, the commercial CDCl$_3$ that is commonly used for NMR samples has 99.8% deuterium and 0.2% protium in its molecules. The residual protons give a small peak (CHCl$_3$) at 7.26 ppm. The residual proton signals for the various solvents are also listed in Table 19.1. **It is important to be aware of the position of these residual signals because you do not want to confuse solvent signals with the signals of your compound.**

TABLE 19.1   Deuterated solvents for NMR spectroscopy

Solvent	Structure	Residual ^{1}H signal (ppm)	^{13}C chemical shift (ppm)
Chloroform-d	CDCl$_3$	7.26 (singlet)	77.0 (triplet)
Acetone-d$_6$	CD$_3$(C=O)CD$_3$	2.04 (quintet)	29.8 (septet), 206.5 (singlet)
Deuterium oxide	D$_2$O	4.6 (broad singlet)	—
Dimethyl sulfoxide-d$_6$	CD$_3$(S=O)CD$_3$	2.49 (quintet)	39.7 (septet)

Polar compounds. Polar chemicals, such as carboxylic acids and poly-hydroxyl compounds, are usually not soluble in $CDCl_3$. However, in most cases these compounds are soluble in deuterium oxide (D_2O). If a carboxylic acid is not soluble in D_2O, it is probably soluble in D_2O containing sodium hydroxide. Adding a drop or two of concentrated sodium hydroxide solution to the sample in D_2O is usually enough to dissolve it.

Problems with the use of D_2O. The use of D_2O presents several problems. There is always a broad peak at approximately 4.6 ppm because of a small amount of HOD present in the original D_2O solvent. This solvent peak can cover important signals from the compound being analyzed. Also, D_2O may exchange with protons in the sample compound, producing HOD. Consider, for example, what happens when a carboxylic acid or alcohol dissolves in D_2O:

$$R-\overset{\overset{\displaystyle O}{\|}}{C}-O-H + D_2O \rightleftharpoons R-\overset{\overset{\displaystyle O}{\|}}{C}-O-D + H-OD$$

Carboxylic acid

$$R-O-H + D_2O \rightleftharpoons R-O-D + H-OD$$

Alcohol

Deuterium nuclei are "invisible" in ^1H NMR spectra, and in an NMR solution there are many more molecules of solvent D_2O than of the sample. The equilibrium positions in these reactions lie well to the right, and the hydroxyl protons do not appear as separate signals but instead merge into the HOD signal.

NMR Reference Calibration

Solvents used for preparing NMR samples often have a small amount of a standard reference substance dissolved in them. However, a reference compound is not really necessary because the residual proton signal of partially deuterated solvent can be used for reference calibration unless sample signals obscure it (see Table 19.1).

Tetramethylsilane. The most common added reference compound is tetramethylsilane, $(CH_3)_4Si$. **Tetramethylsilane, usually referred to as TMS, has been so important as a reference substance in the past that the position of its signal is used to define the 0.0 point on an NMR spectrum.** TMS was chosen because all its protons are equivalent, and they absorb at a magnetic field in which very few other protons in typical organic compounds absorb. TMS is also chemically inert and is soluble in most organic solvents. The amount of TMS in the solvent depends on the type of instrument being used. For a cw NMR spectrometer, the typical concentration of TMS is 1–2%. With a modern FT NMR instrument, the typical concentration of TMS is 0.1%; often TMS is not even added to the NMR solution.

NMR reference for D_2O. Tetramethylsilane is not soluble in deuterium oxide (D_2O), so it cannot be used as a standard with this solvent, and the HOD peak is too broad and variable to be a useful reference standard. The reference substance used for D_2O solutions is the ionic compound sodium 2,2-dimethyl-2-silapentane-5-sulfonate (DSS), $(CH_3)_3SiCH_2CH_2CH_2SO_3^-Na^+$. Its major signal appears at nearly the same position as the TMS absorption. Acetone can also be used as a reference in D_2O solutions as long as its signal does not interfere with signals from the sample. In D_2O solutions, the signal for acetone appears at 2.22 ppm.

Preparing an NMR Sample Solution

The appropriate concentration of the sample solution depends on the type of NMR instrumentation available. A sample mass of 4–20 mg of compound dissolved in approximately 0.5–0.7 mL of solvent is used to prepare a modern high-frequency FT NMR sample solution.

NMR tubes. NMR tubes are delicate, precision pieces of equipment. The most commonly used NMR tubes are made of thin glass and their rims are easily chipped if not handled carefully. **Caution:** Chipping occurs most often during pipetting of the sample into the tube and when trying to remove the plastic cap from the top of the tube.

Check solubility in deuterium-free solvent first. Before using an expensive deuterated solvent for an NMR analysis, be sure that your compound dissolves in the deuterium-free solvent. Prepare a preliminary sample using the necessary amount of solvent in a small vial or test tube. If the preliminary test is satisfactory, place the necessary amount of your sample in another vial or small test tube and add approximately 0.7 mL of deuterated solvent. Agitate the mixture to facilitate dissolution of the sample. If a clear, homogeneous solution is obtained, transfer the sample to the NMR tube with a glass Pasteur pipet.

Particulate matter in the sample solution. If a clear, homogeneous solution is not obtained, the particulate material must be removed before the sample is transferred to the NMR tube. Particulate material may contain paramagnetic metallic impurities that will produce extensive line broadening and poor signal intensity in the NMR spectrum. A convenient filter can be prepared by inserting a small wad of glass wool into the neck of a glass Pasteur pipet [see Technique 8.5]. The narrow end of the filter pipet is placed in the NMR tube and the sample to be filtered is transferred into the filter with a second Pasteur pipet. Pressure from a pipet bulb can be used to force any solution trapped in the filter into the NMR tube.

Height of the NMR solution in the tube. Only a small part of an NMR tube is in the effective probe area of the instrument. Typically, the height of the sample in the tube should be 25–30 mm (Figure 19.8). However, the required height can be 50–55 mm in some NMR instruments. You need to ascertain the required minimum for the instrument you are using. Often a gauge is available in the lab for checking the solution height in the sample tube. If the solution height is slightly short, add a few drops of solvent to bring it to the required level and agitate the NMR tube to thoroughly mix the solution. After capping the NMR tube, wipe off any material on the outside.

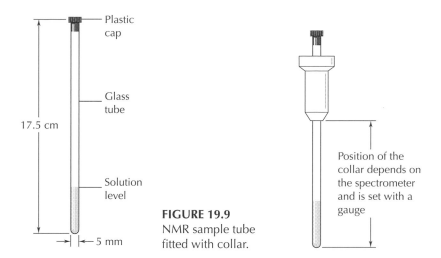

FIGURE 19.8
NMR sample tube filled
to the correct height.

Plastic
cap

Glass
tube

17.5 cm

Solution
level

5 mm

FIGURE 19.9
NMR sample tube
fitted with collar.

Position of the
collar depends on
the spectrometer
and is set with a
gauge

Recovery of the sample. Because none of the sample is destroyed when taking an NMR spectrum, the sample can be recovered if necessary by evaporating the solvent.

Obtaining the NMR Spectrum

Before the NMR sample tube is placed into the magnet of the spectrometer, it is fitted with a collar that is made of a nonmagnetic plastic or ceramic material (Figure 19.9). The collar positions the sample at a precise location within the magnetic field where the RF transmitter/receiver coil is located. A depth gauge provided with the instrument is used to set the position of the collar on the NMR sample tube. A second purpose of the collar can be to enable the sample to spin around its vertical axis once it has been placed in the magnet.

The magnetic field in the RF transmitter/receiver coil region must be homogeneous; that is, the strength and direction of the magnetic field must be exactly the same at every point. A homogeneous magnetic field is achieved through a complex adjustment called *shimming.* Even after shimming, some small magnetic field inhomogeneities may be present. Spinning the sample serves to average out these inhomogeneities, which allows acquisition of spectra with sharp, well-defined peaks. NMR tubes are selected for uniform wall thickness and minimum wobble. Too much sample in the tube is not only a waste of material, it also tends to make the tube top-heavy, often resulting in poor spinning performance and thus poor-quality spectra. With some of the latest NMR spectrometers, the magnet technology has advanced to the point that spinning the sample is not recommended.

Cleaning the NMR Sample Tube

After the spectrum has been obtained, the NMR tube should be cleaned, usually by rinsing with a solvent such as acetone, and then allowing the tube to dry. Solvents cling tenaciously to the inside surface of the long, thin NMR tubes, and a long drying period or passing a stream of dry nitrogen gas through the tube is required to remove all residual solvent. If NMR tubes are not cleaned soon after use, the solvent usually evaporates and leaves a caked or gummy residue that can be difficult to dissolve.

19.3   Summary of Steps for Preparing an NMR Sample

1. Test the solubility of the sample in ordinary, nondeuterated solvents. Select a solvent that dissolves the sample completely.
2. Place 4–20 mg of the sample in a clean, small vial or test tube.
3. Add 0.5–0.7 mL of the appropriate deuterated solvent.
4. Agitate the mixture in the vial to produce dissolution of the sample.
5. Transfer the sample solution into a clean NMR tube using a glass Pasteur pipet. If there are any solids present, filter the solution through a small plug of glass wool.
6. Check the level of the sample in the tube. If needed, add drops of solvent to bring the solution to the recommended level for the instrument and agitate the mixture to produce a homogeneous solution.
7. Cap the NMR tube.
8. Wipe the outside of the tube to remove any material that may impede smooth spinning of the sample in the NMR instrument.

Interpreting ¹H NMR Spectra

19.4   NMR Information

Typically, four types of information can be extracted from a ^1H NMR spectrum. All of them are important in determining the structure of a compound, and all are discussed in the following sections.

- *Number of different kinds of protons* in the molecules of the sample, given by the number of groups of signals [see Technique 19.5]
- Relative number of protons contributing to each group of signals in the spectrum, called the *integration* [see Technique 19.6]
- Positions of the groups of signals along the horizontal axis, called the *chemical shift* [see Techniques 19.7 and 19.8]
- Patterns within a group of signals, called *spin-spin coupling* [see Technique 19.9]

19.5   How Many Types of Protons Are Present?

As the first step in analyzing an NMR spectrum, examine the entire spectrum. A common mistake is to focus on some detail in the spectrum, often a prominent signal, and develop an analysis from an assumption that is consistent with only that detail. Sometimes

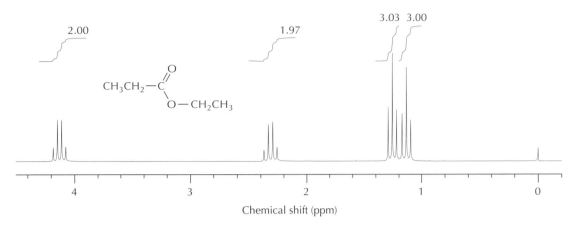

FIGURE 19.10
^{1}H NMR spectrum of ethyl propanoate at 200 MHz.

this method works, but many times it does not. A general method for analysis starts by looking at the entire spectrum and counting the number of groups of signals. A structure consistent with the spectrum is required to have at least this many different kinds of protons. This number is a minimum requirement, and often, as the analysis is refined, it is possible to divide a group of signals into subsets of protons that are subtly different from each other. If you examine the 200-MHz NMR spectrum of ethyl propanoate (Figure 19.10), you will see that four groups of signals are centered at 1.15, 1.26, 2.32, and 4.13 ppm along the horizontal scale. **Note that the horizontal scale is read from right to left, with 0.0 ppm at the far right.**

19.6   Counting Protons (Integration)

Above each group of signals in the 200-MHz NMR spectrum of ethyl propanoate in Figure 19.10 is what looks like a set of steps with a number over it. **The height of each set of steps corresponds to the total signal intensity encompassed by the set.** The numbers correspond to the heights normalized to one of the signals. Software on modern digital NMR spectrometers makes normalization an easy task. Reading from right to left, the normalized integration values for the groups of signals are 3.00, 3.03, 1.97, and 2.00, respectively.

Integration values represent the relative number of each kind of proton in the molecule. If the normalization is not done correctly, the integration values will be a multiple of the true values. Also, integration values are usually not neat, whole-number ratios. Deviations from whole numbers can be as much as 10% and are usually attributed to differences in the amount of time it takes different types of excited hydrogen nuclei to relax back to their lower energy spin states. In acquiring NMR data, it is important to allow enough time for the nuclei to relax. Otherwise, the measured integrals will not accurately reflect the relative number of protons responsible for the signals.

In addition, if the integration is done manually, you must use good judgment about where to start and stop each set of steps.

The integrals for the spectrum in Figure 19.10 are interpreted as 3:3:2:2. The two groups of signals at 1.15 ppm and 1.26 ppm, with three protons each, are produced by two groups of nearly equivalent kinds of protons. The group of signals at 2.32 ppm and the group at 4.13 ppm are each produced by a group of two equivalent protons.

- Primary hydrogens, those on a carbon atom with three hydrogens attached, are called *methyl* protons.
- Secondary hydrogens, on a carbon atom with two hydrogens attached, are called *methylene* protons.
- A tertiary hydrogen, on a carbon atom with only one hydrogen attached, is called a *methine* proton.

EXERCISE

Refer to the structure of ethyl propanoate and identify the set of protons that is responsible for each of the four groups of signals in its NMR spectrum.

$$CH_3CH_2 - C \overset{O}{\underset{O-CH_2CH_3}{\big\backslash}}$$

Ethyl propanoate

Answer: Because there are two groups of two protons and two groups of three protons, we cannot unambiguously assign the signals without more information. But help is on the way. In the next section you will find out how to use the positions of the signals along the horizontal scale to make the necessary assignments.

19.7   Chemical Shift

An NMR spectrum is a plot of the intensity of the NMR signals versus the magnetic field or frequency. Nuclei that are chemically equivalent, such as the four protons in methane (CH_4) or the two protons in dichloromethane (CH_2Cl_2), show only one peak in the NMR spectrum. However, protons that are not chemically equivalent absorb at different frequencies. The local magnetic field experienced by the different protons in a molecule varies with different magnetic environments within the molecule. At 300 MHz, the typical range of these frequencies is about 3500 Hz.

Most important, the positions of the signals along the horizontal scale of an NMR spectrum, called the *chemical shifts,* can be correlated with a molecule's structure. The goal of Techniques 19.7 and 19.8 is to show how the chemical shifts can be used to determine the structures of organic compounds. Arguably, **the use of chemical shifts is the most powerful of all the information available in NMR spectroscopy.**

Chemical Shift Units
(Parts per Million)

Because it is difficult to reproduce magnetic fields exact enough for NMR spectroscopy, an internal standard is used as a reference point. The position of an NMR signal is measured relative to the absorption of the standard. Tetramethylsilane, $(CH_3)_4Si$, is the standard for ^1H and ^{13}C NMR. Chemical shifts are measured at a frequency (Hz) corresponding to a signal's position relative to tetramethylsilane, usually referred to as TMS. It is conventional, however, to convert frequency to a value **δ** **(ppm)** by dividing the chemical shift frequency by the operating frequency of the spectrometer. This conversion produces an important result; **the chemical shift (δ) is independent of the frequency of the spectrometer.**

M = mega = million

$$\delta \text{ (ppm)} = \frac{\text{frequency of the signal (in Hz, from TMS)}}{\text{applied spectrometer frequency (in MHz)}}$$

Because the frequency of an NMR spectrometer is given in megahertz (MHz), the δ values are always given in parts per million (ppm). On the chemical shift scale of an NMR spectrum, the position of the TMS absorption is at the far right and is set at 0.0 ppm. The δ values increase to the left of the TMS peak.*

Consider the NMR spectra of *tert*-butyl acetate shown in Figure 19.11. The *tert*-butyl group, which has nine equivalent protons, and the methyl group, which has three equivalent protons, give a relative integration of $3:1$. In the 60-MHz spectrum (Figure 19.11a), the difference between the signals of TMS and the *tert*-butyl group is 87 Hz. In the 200-MHz spectrum (Figure 19.11b), the difference between these

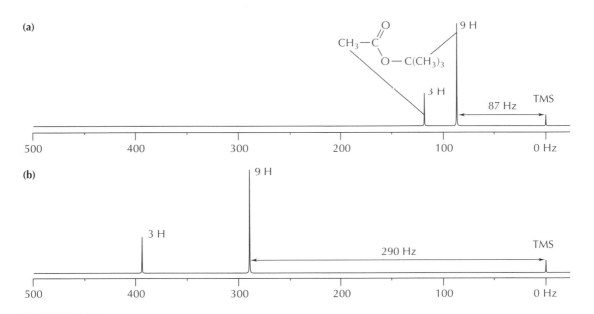

FIGURE 19.11
^1H NMR spectra of *tert*-butyl acetate in the region 0–500 Hz at (a) 60 MHz and (b) 200 MHz.

*In the older NMR literature a τ (tau) scale for chemical shifts was used, in which the TMS absorption signal was given the value of 10.0 ppm and the chemical shift values decreased to the left of TMS on an NMR spectrum. With the τ system, a chemical shift of 2.0 ppm (δ), for example, would be 8.0 ppm.

same signals is 290 Hz. Dividing the signals' frequency difference by the operating frequency of the instrument, we find that the chemical shift (δ) of the *tert*-butyl protons is 1.45 ppm.

$$1.45 \text{ ppm} = 87/60 = 290/200$$

The position of the signal in terms of its chemical shift (δ) is the same, regardless of the magnetic field strength. To compare NMR spectra from different instruments, the chemical shift scales for all NMR spectra are plotted using ppm units.

EXERCISE

On an NMR instrument operating at 60 MHz, the signal for the methyl group of *tert*-butyl acetate is shifted 118 Hz relative to the signal for TMS (see Figure 19.11a).
(a) What is the chemical shift (δ) of the methyl group signal?
(b) What is the frequency difference (in Hz) between the signal for the methyl group and the signal for TMS on an NMR instrument operating at 200 MHz (see Figure 19.11b)?
Answer: (a) δ = 118/60 = 1.97 ppm   (b) ΔHz = 1.97 × 200 = 394 Hz

Figure 19.12 shows the approximate chemical shift regions of signals for different types of protons attached to carbon, oxygen, and nitrogen atoms. A list of chemical shifts for different types of protons is given in Table 19.2.

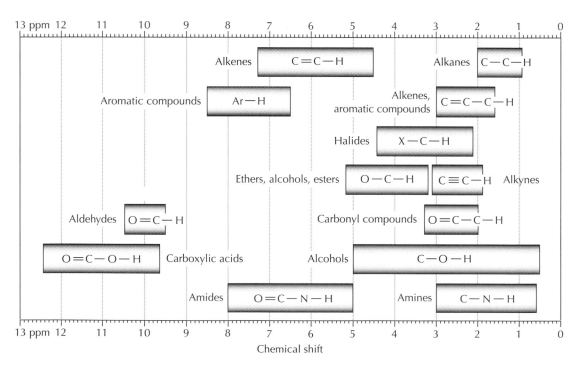

FIGURE 19.12
Approximate regions of chemical shifts for different types of protons in organic compounds.

TABLE 19.2	Characteristic ^{1}H NMR chemical shifts in CDCl$_3$
Compound	**Chemical shift (δ, ppm)**
TMS	0.0
Alkanes (C—C—**H**)	0.9–1.9
Amines (C—N—**H**)	0.6–3.0
Alcohols (C—O—**H**)	0.5–5.0
Alkenes^a (C=C—C—**H**)	1.6–2.5
Alkynes (C≡C—**H**)	1.7–3.1
Carbonyl compounds (O=C—C—**H**)	1.9–3.3
Halides (X—C—**H**)	2.1–4.5
Aromatic compounds^b (Ar—C—**H**)	2.2–3.0
Alcohols, esters, ethers (O—C—**H**)	3.2–5.2
Alkenes (C=C—**H**)	4.5–7.3
Phenols (Ar—O—**H**)	4.0–8.0
Amides (O=C—N—**H**)	5.0–8.0
Aromatic compounds (Ar—**H**)	6.5–8.5
Aldehydes (O=C—**H**)	9.5–10.5
Carboxylic acids (O=C—O—**H**)	9.7–12.5

a. Allylic protons.

b. Benzylic protons.

Diamagnetic Shielding

The chemical shift of a hydrogen nucleus is strongly influenced by the electron density surrounding it. Under the influence of an applied magnetic field, circulating electrons in the spherical electron cloud induce a small magnetic field opposed to the applied field, as illustrated in Figure 19.13. Thus, the effective magnetic field that a proton feels is a little less than the applied field. The electron cloud is said to shield the nucleus from the applied magnetic field and the effect is called *local diamagnetic shielding.*

If the electron density around a proton is decreased, the opposing induced magnetic field will be smaller. Therefore, the nucleus is less shielded from the applied magnetic field, and the proton is said to be *deshielded.* With greater deshielding, the effective magnetic field felt by the proton increases, and the chemical shift of its signal increases. For example, the protons of methane resonate at 0.23 ppm. Attaching an electron-withdrawing chlorine atom to the carbon atom pulls electron density away from the electron cloud surrounding the nearby protons. Thus, the chlorine deshields the protons. The protons of chloromethane resonate at 3.1 ppm.

Magnitude of the deshielding effect. The magnitude of the deshielding effect decreases rapidly as the distance from the electron-withdrawing substituent increases. This effect is demonstrated by the decrease in the chemical shifts of methyl protons as their distance from a bromine atom increases.

CH$_3$Br	CH$_3$CH$_2$Br	CH$_3$CH$_2$CH$_2$Br	CH$_3$CH$_2$CH$_2$CH$_2$Br
2.69 ppm	1.66 ppm	1.06 ppm	0.93 ppm

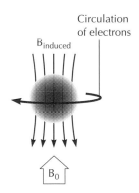

Circulation of electrons

B$_{induced}$

B$_0$

FIGURE 19.13
The opposing magnetic field induced by circulation of electrons around a nucleus in an applied magnetic field, B$_0$. The nucleus is partially shielded from the applied magnetic field by the opposing magnetic field.

Deshielding and shielding effects are additive. For instance, the chemical shift of the protons in substituted methane derivatives increases as the number of attached electron-withdrawing bromine atoms increases.

$$CH_4 \quad CH_3Br \quad CH_2Br_2 \quad CHBr_3$$
$$0.23\ ppm \quad 2.69\ ppm \quad 4.94\ ppm \quad 6.82\ ppm$$

The additive nature of the deshielding effect is also seen as the carbon atom bearing the proton becomes more highly substituted. With the same electron-withdrawing groups nearby, tertiary hydrogen atoms have a greater chemical shift than do secondary hydrogens. Likewise, secondary hydrogen atoms have a greater chemical shift than do primary hydrogens with the same electron-withdrawing groups nearby. This trend is illustrated by the chemical shifts of the proton(s) attached to the carbon atom adjacent to a bromine atom in bromomethane, bromoethane, and 2-bromopropane.

$$CH_3Br \quad CH_3CH_2Br \quad (CH_3)_2CHBr$$
$$2.69\ ppm \quad 3.37\ ppm \quad 4.21\ ppm$$

Deshielding effects and electronegativity. The position of the signal for a proton attached to a carbon atom also depends on the electronegativity (χ) of the other atoms attached to carbon. The periodic trends seen in the electronegativities of elements are mirrored in the chemical shifts of methyl groups attached to these elements.

	CH_3-I	CH_3-Br	CH_3-Cl	CH_3-F
δ	2.2 ppm	2.7 ppm	3.1 ppm	4.3 ppm
χ	2.66	2.96	3.16	3.98

Similarly, as you move from left to right along a row of elements in the periodic table, the electronegativities increase and the chemical shifts of attached methyl groups also increase.

	$(CH_3)_4C$	$(CH_3)_3N$	$(CH_3)_2O$	CH_3F
δ	0.9 ppm	2.2 ppm	3.2 ppm	4.3 ppm
χ	2.50	3.04	3.44	3.98

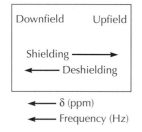

FIGURE 19.14
When the opposing induced magnetic field decreases, the effective magnetic field felt by the nucleus increases. Thus, as a nucleus becomes more deshielded, the chemical shift of the resonance frequency increases.

Summary of shielding and deshielding effects. Let's briefly summarize the effect of shielding and deshielding on the chemical shift of protons and introduce some commonly used terms (Figure 19.14). Increasing the electron density around a nucleus shields it from the applied field, making the effective field experienced by the nucleus smaller. The value of the observed chemical shift of the signal therefore decreases, and, on a typical NMR spectrum, the signal moves to the **right,** which is called an *upfield* shift because at a constant frequency, a slightly higher applied magnetic field is needed for resonance to occur. Decreasing the electron density around a nucleus deshields it, causing the chemical shift to increase and moving the signal to the **left,** resulting in a *downfield* shift.

Anisotropy

In molecules with π-orbitals, local diamagnetic shielding does not completely account for the chemical shifts observed for different protons. The shielding effect on a proton depends in part on the location

(a)

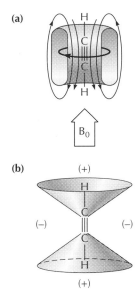

(b)

FIGURE 19.15
(a) Circulation of π-electrons in an applied magnetic field induces an opposing magnetic field that shields the acetylenic proton. (b) Regions of shielding (+) and deshielding (−) for acetylene.

of the proton relative to the induced magnetic field because a π-orbital is not spherically symmetrical. This effect is called *anisotropy*, a term that means having a different effect along a different axis.

Consider acetylene, H—C≡C—H, for example. Although acetylene molecules are oriented more or less randomly because of rapid tumbling in solution, at any one time some of these linear molecules are lined up with the applied magnetic field. In the aligned molecules, the circulation of the electrons in the cylindrical π-orbital system of the triple bond induces a diamagnetic field, as illustrated in Figure 19.15a. This induced magnetic field opposes the applied magnetic field, shielding the acetylene proton and moving its NMR signal upfield. Regions of shielding are often represented as cones with (+) signs as shown in Figure 19.15b. The chemical shift of the protons in acetylene is 1.80 ppm; it is affected both by local diamagnetic shielding effects and by anisotropic shielding.

Alkenes (olefins) and aldehydes also exhibit strong anisotropic effects. When a π-orbital of a double bond is aligned with an applied magnetic field, the circulation of the two π-electrons induces a diamagnetic field perpendicular to the plane of the double bond. Anything in the region above the π-orbital is shielded. However, at the sides of the double bond the flux lines of the induced magnetic field add to the applied magnetic field, which creates a deshielding region in the plane perpendicular to the π-orbital. The shielding (+) and deshielding (−) regions of ethylene and formaldehyde are shown in Figures 19.16a and b. Strong anisotropic effects are demonstrated by the strongly deshielded protons of ethylene and formaldehyde.

$$H_2C{=}CH_2 \qquad\qquad H_2C{=}O$$
$$5.28 \text{ ppm} \qquad\qquad 9.60 \text{ ppm}$$

Because of anisotropic deshielding, protons of methyl groups attached to the carbon atoms of C=C or C=O bonds appear near 2.0 ppm, whereas protons of methyl groups attached to carbon atoms of C—C or C—O bonds appear closer to 1.0 ppm.

Protons attached to benzene rings absorb at a position even farther downfield from that of the vinyl protons in alkenes. The interactions of the six π-electrons of the aromatic ring produce a stronger anisotropic effect than that found with simple alkenes. The ring current created by the movement of these electrons induces a magnetic field, as illustrated in Figure 19.16c. The regions above and below the aromatic ring are shielded (+), whereas the protons at the edge of the ring are deshielded (−). The signal for the protons in benzene appears at 7.36 ppm, about 2 ppm downfield from the signal produced by the protons in ethylene.

FIGURE 19.16
Regions of shielding (+) and deshielding (−) for (a) ethylene, (b) formaldehyde, and (c) benzene.

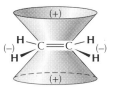

(a)

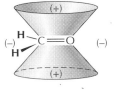

(b)

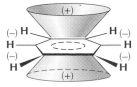

(c)

| **19.8** | ## Quantitative Estimation of Chemical Shifts |

Much of the power of NMR spectroscopy comes from the correlation of molecular structure with positions of signals along the chemical shift scale. As you have already seen, the type of bonding and the proximity of electronegative atoms influence the chemical shift position of protons.

Signals for different types of protons attached to carbon appear in well-defined regions. Tables cataloging these relationships, constructed by compiling large numbers of NMR signals from many organic compounds, contain much data, too much to memorize. However, to master the use of NMR spectroscopy for determining molecular structures, you must be able to use Tables 19.3–19.5, which allow you to calculate estimated chemical shifts. The calculated chemical shifts can then be compared to the signals in the spectrum you are analyzing.

From the empirical correlations in Tables 19.3–19.5, it is possible to calculate the chemical shift of a hydrogen nucleus in a straightforward, additive way. The ability to add the individual effects of nearby functional groups is extremely useful because it allows an estimation of the chemical shifts for most of the protons in organic compounds.

Chemical Shifts of Alkyl Protons

The aggregate effect of multiple functional groups on the chemical shift of the proton(s) of an alkyl group can be determined from Table 19.3.

Base values.   Begin with the base values at the top of Table 19.3. In any proposed molecular structure, primary hydrogen atoms (methyl groups) have a base value of 0.9 ppm. Secondary hydrogen atoms (methylene groups) are somewhat more deshielded, as shown by their chemical shift base value of 1.2 ppm. Tertiary (methine) hydrogen atoms have an even greater chemical shift; their base value is 1.5 ppm.

Effects of nearby substituents.   The effect of each nearby substituent is added to the base value to arrive at the chemical shift of a particular proton in a molecule. If the substituent is directly attached to the carbon atom to which the proton is attached, it is called an *α (alpha) substituent.* If the group is attached to a carbon atom once removed, it is a *β (beta) substituent.* And if the group is attached to a carbon atom twice removed, it is a *γ (gamma) substituent.*

The effect of an α substituent on the chemical shift of the proton is found by using a value from the first column in Table 19.3, and the effects of β and α groups are found in the second and third columns, respectively. Notice that the topmost group, —R, an alkyl group, has no effect on the chemical shift other than changing the base values. When the carbon atom bearing the proton is farther away from the functional group, its effect on the chemical shift of the proton is smaller. The effect of a group more than three carbon atoms away

TABLE 19.3   **Additive parameters for predicting NMR chemical shifts of alkyl protons in CDCl₃ᵃ**

	Base values	
	Methyl	0.9 ppm
	Methylene	1.2 ppm
	Methine	1.5 ppm

Group (Y)	Alpha (α) substituent	Beta (β) substituent	Gamma (γ) substituent
	$\underset{\displaystyle |}{\overset{\displaystyle |}{H-C}}-Y$	$\underset{\displaystyle |}{\overset{\displaystyle |}{H-C}}-\underset{\displaystyle |}{\overset{\displaystyle |}{C}}-Y$	$\underset{\displaystyle |}{\overset{\displaystyle |}{H-C}}-\underset{\displaystyle |}{\overset{\displaystyle |}{C}}-\underset{\displaystyle |}{\overset{\displaystyle |}{C}}-Y$
—R	0.0	0.0	0.0
—C=C	0.8	0.2	0.1
—C=C(C=O)OR	1.0	0.3	0.1
—C≡C—R	0.9	0.3	0.1
—C≡C—Arᵇ	1.2	0.4	0.2
—Ar	1.4	0.4	0.1
—(C=O)OH	1.1	0.3	0.1
—(C=O)OR	1.1	0.3	0.1
—(C=O)H	1.1	0.4	0.1
—(C=O)R	1.2	0.3	0.0
—(C=O)Ar	1.7	0.3	0.1
—(C=O)NH₂	1.0	0.3	0.1
—(C=O)Cl	1.8	0.4	0.1
—C≡N	1.1	0.4	0.2
—Br	2.1	0.7	0.2
—Cl	2.2	0.5	0.2
—OH	2.3	0.3	0.1
—OR	2.1	0.3	0.1
—OAr	2.8	0.5	0.3
—O(C=O)R	2.8	0.5	0.1
—O(C=O)Ar	3.1	0.5	0.2
—NH₂⁻	1.5	0.2	0.1
—NH(C=O)R	2.1	0.3	0.1
—NH(C=O)Ar	2.3	0.4	0.1

a. There may be differences of 0.1–0.5 ppm in the chemical shift values calculated from this table and those measured from individual spectra.

b. Ar = aromatic group.

from the carbon bearing the proton of interest is small enough to be safely ignored. There may be a difference of 0.1–0.5 ppm between the chemical shift value calculated from Table 19.3 and the measured value, but the difference is usually no greater than 0.2–0.3 ppm, close enough to figure out if a proposed structure fits the spectrum.

Identifying α, β, and γ substituents. It is important in calculating estimated chemical shifts to use a systematic methodology. A good way not to forget to include all α, β, and γ substituents for each type of proton in a target molecule is to **write down all the α groups first, then all the β groups, and last the γ groups.** Only then go to Table 19.3, look up the base value and the value for each α, β, and γ substituent from the correct column, and do the necessary addition.

EXERCISE

Identify the α, β, and γ substituents for the two methylene protons and for the methyl group attached to C—4 of 4-methoxy-4-methyl-2-pentanone.

$$H_3CO-\underset{\underset{H_3C}{|}}{\overset{\overset{H_3C}{|}}{C}}-CH_2-\overset{\overset{O}{\parallel}}{C}\diagdown CH_3$$

4-Methoxy-4-methyl-2-pentanone

Answer: The methylene protons have one α substituent, a —(C=O)CH₃ group, listed in Table 19.3 as —(C=O)R. The methylene group also has one β substituent, a methoxy group, listed in Table 19.3 as —OR. The C—4 methyl groups have no α substituents other than an alkyl group, but they do have a β substituent, the methoxy group, as well as a γ substituent, the —(C=O)CH₃ group.

FOLLOW-UP ASSIGNMENT

Identify the α and β substituents for the *tert*-butyl protons of the compound $(CH_3)_3C(C=O)OCH_3$.

Calculating estimated chemical shifts. Table 19.3 is laid out with carbon substituents at the top, followed by the heteroatoms—halogens, and oxygen and nitrogen substituents. To illustrate its use, let us return to the example of ethyl propanoate, whose 360-MHz NMR spectrum is shown in Figure 19.17.

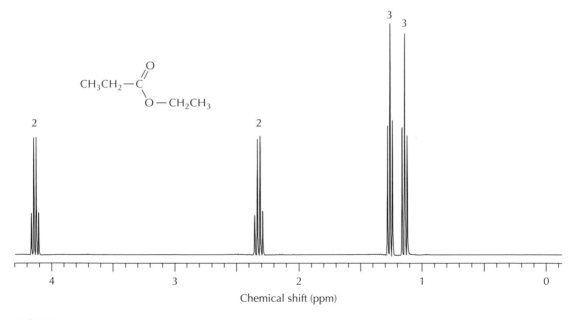

FIGURE 19.17
¹H NMR spectrum of ethyl propanoate at 360 MHz.

WORKED EXAMPLE

The integration in Figure 19.17 shows that the signal at 4.13 ppm comes from a methylene group. Referring to Figure 19.12 or Table 19.2, the best correlation is with a methylene group directly attached (α) to the oxygen atom of the ester group. Using Table 19.3, we can determine with more accuracy if this correlation is valid. Scan down the table to the entry fifth from the bottom to find the $-O(C=O)R$ group. Here is the calculation:

Base value for a methylene group	1.2 ppm
Presence of the α $-O(C=O)R$ group	2.8 ppm
Calculated chemical shift of the methylene protons	4.0 ppm

The calculated value is within 0.13 ppm of the methylene group attached to the oxygen atom, close enough to be consistent with the assignment.

The second methylene group at 2.32 ppm must be attached to the carbonyl carbon. Repeat the calculation for that methylene group, again using Table 19.3 and scanning down to the eighth entry.

Base value for a methylene group	1.2 ppm
Presence of the α $-(C=O)OR$ group	1.1 ppm
Calculated chemical shift of the methylene protons	2.3 ppm

The estimated value of the chemical shift of the second methylene group is within 0.02 ppm of the measured value.

EXERCISE

Estimate the chemical shifts of the two different methyl groups in ethyl propanoate. Are these values consistent with the observed chemical shifts of 1.13 ppm and 1.26 ppm?

Answer: From Table 19.3, the value of the chemical shift for the methyl protons β to the $-(C=O)OR$ group is $0.9 + 0.3 = 1.2$ ppm. The chemical shift of the methyl protons β to the oxygen atom of the $-O(C=O)R$ group is $0.9 + 0.5 = 1.4$ ppm. These calculated values compare reasonably well with the measured chemical shifts of 1.13 and 1.26 ppm. Even though the estimates of the chemical shifts differ by 0.07 and 0.14 ppm from the measured values, their relative order of increasing chemical shift adds to our confidence in the assignments.

FOLLOW-UP ASSIGNMENT

A compound, $C_7H_{14}O_2$, has the structure $CH_3(C=O)CH_2C(OCH_3)(CH_3)_2$. In its NMR spectrum are four separate signals, at 1.21 ppm, 2.08 ppm, 2.46 ppm, and 3.16 ppm, with the relative integrations of $6:3:2:3$. This integration pattern is consistent with the structure of the compound having three kinds of methyl groups, one of which is duplicated, and one methylene group. Calculate the chemical shifts for each of the four kinds of protons in $C_7H_{14}O_2$ using Table 19.3 and assign them to their correct NMR signals.

TABLE 19.4	**Additive parameters for predicting NMR chemical shifts of aromatic protons CDCl$_3$**		
	Base value	7.36 ppm[a]	
Group	*ortho*	*meta*	*para*
—CH$_3$	−0.18	−0.11	−0.21
—CH(CH$_3$)$_2$	−0.14	−0.08	−0.20
—CH$_2$Cl	0.02	−0.01	−0.04
—CH=CH$_2$	0.04	−0.04	−0.12
—CH=CHAr	0.14	−0.02	−0.11
—CH=CHCO$_2$H	0.19	0.04	0.05
—Ar	0.23	0.07	−0.02
—(C=O)H	0.53	0.18	0.28
—(C=O)R	0.60	0.10	0.20
—(C=O)Ar	0.45	0.12	0.23
—(C=O)OCH$_3$	0.68	0.08	0.19
—(C=O)OCH$_2$CH$_3$	0.69	0.06	0.17
—(C=O)OH	0.77	0.11	0.25
—(C=O)Cl	0.76	0.16	0.33
—(C=O)NH$_2$	0.46	0.09	0.17
—C≡N	0.29	0.12	0.25
—Cl	−0.02	−0.07	−0.13
—Br	0.13	−0.13	−0.08
—OH	−0.53	−0.14	−0.43
—OR	−0.45	−0.07	−0.41
—OAr	−0.36	−0.04	−0.28
—O(C=O)R	−0.27	0.02	−0.13
—O(C=O)Ar	−0.14	0.07	−0.09
—NH$_2$	−0.71	−0.22	−0.62
—N(CH$_3$)$_2$	−0.68	−0.15	−0.73
—NH(C=O)R	0.14	−0.07	−0.27
—NO$_2$	0.87	0.20	0.35

a. Base value is the measured chemical shift of benzene in CDCl$_3$ (1% solution).

Chemical Shifts of Aromatic Protons

The chemical shifts of protons on substituted benzene rings can also be calculated. To estimate the chemical shifts, the contributions of substituents shown in Table 19.4 are added to a base value of 7.36 ppm, the chemical shift for the protons of benzene dissolved in CDCl$_3$.

WORKED EXAMPLE

Using Table 19.4, estimate the chemical shift of H$_a$ in the structure of methyl 3-nitrobenzoate.

Methyl 3-nitrobenzoate

There are two functional groups that affect the chemical shift of H_a, an *ortho-*(C=O)OCH$_3$ group and an *ortho-*nitro group. The contribution of the *ortho-*(C=O)OCH$_3$ group to the chemical shift of H_a is the eleventh item of Table 19.4; it is 0.68 ppm. The contribution of the *ortho-*NO$_2$ group, at the end of the table, is an additional 0.87 ppm. Adding these values to the base value of 7.36 ppm gives an estimated chemical shift of 8.91 ppm for H_a, compared to a measured value of 8.87 ppm.

Base value for a benzene ring	7.36 ppm
Presence of the *ortho-*(C=O)OCH$_3$ group	0.68 ppm
Presence of the *ortho-*NO$_2$ group	0.87 ppm
Calculated chemical shift for H_a	8.91 ppm
Measured chemical shift for H_a	8.87 ppm

FOLLOW-UP ASSIGNMENT

The measured chemical shifts for the remaining three aromatic protons of methyl 3-nitrobenzoate are 7.67 ppm, 8.38 ppm, and 8.42 ppm. The chemical shifts have been assigned to H_c, H_d, and H_b, respectively. Using Table 19.4, calculate the estimated chemical shifts of these three aromatic protons and justify their assignments.

Chemical Shifts of Vinyl Protons

Chemical shifts of protons attached to C=C bonds, called *vinyl* protons, can be estimated using Table 19.5. The estimated chemical shift for a vinyl proton is the sum of the base value of 5.28 ppm, the chemical shift for H_2C=CH_2, and the contributions for all *cis, trans,* and *geminal (gem)* substituents. A **geminal group** is the one that is attached to the same carbon atom as the vinyl proton whose estimated chemical shift is being calculated.

$$cis \diagdown \quad \diagup H$$
$$C=C$$
$$trans \diagup \quad \diagdown gem$$

WORKED EXAMPLE

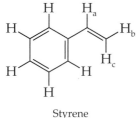

Styrene

Styrene, the monomer from which polystyrene is made, has the formula C_6H_5CH=CH_2. In addition to the three signals of the protons attached directly to the benzene ring, there are separate NMR signals for the three vinyl protons at 5.25 ppm, 5.75 ppm, and 6.70 ppm. Calculate the expected chemical shifts for H_a, H_b, and H_c of styrene and assign the three measured signals to the correct protons.

Table 19.5 has a base value of 5.28 ppm, which will be part of the calculation for all three vinyl protons. H_a has the phenyl group (C_6H_5—) on the same carbon atom; it is a *geminal* group. The phenyl group is about halfway down Table 19.5 and its *gem* parameter is 1.38 ppm. Here is the calculation for the chemical shift of H_a.

Base value for a vinyl proton	5.28 ppm
Presence of the *gem* C$_6$H$_5$— group	1.38 ppm
Calculated chemical shift of H_a	6.66 ppm
Measured chemical shift of H_a	6.70 ppm

In the same manner, we can calculate the chemical shifts for H_b and H_c. The phenyl group is *trans* to H_b, so −0.07 ppm must be added to the base

TABLE 19.5 Additive parameters for predicting NMR chemical shifts of vinyl protons in $CDCl_3$[a]

$$cis \diagdown \qquad H \diagup$$

Base value   5.28 ppm

Group	gem	cis	trans
—R	0.45	−0.22	−0.28
—CH=CH$_2$	1.26	0.08	−0.01
—CH$_2$OH	0.64	−0.01	−0.02
—CH$_2$X (X = F, Cl, Br)	0.70	−0.11	−0.04
—(C=O)OH	0.97	1.41	0.71
—(C=O)OR	0.80	1.18	0.55
—(C=O)R	1.10	1.12	0.87
—(C=O)H	1.02	0.95	1.17
—Ar	1.38	0.36	−0.07
—Br	1.07	0.45	0.55
—Cl	1.08	0.18	0.13
—OR	1.22	−1.07	−1.21
—OAr	1.21	−0.60	−1.00
—O(C=O)R	2.11	−0.35	−0.64
—NH$_2$, —NHR, —NR$_2$	0.80	−1.26	1.21
—NH(C=O)R	2.08	−0.57	−0.72

a. There may be small differences in the chemical shift values calculated from this table and those measured from individual spectra.

value: 5.28 + (−0.07) = 5.21 ppm. This value fits well with the signal at 5.25 ppm in the NMR spectrum of styrene. The phenyl group is *cis* to H$_c$, so 0.36 ppm must be added to the base value: 5.28 + 0.36 = 5.64 ppm. It seems clear that the 5.75-ppm signal must be H$_c$.

EXERCISE

Consider the structure of ethyl *trans*-2-butenoate, $C_6H_{10}O_2$. Estimate the chemical shift for each of the two different vinyl protons in the molecule using Table 19.5 and then assign H$_a$ and H$_b$. The measured chemical shift values are 5.80 ppm and 6.90 ppm.

Ethyl *trans*-2-butenoate

Answer: Calculating the estimated chemical shift of H$_a$, we find

Base value for a vinyl proton	5.28 ppm
Presence of the *gem* —(C=O)OR group	0.80 ppm
Presence of the *cis* R group	−0.22 ppm
Calculated chemical shift of H$_a$	5.86 ppm
Measured chemical shift of H$_a$	5.80 ppm

For the estimated chemical shift of H_b, we find

Base value for a vinyl proton	5.28 ppm
Presence of the *cis* —(C=O)OR group	1.18 ppm
Presence of the *gem* R group	0.45 ppm
Calculated chemical shift of H_b	6.91 ppm
Measured chemical shift of H_b	6.90 ppm

Using Tables 19.3–19.5 in Combination

Now you can test your skills in the use of Tables 19.3–19.5, as well as Figure 19.12.

PROBLEM ONE

Figure 19.18 is the ^{1}H NMR spectrum of a compound with the molecular formula $C_6H_{12}O_2$. It is an ester, which is one of the two following isomers,

$$(CH_3)_3C(C=O)OCH_3 \text{ or } CH_3(C=O)OC(CH_3)_3$$

Calculate the chemical shifts for the two different kinds of methyl groups in each structure and then assign the NMR signals in Figure 19.18 to the appropriate methyl groups in the correct isomer.

Hint: Look first at the whole spectrum, paying attention to the integrals that are associated with the two signals. Use the spectrum to measure each of the chemical shift values. Consider each isomer and think about the proximity of the two kinds of protons to electronegative atoms. Make a hypothesis as to which isomer seems correct. Then calculate the estimated chemical shifts for each of the two possible isomers using Table 19.3. Decide which molecular structure is correct and assign the NMR signals to the appropriate protons.

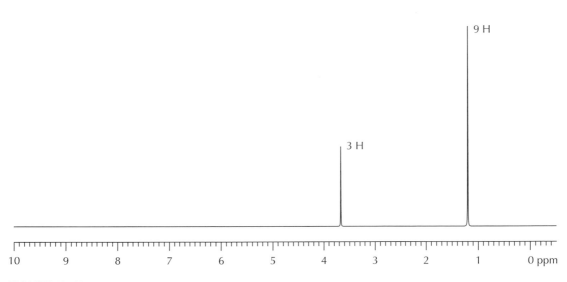

FIGURE 19.18
^{1}H NMR spectrum of a compound with the molecular formula $C_6H_{12}O_2$, at 300 MHz.

PROBLEM TWO

Calculate the estimated chemical shifts of each of the five types of protons in 4-bromo-1-butene and assign each of them to their respective NMR signals. Briefly discuss any ambiguities in your assignments. The observed chemical shift values are 2.62 ppm, 3.41 ppm, 5.10 ppm, 5.15 ppm, and 5.80 ppm.

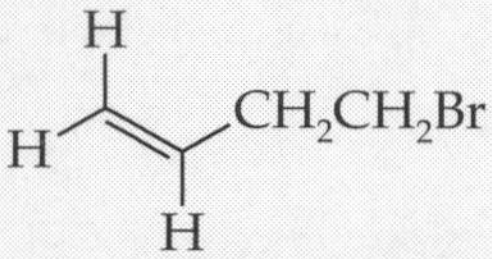

4-Bromo-1-butene

Hint: Decide which of the protons are alkyl protons and which are vinyl protons and then use Tables 19.3 and 19.5 to calculate the estimated chemical shifts. Assign each observed NMR signal to the appropriate proton(s).

PROBLEM THREE

---

trans-1-(*para*-Methoxyphenyl)propene has the following structure.

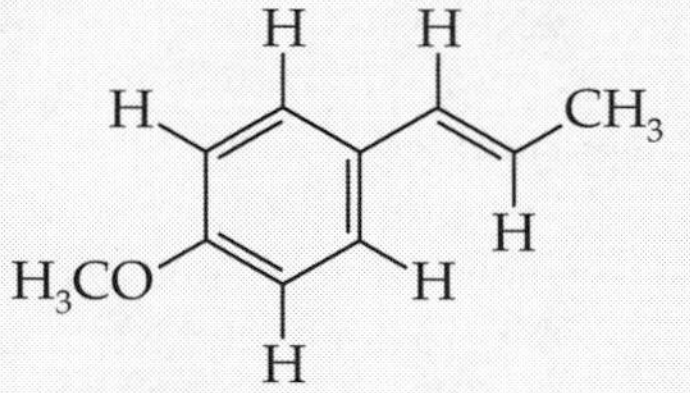

trans-1-(*para*-Methoxyphenyl)propene

Its measured NMR signals are observed at 1.83 ppm, 3.75 ppm, 6.07 ppm, 6.33 ppm, 6.80 ppm, and 7.23 ppm. Their respective integrations are 3:3:1:1:2:2. Which NMR signals are produced by alkyl protons, aromatic protons, and vinyl protons? Using Tables 19.3–19.5, calculate the estimated chemical shift for each type of proton in *trans*-1-(*para*-methoxyphenyl)-propene and assign the observed chemical shifts to the correct protons.

A Final Word on Calculating Estimated Chemical Shifts

In general, Tables 19.3–19.5 provide good estimates. However, a word of caution is in order. It is important to remember that the simple acyclic compounds used to generate the tables incorporate some structural features that may not be present in every situation. Anisotropic effects of rings, rigid stereochemistry, multiple deshielding groups, hindered rotation, and hydrogen bonding can lead to estimated chemical shifts that are different from the actual chemical shifts.

Rings. The signals produced by methylene protons in rings are slightly deshielded compared to the signals produced by methylene protons in acyclic compounds. Signals produced by methylene protons in cyclopentanes and cyclohexanes typically appear near 1.5 ppm compared to 1.2 ppm for open-chain compounds.

Cyclopropanes and oxiranes are unique in that the σ-bonding in three-membered rings has some π-orbital character, which produces an anisotropic effect and shielding above and below the plane of the ring. Chemical shifts of protons on cyclopropane and epoxide rings are approximately 1.0 ppm upfield from their acyclic counterparts.

Rigid stereochemistry. In rigid molecules, protons can be held in positions within an anisotropic shielding or deshielding region of a functional group. These effects can shift a proton to a position that can differ significantly from the estimated chemical shift. For example, the chemical shift of the methyl group *trans* to the carbomethoxy group of methyl 3-methyl-2-butenoate is 1.90 ppm, as predicted using Table 19.3. However, the *cis*-methyl group appears at 2.18 ppm. The *cis*-methyl group is in the deshielding region of the carbonyl group, which causes it to be shifted downfield.

1.90 ppm

2.18 ppm

Methyl 3-methyl-2-butenoate

Multiple deshielding groups. If there are multiple deshielding α substituents, especially alkoxy groups and halogens attached to a single carbon atom, the calculated chemical shift value of its protons can sometimes differ from the measured chemical shift by more than 1 ppm. The divergence between the actual chemical shifts and the estimates can be seen in the following series, as more methoxy groups are attached to the carbon atom of methane.

	CH_3OCH_3	$CH_2(OCH_3)_2$	$CH(OCH_3)_3$
Measured	3.24 ppm	4.58 ppm	4.97 ppm
Estimation	3.0 ppm	5.4 ppm	7.8 ppm

Hindered rotation. The estimates of chemical shifts of protons *ortho* to bulky groups on benzene rings can differ considerably from the measured values. This difference is evident in the calculation of chemical shift values for acetanilides (Ar—NH—(C=O)CH$_3$) that are substituted in the *ortho* position with substituents such as bromine, chlorine, or a nitro group. In these compounds, the chemical shift of the *ortho* proton is nearly 1 ppm downfield from the estimated chemical shift calculated from Table 19.4. Hydrogen bonding between the amide hydrogen and the *ortho* substituent impedes rotation about the C—N bond, freezing the conformation of the molecule so that the carbonyl group is always located in the vicinity of the *ortho* hydrogen.

| 19.9 | Spin-Spin Coupling (Splitting) |

The chemical shifts and integrals of NMR signals provide a great deal of information about the structure of a molecule. However, this information is often not enough to determine the structure. Closer examination of NMR signals reveals that they are generally not shapeless blobs but highly structured patterns with a multiplicity of lines. Reexamine the spectrum of ethyl propanoate shown in Figure 19.17 (page 286). The signal at 4.1 ppm is actually a group of four peaks, as is the signal at 2.3 ppm. The signals at 1.2 and 1.1 ppm are groups of three peaks. The fine structure of these patterns is caused by interactions between the proton(s) producing the signals and neighboring nuclei, particularly other protons. The effects are small compared with those of shielding and deshielding, but analysis of the patterns provides valuable information about the local environments of protons in a molecule.

Vicinal Coupling ($^3J_{HH}$):

The interactions that cause the fine structure of NMR signals are transmitted through the bonding framework of the molecules. They are usually observable only when the interacting nuclei are near one another. The most commonly observed effects are produced by the interaction between protons attached to adjacent carbon atoms. These protons, which are separated by three bonds, are called *vicinal*, or nearby, protons.

A proton that is affected by the spin states of another nucleus is *coupled* to that nucleus and its signal is split into multiple signals. A simple example of coupling between two vicinal hydrogen atoms can be seen in the NMR spectrum of 1,1,2-tribromo-2-phenylethane shown in Figure 19.19. Both H_a and H_b have a spin of $\frac{1}{2}$ and therefore have two spin states, one aligned with the applied magnetic field and one opposed to it. In the absence of H_b, H_a would exhibit a single peak at 5.97 ppm. However, in the presence of the neighboring H_b, H_a is affected by the spin state of H_b.

The effective magnetic field felt by H_a increases a little when the magnetic field of H_b is aligned with the applied magnetic field. The aligned orientation leads to a slight deshielding effect, and the position of the H_a signal moves slightly downfield. The magnetic field of the spin state of H_b opposed to the applied magnetic field decreases the effective magnetic field felt by H_a, moving the H_a position slightly upfield. Thus, the signal for H_a is split into a *doublet* because the number of nuclei in each spin state is nearly equal—two peaks of nearly equal intensity are observed.

Vicinal Protons

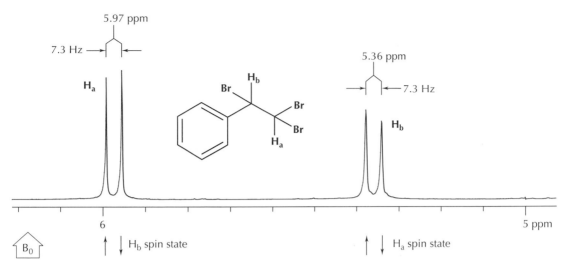

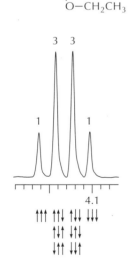

FIGURE 19.19
A section of the ^{1}H NMR spectrum of 1,1,2-tribromo-2-phenylethane at 360 MHz.

The distance between the signals of the doublet is called the *coupling constant (J).* When it is a vicinal (three-bond) coupling constant that involves two protons, the notation is $^3J_{HH}$. Coupling constants are measured in Hz (cycles per second), and their values are independent of the spectrometer operating frequency. In Figure 19.19, the value of the coupling constant is 7.3 Hz. In an analogous manner, proton H_a interacts with proton H_b, and H_b also appears as a doublet with the same coupling constant. The fact that interacting protons have coupling constants of exactly the same value is very useful for identifying which protons are coupled to each other.

Now consider a slightly more complicated pattern. An expanded section of the 360-MHz NMR spectrum of ethyl propanoate near 4.1 ppm is shown in Figure 19.20. This set of NMR signals is produced by the methylene group **b**, which has a relative integration of two protons. The protons of the methylene group are coupled to the protons of the adjacent methyl group, and they split into a four-peak pattern called a *quartet.* The four peaks, which are due to the three protons of the methyl group, are produced by the four spin states shown in Figure 19.20:

1. The three spins of the methyl protons aligned with the applied magnetic field (left signal)
2. Two spins of the methyl protons aligned with and one opposed to the applied magnetic field
3. One spin aligned with and two spins of the methyl protons opposed to the applied magnetic field
4. The three spins of the methyl protons opposed to the applied magnetic field

FIGURE 19.20
Signal at 4.13 ppm in the ^{1}H NMR spectrum of ethyl propanoate at 360 MHz.

Statistically, there are three possible combinations that lead to spin states 2 and 3. Because every combination of spins has the same probability of occurring, the relative intensities of the four peaks in the pattern are 1:3:3:1. The coupling constant turns out to be 7.1 Hz.

EXERCISE

Analyze the **_triplet_** (three-peak) pattern of methyl group **a** of ethyl propanoate at 1.26 ppm, shown in Figure 19.21. The protons of this methyl group are coupled to the protons of methylene group **b** (see Figure 19.20).

Answer: The key to the splitting pattern of the methyl group is the number of spin states of the methylene group to which it is coupled. As usual, the spins of the two methylene protons have an equal probability of being aligned or opposed to the applied magnetic field. Three combinations are possible. The two spins can be aligned, one can be aligned and the other opposed, or the two spins can be opposed to the applied magnetic field. There is twice the probability of one spin aligned and one opposed. This produces a triplet pattern for the nearby methyl group, with the relative intensities 1:2:1.

We can check to make sure that methyl group **a** is coupling with methylene group **b** by calculating the coupling constant, J, between them. If they are coupled, J must be 7.1 Hz, which is the value measured from the 4.13 ppm set of peaks. Figure 19.21 gives the positions of the peaks for the methyl group in Hz, and the distance between the peaks must be the same.

$$459.9 \text{ Hz} - 452.8 \text{ Hz} = 7.1 \text{ Hz}$$
$$452.8 \text{ Hz} - 416.8 \text{ Hz} = 7.2 \text{ Hz}$$

Within experimental error, the coupling constants are the same.

Singlet _One peak_
Doublet _Two peaks_
Triplet _Three peaks_
Quartet _Four peaks_

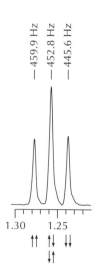

FIGURE 19.21
Signal at 1.26 ppm in the ^{1}H NMR spectrum of ethyl propanoate at 360 MHz.

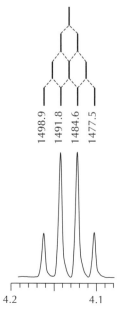

FIGURE 19.22
Splitting tree for the signal at 4.13 ppm in the ^{1}H NMR spectrum of ethyl propanoate at 360 MHz.

***Splitting Trees and
the N+1 Rule***

A common device for predicting and analyzing the fine structure of coupling patterns is a ***splitting tree,*** constructed by mapping the effect of each spin-spin coupling on a signal. The splitting tree for the methylene group **b** of ethyl propanoate is shown in Figure 19.22. Notice that there are three branching sites in the tree—one set of branches for each proton in methyl group **a**, whose coupling produces the splitting tree.

Splitting of signals.   Because the three adjacent methyl protons are equivalent to one another, the methylene signal can be thought of as splitting into doublets three times. The signal is split into a doublet by the first methyl proton. The coupling with a second methyl proton splits each signal of the doublet into two signals. Because the coupling constants of these two interactions are exactly the same, the position of the high-field signal of one doublet reinforces the position of the low-field signal of the second doublet. If no more splitting occurred, the pattern would consist of three equally spaced signals, the center signal having twice the intensity of the two outer ones. Coupling with the third methyl proton, however, again splits each of the signals of the triplet into two signals. Because the coupling constant is the same, the signals again reinforce each other. As seen in the splitting tree in Figure 19.22, the resulting pattern is a ***quartet,*** a group of four equally spaced signals. The ratio of the intensities is $1:3:3:1$.

The presence of doublets, triplets, and quartets in NMR spectra has led to the N + 1 rule for multiplicity. This rule states: **A proton that has N equivalent protons on adjacent carbon atoms will be split into N + 1 signals.**

Pascal's triangle.   The ratio of the intensities of the multiplet signals can be obtained from Pascal's triangle (Figure 19.23), a triangular arrangement of the mathematical coefficients obtained by a binomial expansion. The N + 1 rule assumes that all N protons are equivalent, with equal coupling constants. If the protons are not all equivalent, the coupling constants will probably not all be equal. In that case, the total number of peaks will be greater than N + 1. Multiple couplings are discussed later in this section (page 298.).

Returning to the 360-MHz NMR spectrum of ethyl propanoate (see Figure 19.17), you will see that the methylene signal at 4.13 ppm appears as a quartet because of the splitting by the three hydrogen nuclei of the adjacent methyl group (N = 3; N + 1 = 4). In turn, the three-proton signal at 1.26 ppm appears as a triplet because of the two hydrogen nuclei of the adjacent methylene group (N = 2, N + 1 = 3). **This triplet-quartet pattern is seen quite often and is diagnostic for an ethyl group.** A second triplet-quartet pattern occurs in the NMR spectrum of ethyl propanoate, indicative of the presence of a second ethyl group. In this case, the quartet is located at 2.32 ppm because the methylene component of the ethyl group is attached to a carbonyl group.

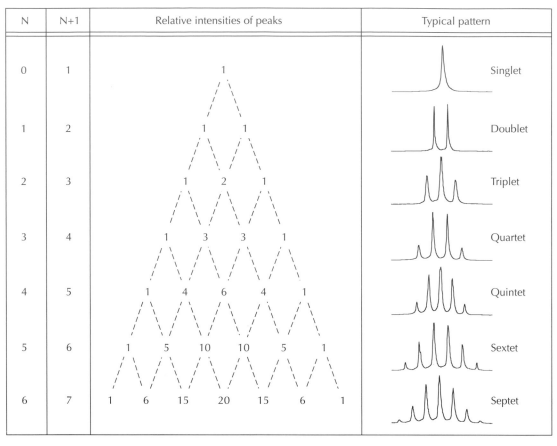

N	N+1	Relative intensities of peaks	Typical pattern
0	1	1	Singlet
1	2	1 1	Doublet
2	3	1 2 1	Triplet
3	4	1 3 3 1	Quartet
4	5	1 4 6 4 1	Quintet
5	6	1 5 10 10 5 1	Sextet
6	7	1 6 15 20 15 6 1	Septet

FIGURE 19.23
Pascal's triangle can be used to predict the multiplicity and relative intensities of the signal of any magnetically active nucleus coupled to N equivalent nuclei of spin $\frac{1}{2}$. It applies to both ^1H and ^{13}C spectra.

Multiple Couplings
In most organic compounds will be coupling constants that are similar in magnitude as well as coupling constants that are quite different. This situation creates patterns that are more complicated than the ones shown in Figures 19.17 and 19.19. The 360-MHz spectrum of ethyl *trans*-2-butenoate shown in Figure 19.24 is an example. The spectrum shows five groups of signals at 1.2 ppm, 1.8 ppm, 4.1 ppm, 5.8 ppm, and 6.9 ppm, with integral values of 3:3:2:1:1, respectively. Notice that Figure 19.24 has an expanded inset near every set of peaks. It is often necessary to expand regions in your NMR spectra to more clearly reveal the detail of the splitting patterns.

The quartet pattern at 4.1 ppm, which integrates to two protons, and the triplet pattern at 1.2 ppm, which integrates to three protons, indicates that an ethyl group is incorporated into the structure of the molecule. The chemical shift of the methylene group suggests that the ethyl group is bonded to an oxygen atom. Taken together, the partial structure producing the signals at 4.1 ppm and 1.2 ppm must be $—OCH_2CH_3$.

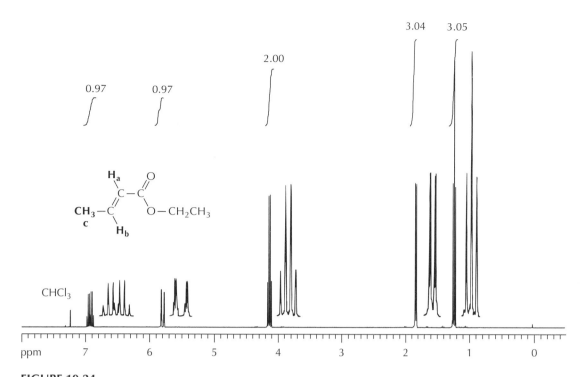

FIGURE 19.24
^{1}H NMR spectrum of ethyl *trans*-2-butenoate at 360 MHz, with superimposed expanded (4×) insets adjacent to the signals.

The three-proton signal at 1.8 ppm is the result of a second methyl group. Its position is consistent with the chemical shift of a methyl group attached to a C=C bond (see Figure 19.12). The signal appears to be a doublet, indicating that there is only one proton on the adjacent carbon atom, which confirms CH_3—CH=C— as a component part of the molecule. From Table 19.2, we know that the two signals at 5.8 ppm and 6.9 ppm are produced by protons attached to the carbon atoms of a C=C bond. Because the signal at 5.8 ppm is a doublet, there is only one other proton on an adjacent carbon. Therefore, we can assign it to the vinyl proton (H_a) next to the carbonyl group. Taking all these observations into consideration leads us to the conclusion that the partial structure that produces the signals at 5.8 ppm, 6.9 ppm, and 1.8 ppm must be —(C=O)—CH=CH—CH_3.

EXERCISE

What coupling pattern accounts for the complex set of peaks centered at 6.93 ppm in Figure 19.25, which is an expanded section of the H_b signal in the NMR spectrum of ethyl *trans*-2-butenoate in Figure 19.24?

Answer: The signal at 6.93 ppm is produced by the CH_3—CH=CH— proton. This set of peaks is a good deal more complex than the simple N + 1 pattern we have seen previously. It seems to have eight peaks. If the coupling constants between the vinyl proton and the four protons on adjacent carbon atoms were the same, the signal should appear as a quintet (N = 4,

N + 1 = 5); however, it clearly is not a five-peak pattern. If there were only coupling with the other vinyl proton (H_a) signal at 5.8 ppm, the N + 1 rule predicts that the signal at 6.9 ppm would appear as a doublet. However, because there is also coupling with the adjacent methyl group (H_c), each peak of the doublet is split into a quartet. Two overlapping quartets produce the observed eight-peak pattern in Figure 19.25.

By accurately measuring the distances between the signals, it is possible to determine the coupling constants. One coupling constant can be determined by measuring the distance from the outermost signal of the pattern to the adjacent signal. Using the leftmost signals, we can calculate this coupling constant to be 6.9 Hz:

$$2513.5 \text{ Hz} - 2506.6 \text{ Hz} = 6.9 \text{ Hz}$$

This value is the coupling constant between the three hydrogen atoms of the methyl group at 1.8 ppm and the vinyl proton at 6.9 ppm.

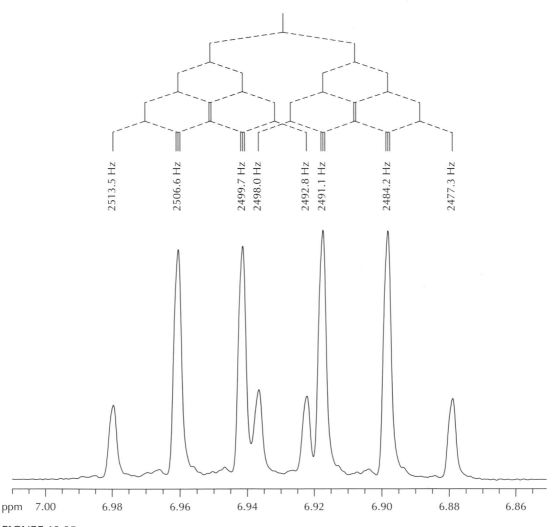

FIGURE 19.25
Splitting tree for the vinyl proton signal at 6.9 ppm in the ^{1}H NMR spectrum of ethyl *trans*-2-butenoate.

The distance in Hz between the outermost signals of the pattern (2513.5 Hz − 2477.3 Hz = 36.2 Hz) is the sum of all the coupling constants. The coupling constant between the two vinyl protons can be calculated by subtracting the coupling constants of each proton in the methyl group from this sum:

$$36.2 \text{ Hz} - (3 \times 6.9 \text{ Hz}) = 15.5 \text{ Hz}$$

The splitting tree for the proton appearing at 6.93 ppm is shown at the top of Figure 19.25.

Other Types of Coupling

Most of the observed coupling in NMR spectroscopy is a result of vicinal coupling through three bonds, H_a—C—C—H_b. However, coupling through one, two, and four bonds can also be observed. One-bond coupling occurs between ^{13}C and ^{1}H ($^{1}J_{CH}$). Because the relative abundance of ^{13}C is so small, the signals produced by this splitting are usually negligible in a ^{1}H NMR spectrum. With concentrated samples, it is possible to observe this splitting by turning up the amplitude, as demonstrated by the ^{1}H NMR spectrum of chloroform shown in Figure 19.26. This type of coupling is a major consideration when observing ^{13}C signals, as you will see in the section on ^{13}C NMR spectroscopy.

Geminal coupling. Coupling through two bonds ($^{2}J_{HH}$), or geminal coupling, occurs between two protons attached to the same carbon atom, H_a—C—H_b. In many molecules, these two protons are equivalent and coupling is not observed. However, geminal cou-

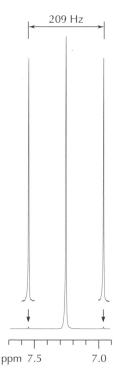

209 Hz

ppm 7.5   7.0

FIGURE 19.26 Expanded and amplified section of the ^{1}H NMR spectrum of chloroform at 360 MHz, showing ^{13}C splitting.

pling is frequently observed in compounds with vinyl methylene groups, $H_2C=C-$, where the two geminal protons can be non-equivalent. Other examples where methylene protons are not equivalent are discussed in the advanced NMR topics section [see Technique 19.12].

Allylic coupling. Coupling through four bonds ($^4J_{HH}$) is often observed in compounds containing carbon-carbon double bonds ($H_a-C=C-C-H_c$) and is called *allylic coupling.* When the NMR spectrum of ethyl *trans*-2-butenoate is expanded, its allylic coupling can be seen. Expansion of the signals at 5.8 ppm and 1.84 ppm (H_a and H_c in Figure 19.24) reveals further fine structure (Figure 19.27a and b). At 5.8 ppm, the pattern of the NMR signal is a doublet of quartets and the signal at 1.84 ppm is a doublet of dou-

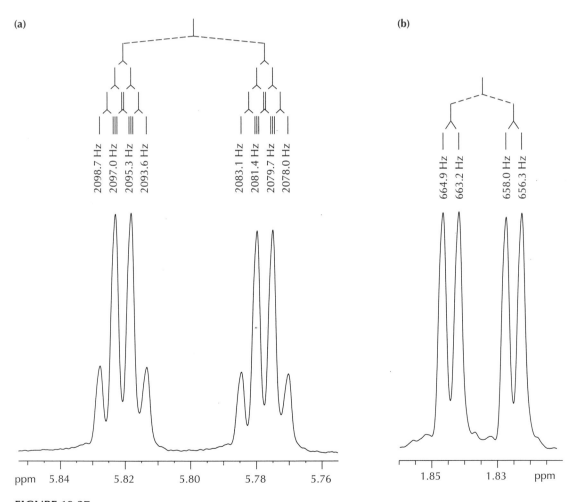

FIGURE 19.27
Expansions of signals at (a) 5.8 ppm and (b) 1.8 ppm in the ^1H NMR spectrum of ethyl *trans*-2-butenoate.

blets. The coupling constant for this $^4J_{HH}$ coupling is quite small, only 1.7 Hz. Until the signals are expanded, it is hardly noticeable.

Magnitude of Coupling Constants

The magnitude of coupling constants can reveal valuable information about the structure of a molecule. The magnitude is related to the number of bonds between the interacting protons; the more bonds, the smaller the coupling constant. The size of vicinal coupling for alkyl protons ranges from about 2 to 15 Hz. The size of geminal coupling depends on bond angles and hybridization; for alkyl protons it is generally on the order of 10–16 Hz. The geminal coupling of vinyl protons is much smaller (0–3 Hz). Coupling through four or more bonds is also very small, 0–3 Hz. Typical coupling constants for various arrangements of protons are listed in Table 19.6.

TABLE 19.6 Typical proton-proton coupling constants

Arrangement of protons	J_{ab}(Hz)	Arrangement of protons	J_{ab}(Hz)	Arrangement of protons	J_{ab}(Hz)
Free rotation	7		10 to 16		0 to 3
Anti	8 to 13		11 to 14		12 to 18
Gauche	2 to 4		8 to 13		6 to 12
	6 to 9		2 to 6		4 to 10
	1 to 3		2 to 5		0.5 to 2
	0 to 1				0

FIGURE 19.28
Dependence of the coupling constant on dihedral angle, ϕ, formed by two vicinal CH bonds (Karplus relationship). Coupling constants usually fall between the two curves, which result from different assumptions.

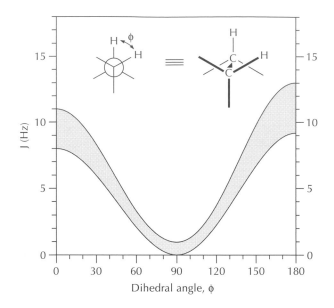

Our analysis of Figure 19.25 showed that the coupling constant for the splitting of the two vinyl protons is 15.5 Hz. This large coupling constant indicates that the protons are *trans* to one another; therefore, the molecule must be a *trans* (E) alkene.

If there is free rotation about a carbon-carbon single bond connecting the coupled protons, the vicinal or three-bond coupling constants are usually about 7 Hz. If rotation about the carbon-carbon bond is restricted, the coupling constant can range from 0 to 13 Hz. The size of a vicinal coupling constant is related to the angle ϕ on a Newman projection of the interacting protons (Figure 19.28). This angle is called the **dihedral angle.**

In the early days of NMR spectroscopy, Martin Karplus at Harvard studied the relationship between the size of a coupling constant and the dihedral angle. His conclusions are now widely accepted and are often presented as a plot of the coupling constant versus dihedral angle. This plot is called a **Karplus curve** and is shown in Figure 19.28. The important characteristics of the Karplus relationship are the minimum value of the coupling constant at a dihedral angle of 90° and the large values of the vicinal coupling constant at dihedral angles of 0° and 180°.

19.10   Sources of Confusion

Using NMR spectroscopy to analyze the structures of organic compounds is a logical process. However, sometimes the process of interpreting a spectrum becomes complicated beyond the factors of chemical shift and spin-spin splitting that we have discussed. It is important to be aware of some of the complicating factors so that you can make rational choices when confronted with unexpected, confusing, or poorly defined signals in a spectrum.

In this section we will briefly discuss four common, potentially confusing areas in the interpretation of NMR spectra:

- History of the NMR sample: Mixtures of compounds
- Overlap of NMR signals
- NMR sample preparation and data acquisition
- O—H and N—H protons

History of the NMR Sample: Mixtures of Compounds

Extra signals in an NMR spectrum are often a product of a mixture of the compound whose structure you want to ascertain, with solvents, starting materials, reaction side products, and residual proton signals from the deuterated solvent. Much puzzlement and frustration can be avoided by careful consideration of what might be present in the NMR tube.

Sources of extra signals.   To determine the source of extra signals, it is important to know the history of the NMR sample. If you prepared it, you already know the solvents and reagents that could be present. What solvent did you use for the reaction mixture? If you purified your compounds by extraction, recrystallization, or chromatography, what solvents did you use? What does the NMR spectrum of the starting material look like? What solvent did you use to clean the NMR sample tube?

Table 19.7 lists some organic solvents that are common impurities in NMR samples and the chemical shift positions and multiplicity of their NMR signals.*

TABLE 19.7   NMR signals of common solvents

Solvent	NMR signals[a]
Acetone	2.1 (s)
Benzene	7.4 (s)
Chloroform	7.3 (s)
Cyclohexane	1.4 (s)
Dichloromethane	5.3 (s)
Diethyl ether	1.2 (t), 3.5 (q)
Dimethyl sulfoxide	2.5 (s)
Ethanol (anhydrous)	1.2 (t), 3.0 (t) or (s), 3.7 (m) or (q)
Ethyl acetate	1.3 (t), 2.0 (s), 4.1 (q)
Hexane	0.9 (t), 1.3 (m)
2-Propanol	1.2 (d), 2.6 (d) or (s), 4.0 (m)
Methanol (anhydrous)	2.3 (q) or (s), 3.4 (d) or (s)
Tetrahydrofuran	1.9 (m), 3.8 (m)
Toluene	2.3 (s), 7.2 (m) or (s)
Water (dissolved)	1.6 (s)
Water (bulk)	4.5 (br s)

a. In $CDCl_3$. Multiplicity of signal is shown in parentheses.

*An extensive, useful list of impurity peaks has been collected by Gottlieb, Kotlyar, and Nudelman (*J. Org. Chem.* **1997**, *62*, 7512–7515).

Example of a typical mixture. The NMR spectrum shown in Figure 19.29 is a typical example of a mixture encountered in the laboratory. The material for the sample was obtained from the base-catalyzed dehydrochlorination of 2-chloro-2-methylpentane.

The expected products of the reaction are 2-methyl-1-pentene and 2-methyl-2-pentene, and it is quite possible that they will be contaminated with a small amount of 1-propanol, the solvent used in the reaction. The singlet at 2.2 ppm and the triplet at 3.6 ppm are produced by the hydroxyl proton of 1-propanol and the methylene group attached to the hydroxyl group. The other signals of 1-propanol, at 1.55 ppm and 0.92 ppm, are obscured by signals of the two alkenes.

Even though every signal in the spectrum is not distinct, much useful information can be obtained because each component of the mixture exhibits unique features. For example, the ability of NMR to count each of the protons in every component of a mixture can be put to good use. This information allows you to tell which sets of NMR peaks relate to the same compound, because they must have integral proton ratios. It also allows you to determine the molar composition of the mixture.

Calculating the ratio of products. The integrals of the vinyl proton signals at 5.1 ppm and 4.7 ppm can be used to determine the ratio of

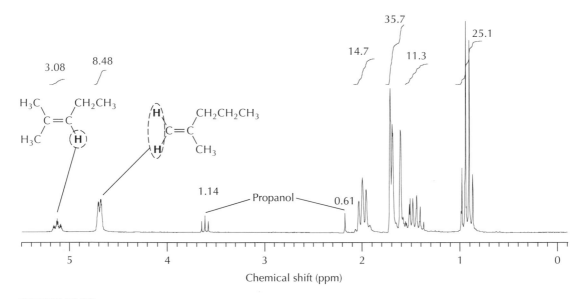

FIGURE 19.29
200-MHz ^{1}H NMR spectrum of the reaction product from dehydrochlorination of 2-chloro-2-methylpentane, a mixture of 2-methyl-1-pentene and 2-methyl-2-pentene.

2-methyl-2-pentene to 2-methyl-1-pentene in the mixture. The broad triplet at 5.1 ppm is caused by the vinyl proton attached to C-3 in 2-methyl-2-pentene. The two broad signals at 4.7 ppm are produced by the two protons attached to C-1 in 2-methyl-1-pentene. To calculate the molar ratio of the two alkenes in the product mixture, the integrals need to be normalized by dividing them by the number of protons causing the signals.

$$\frac{\text{moles of 2-methyl-1-pentene}}{\text{moles of 2-methyl-2-pentene}} = \frac{8.48/2}{3.08/1} = \frac{1.38}{1}$$

The calculation shows that the dehydrochlorination of 2-chloro-2-methylpentane produces 58% 2-methyl-1-pentene and 42% 2-methyl-2-pentene.

Overlap of NMR Signals

Signals from protons with similar chemical shifts may overlap with one another, leading to broad and poorly defined patterns. Often these patterns are so poorly defined that analyzing the coupling is not even tempting. An example of poorly defined peaks is shown in the 60-MHz NMR spectrum of 1-butanol (Figure 19.30a). The region from 1–2 ppm exhibits a complex multiplet integrating to four protons, which includes both of the two similar methylene groups of 1-butanol. In addition, little can be learned from the methyl protons that appear at 0.7–1.0 ppm.

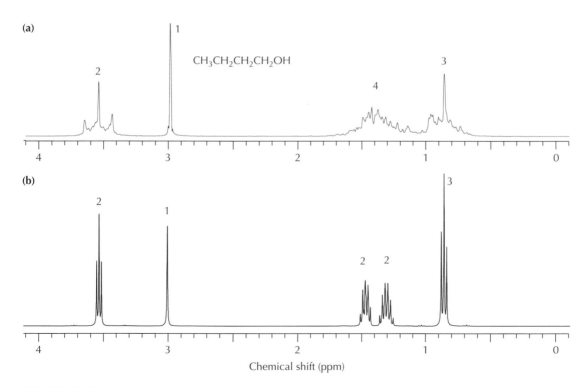

FIGURE 19.30
^{1}H NMR spectra of 1-butanol at (a) 60 MHz and (b) 360 MHz.

Sometimes it is possible to unravel complex multiplets with an NMR spectrum obtained using a higher-field instrument. The 360-MHz NMR spectrum of 1-butanol is shown in Figure 19.30b. In this spectrum, the chemical shifts of the two methylene groups are different enough so that their signals are separated into two well-defined multiplets; one signal at 1.3 ppm is a six-peak multiplet and the other signal at 1.47 ppm is a quintet. Also, notice that the methyl group at 0.86 ppm is a well-defined triplet.

There are also cases where two or more well-defined patterns overlap, producing what at first glance may resemble a single pattern but which, on closer examination, has coupling constants and/or signal intensities that do not make sense. A good example of such a deceptive pattern is the apparent quartet at 1.15 ppm in the 60-MHz NMR spectrum of ethyl propanoate. An expanded section of this 60-MHz spectrum, containing the four-line pattern, is shown in Figure 19.31a. Figure 19.31b shows the same section of the spectrum obtained on a 360-MHz NMR instrument; at the higher magnetic field the pattern separates into two triplets, one centered at 1.13 ppm and the other centered at 1.26 ppm.

NMR Sample Preparation and Data Acquisition

Broad and distorted signals are often the result of either poor sample preparation or improper operation or adjustment of the NMR instrument.

Spinning sidebands. Reflections of a signal, called *spinning sidebands,* that appear symmetrically around the signal are evidence of problems with NMR instrument settings, especially those that control the homogeneity of the magnetic field. Sometimes the spinning sidebands are large enough that a quick glance suggests that a singlet peak may be a triplet. However, it is rare that the sidebands are large enough to approach the 1:2:1 peak heights necessary for a

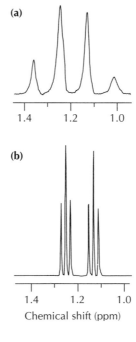

FIGURE 19.31
^{1}H NMR spectra of the methyl group signals of ethyl propanoate at (a) 60 MHz and (b) 360 MHz.

triplet pattern that is caused by spin-spin coupling. You can always tell if the set of peaks is a real triplet by examining the relative heights of the three peaks. In addition, spinning sidebands will appear around all of the peaks in the NMR spectrum, not just one or two of them.

Poor sample preparation. Following are typical examples of poor sample preparation:

- The sample is not completely dissolved or there are insoluble impurities present.
- The sample solution is too concentrated.
- The height of the solution in the NMR tube is not correct.

Problems during data acquisition. Following are examples of problems that occur during data acquisition:

- The sample tube is not positioned properly in the spin collar.
- The sample is not spinning evenly during data acquisition.
- The NMR sample tube is spinning too rapidly during the data acquisition, leading to a vortex in the sample solution.
- The spectrometer is not tuned properly.

Some of these sources of confusion cannot be avoided, but many of them can be minimized with careful sample preparation and conscientious data acquisition.

O—H and N—H Protons

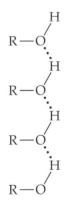

Intermolecular
hydrogen
bonding

Hydrogen bonding. Hydrogen bonding involving oxygen or nitrogen atoms draws electron density away from the O—H and N—H protons, deshielding them and shifting their signals downfield. Because concentration and temperature affect the extent of intermolecular hydrogen bonding, the chemical shift of protons attached to oxygen and nitrogen atoms in alcohols, amines, and carboxylic acids can appear over a wide range. In fact, O—H and N—H signals often vary in two NMR spectra of the same compound in the same solvent because the concentration differs. In dilute samples, there is little or no intermolecular hydrogen bonding and the signals may have small chemical shift values. Intermolecular hydrogen bonding in concentrated samples shifts O—H or N—H peaks downfield. The extreme case where hydrogen bonding causes deshielding occurs with carboxylic acids, which have a chemical shift of 10–13 ppm for the O—H proton.

Proton exchange. Another potential source of confusion is the chemical exchange of O—H and N—H protons, which has two consequences for NMR spectroscopy. First, a proton may not be attached to the heteroatom long enough for coupling to occur with nearby protons. Second, the signal for the exchangeable protons can merge into a single peak with an unpredictable chemical shift.

   If a proton on the oxygen atom of an alcohol exchanges rapidly, which it usually does in CDCl$_3$ that contains a small amount of water or acid, no coupling between the hydroxyl proton and protons on the adjacent carbon is observed. The hydroxyl proton signal becomes a broadened singlet. However, if the sample solution used to obtain the NMR spectrum is anhydrous and acid-free, splitting is often observed because exchange of the O—H proton is relatively slow under these conditions, and the hydroxyl proton signal is split by the protons attached to the adjacent carbon atom. The choice of solvent can affect the likelihood of proton exchange. Samples of alcohols prepared in deuterated chloroform almost always show proton exchange, whereas the use of dimethyl sulfoxide-d$_6$ suppresses it.

The NMR spectrum of methanol dissolved in CDCl$_3$ is shown in Figure 19.32a. Proton exchange is evident because both the signal produced by the methyl protons and the signal produced by the hydroxyl proton appear as singlets. In the NMR spectrum of methanol dissolved in CD$_3$SOCD$_3$, shown in Figure 19.32b, the signal produced by the methyl protons appears as a doublet and the signal produced by the hydroxyl proton appears as a quartet, as predicted by the N + 1 rule. The protons are coupled because there is no chemical exchange of the hydroxyl proton in DMSO-d$_6$. Notice also that the differing amounts of intermolecular hydrogen bonding in the two solvents cause very different chemical shifts for the O—H proton of methanol.

The second consequence of chemical exchange is that O—H and N—H protons can exchange so quickly that they merge into a common environment and become combined into a single "averaged" NMR peak, whose chemical shift depends on concentration, solvent, temperature, and the presence of water or acid. When NMR samples are dissolved in D$_2$O, all O—H and N—H protons in the compounds merge into a broadened peak at the chemical shift of H—O—D.

Using chemical exchange as a diagnostic probe. Chemical exchange can be used as a diagnostic probe for protons of alcohols and amines dissolved in a deuterated organic solvent. The experiment is carried out by obtaining a second NMR spectrum after addition of a drop of D$_2$O to the solution. Hydroxyl and amine protons in

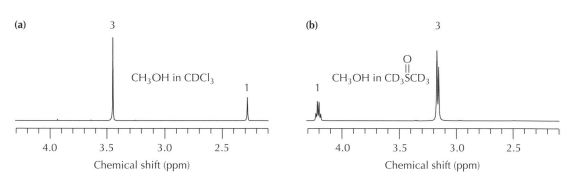

FIGURE 19.32
360-MHz ^{1}H NMR spectra of methanol in (a) deuterochloroform and (b) d$_6$-dimethyl sulfoxide.

the molecule are replaced by deuterons through chemical exchange. If the compound is an alcohol or amine, the signal resulting from the exchangeable proton disappears and a new signal produced by HOD appears at approximately 4.6 ppm.

19.11   Two Case Studies

NMR spectroscopy is the principal tool used by organic chemists for determining the structures of organic compounds. In this section we look at the ^{1}H NMR spectra of two organic molecules and show how the information derived from their spectra can help determine their molecular structures.

Four Major Pieces of Information from an NMR Spectrum

To start, it may be useful to recap the four major pieces of information that are used in the interpretation of a ^{1}H NMR spectrum of a pure compound:

- The *number of signals* tells us how many kinds of nonequivalent protons are in the molecule.
- The *integration* determines the relative number of nonequivalent protons in a ^{1}H NMR spectrum.
- The *chemical shift* provides important information on the environment of a proton. Downfield signals (larger ppm values) suggest nearby deshielding oxygen atoms, halogen atoms, or π-systems. Tables 19.2–19.5 and Figure 19.12 are useful aids for correlating chemical shifts with molecular structure.
- The *splitting* of signals, caused by spin-spin coupling of protons to other protons, reveals the presence of nearby protons that produce the coupling. Values of coupling constants can establish coupling connections between protons and can reveal stereochemical relationships (see Table 19.6).

Analysis of the Spectrum

First you should examine the entire spectrum without being too eager to focus on a prominent signal or splitting pattern. To ensure success, a structured and logical approach to the interpretation of an NMR spectrum is necessary. In time, after you have interpreted numerous spectra, you can replace the structured approach with a less formal one. The following approach will assist you in learning this skill:

1. Make inferences and deductions based on the spectral information.
2. Build up a collection of structure fragments.
3. Put the pieces together into a molecular structure that is consistent with the data and with the rules of chemical bonding.
4. Confirm the chemical shift assignments with calculated chemical shifts based on Tables 19.3–19.5. Any inconsistencies between the values should be examined and resolved. Explanations of minor inconsistencies usually hinge on subtle structural features of the molecule.

Double-Bond Equivalents

In NMR problem sets, the molecular formula of a pure compound is often provided. If it is available, the molecular formula can be used to determine the **double-bond equivalents (DBE)**, which provide the number of double bonds and/or rings present in the molecule.

$$\text{double-bond equivalents (DBE)} = C + \frac{N - H - X}{2} + 1$$

where C is the number of carbon atoms, H is the number of hydrogen atoms, X is the number of halogen atoms, and N is the number of nitrogen atoms. Other names for double-bond equivalents are *degree of unsaturation* and *index of unsaturation.*

Organizing the Spectral Information

We suggest the following method for organizing information from an NMR spectrum. First, prepare an informal table with the following headings:

- Chemical Shift (ppm)
- ^1H Type
- Integration
- Splitting Pattern
- Possible Structure Fragment(s)

Chemical Shift. In the Chemical Shift column, list the positions (or ranges) of all the signals in the spectrum.

^1H Type. Based on the chemical shifts, use Figure 19.12 and Table 19.2 as guides for entering likely structural assignments in the ^1H Type column for each NMR signal, for example, Ar—H, =C—H, —O—C—H, and so on. Consider any reasonable structure within the chemical shift range. A few types of protons can appear over a wide range of chemical shifts, but at this early juncture it is better to err on the side of being too inclusive. As the analysis is refined the possibilities can usually be narrowed down.

Integration. Enter the value of each integral, rounded to whole numbers, in the Integration column. Remember that the integrals must add up to the total number of hydrogen atoms in the molecular formula.

Splitting Pattern. In the Splitting Pattern column, enter a description of the splitting pattern. Be as precise as possible, using standard descriptive terms, such as doublet, triplet, quartet, and combinations of these terms, such as doublet of triplets. If you have processed the FID to obtain the NMR spectrum or if you have access to the actual NMR data collected on the spectrometer, rather than just being given the spectrum, you should consider expanding regions in your NMR spectrum to reveal the detail of important splitting patterns. Later in the NMR analysis, when you have a complete structure proposal to consider, you may also wish to measure some of the coupling constants. Coupling constants can be used to determine which signals are coupled to one another and in some cases to assign stereochemistry.

Possible Structure Fragment(s). The Possible Structure Fragment(s) column is where you pull all the information together and enter all of the structural fragments that are consistent with the data for each NMR signal. Be flexible and consider all reasonable possibilities. For example, in the spectrum you may have a quartet that integrates to two protons. The quartet is probably the result of coupling to three protons with equal coupling constants. Two arrangements are consistent with this pattern, —**CH₂**—CH₃ and perhaps —CH—**CH₂**—CH₂ (boldface indicates the protons exhibiting the quartet).

Proposed Structure

Once you have constructed the table, the analysis is a matter of eliminating proposed structure fragments that are inconsistent and then putting the remaining structure fragments together into reasonable proposals for the structure of the compound. The final structure must be consistent with all the NMR data. Of particular importance is calculating the quantitative estimation of the chemical shifts (Tables 19.3–19.5). In all but the simplest cases, it will be important to estimate the chemical shift for each type of proton. The estimated chemical shifts will allow you to eliminate proposed structures inconsistent with the chemical shift data.

PROBLEM ONE

An organic compound has a molecular formula of $C_5H_{12}O$. Its 200-MHz ^1H NMR spectrum is shown in Figure 19.33. Determine its structure.

Double-bond equivalents. First, determine the compound's double-bond equivalents (DBE).

$$DBE = C + \frac{N - H - X}{2} + 1 = 5 + \frac{0 - 12 - 0}{2} + 1 = 0$$

Because DBE = 0, we know that the compound contains no rings or double bonds. Because the molecule has no double bonds and contains an oxygen atom, it must be either an alcohol or an ether.

Table of data from the spectrum. The data from the spectrum are summarized in Table 19.8. An assignment for the signal at 2.36 ppm is tentative because that region is normally where protons on carbons adjacent to alkenes and carbonyl groups appear, and we know from the DBE calculation that there are no double bonds in the molecule. However, the proton on the oxygen atom of an alcohol could also appear in this chemical shift region.

Assembling structure fragments. There are two possible fragments that could explain the splitting of the signal at 1.46 ppm:

$$CH_3-CH_2- \quad \text{or} \quad -CH_2-CH_2-CH-$$

The ethyl fragment has to be eliminated because there is no signal exhibiting a pattern consistent with the methyl portion of that fragment—a three-proton triplet at approximately 1.0 ppm.

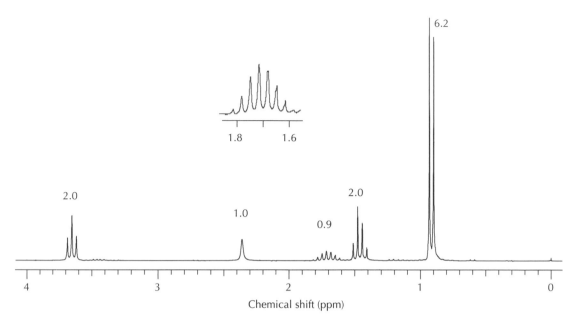

FIGURE 19.33
200-MHz ^{1}H NMR spectrum of $C_5H_{12}O$.

The splitting pattern at 1.71 ppm is difficult to know with certainty. Because the outermost signals of highly split patterns are very small relative to the other signals, the multiplet could be either an octet or a nonet. In either case, there must be two methyl groups attached to a methine group. The downfield triplet signal at 3.65 ppm is the result of a methylene group attached to an oxygen atom.

From the analysis so far, we can propose that the compound $C_5H_{12}O$ is an alcohol with an isopropyl group ((CH_3)$_2$CH—), as well as a methylene group flanked by methine and methylene groups (—CH—CH_2—CH_2—). In addition, there seems to be a methylene group attached to an oxygen atom and a second methylene group (—CH_2—CH_2—O—). This array of fragments consists of eight carbon atoms, sixteen hydrogen atoms, and one oxygen atom. Obviously, there are some atoms common to more than one fragment.

TABLE 19.8 Interpreted data from ^{1}H NMR spectrum (200 MHz) of $C_5H_{12}O$

Chemical shift (ppm)	^{1}H type	Integration	Splitting pattern	Possible structure fragment(s)
0.91	C—C—**H**	6	Doublet	—CH(CH$_3$)$_2$
1.46	C—C—**H**	2	Quartet	CH$_3$CH$_2$—
				or
				—CH$_2$CH$_2$CH—
1.71	C—C—**H**	1	Multiplet	—CH$_2$CH(CH$_3$)$_2$
				or
				—CHCH(CH$_3$)$_2$
2.36	Perhaps C—O—**H**	1	Broad singlet	R—O—H
3.65	O—C—**H**	2	Triplet	—CH$_2$CH$_2$O—

To solve this puzzle, set the three structure fragments out side-by-side to look for possible overlap:

$(CH_3)_2CH-$ $-HC-CH_2-CH_2-$ $-CH_2-CH_2-O-H$
    1   **2**   **3**

Possible structures. It looks as if fragment 2 overlaps both fragment 1 and fragment 3. Combining fragments 1 and 3 produces a five-carbon alcohol, $(CH_3)_2CH-CH_2-CH_2-OH$, 3-methyl-1-butanol. The estimated chemical shifts from Table 19.3 are shown in the following structure; as you can see, the correspondence is very good.

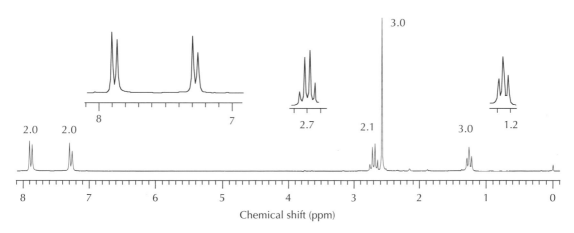

Observed chemical shifts of 3-methyl-1-butanol
(calculated estimates of chemical shifts)

FOLLOW-UP ASSIGNMENT

Using Table 19.3, calculate the chemical shifts for each of the four kinds of protons attached to carbon atoms in 3-methyl-1-butanol. Do your answers correspond to the shifts shown in the structure at the end of the problem?

PROBLEM TWO

An organic compound has a molecular formula of $C_{10}H_{12}O$. Its 200-MHz ^1H NMR spectrum is shown in Figure 19.34. Determine the structure of this compound.

FIGURE 19.34
200-MHz ^1H NMR spectrum of $C_{10}H_{12}O$.

Double-bond equivalents. The double-bond equivalent calculation indicates that the compound contains a combination of five double bonds and/or rings.

$$DBE = C + \frac{N - H - X}{2} + 1 = 10 + \frac{0 - 12 - 0}{2} + 1 = 5$$

Whenever there is a large DBE value, it is likely that the molecular structure of the compound incorporates one or more benzene rings, because each benzene ring accounts for four DBEs, three double bonds and one ring. The NMR spectrum confirms this assumption by the presence of signals in the aromatic proton region (6.5–8.5 ppm). The total integration of the aromatic protons is four, implying that the benzene ring is disubstituted. Moreover, the symmetry of the two signals in the aromatic region, a pair of doublets, indicates two groups of equivalent protons. This pattern is possible only if the two substituents are attached to the 1 and 4 positions of the benzene ring (*para* substitution). Several **signature patterns** are observed in NMR spectra, and a pair of symmetrical doublets in the aromatic region is one of the more frequently encountered ones.

1,4 disubstitution with
two types of
aromatic protons

Table of data from the spectrum. The NMR data from Figure 19.34 are summarized in Table 19.9.

Possible structure fragments. The three-proton triplet at 1.25 ppm and two-proton quartet at 2.70 ppm are another signature pattern, which indicates an ethyl group. The 2.70-ppm chemical shift of the two-proton quartet indicates the environment of the methylene protons, suggesting that the ethyl group could be a part of four possible structure fragments:

$CH_3CH_2C{=}O$ CH_3CH_2Ar $CH_3CH_2C{=}C$ $CH_3CH_2C{\equiv}C$

A decision to eliminate two of these possibilities can be made reasonably easily. The compound has the molecular formula $C_{10}H_{12}O$. The benzene ring has six carbons, and the latter two possible structure fragments have four carbons each. They would leave no room for the methyl group appearing at 2.58 ppm, nor would they indicate how the oxygen atom could be incorporated in the structure. Therefore, the only viable structure fragment options for the ethyl group are $CH_3CH_2C{=}O$ and $CH_3CH_2{-}Ar$.

The only NMR signal left to analyze is the methyl group at 2.58 ppm, which must be in one of two possible structure fragments, $CH_3C{=}O$ or CH_3Ar. To summarize, $C_{10}H_{12}O$ includes a methyl group ($CH_3{-}$), an ethyl group ($CH_3CH_2{-}$), and a *para*-disubstituted benzene ring (C_6H_4). The atom

TABLE 19.9 Interpreted data from ^1H NMR spectrum (200 MHz) of $C_{10}H_{12}O$

Chemical shift (ppm)	^1H type	Integration	Splitting pattern	Possible structure fragment(s)
1.25	C—C—H	3	Triplet	CH_3CH_2—
2.58	O=C—C—H	3	Singlet	CH_3—Cb=O
	or			or
	Ar—C—H			CH_3—Ar
	or			or
	C=C—C—H			CH_3—C=C
	or			or
	C≡C—H			CH_3—C≡C
2.70	O=C—C—H	2	Quartet	CH_3CH_2C=O
	or			or
	Ar—C—H			CH_3CH_2—Ar
	or			or
	C=C—C—H			CH_3CH_2—C=C
	or			or
	C≡C—H			CH_3CH_2—C≡C
7.28	Ar—H	2	Doublet	
7.88	Ar—H	2	Doublet	

count in the fragments is nine carbon atoms and twelve hydrogen atoms, leaving only one carbon atom and one oxygen atom to be accounted for. Because one more double-bond equivalent is required, the last fragment for the molecule is a carbonyl group, which is also consistent with the possible structure fragments.

Possible structures. Two possible structures are consistent with the data: 4'-methylphenyl-1-propanone and 4'-ethylphenyl-1-ethanone.

4'-Methylphenyl-1-propanone   4'-Ethylphenyl-1-ethanone

| 19.12 | Advanced Topics in ^{1}H NMR |

Second-Order Effects

Observed splitting patterns may differ from patterns predicted by
simple coupling rules as a result of *second-order effects.* As the
chemical shifts of the coupled protons become closer to one
another, second-order effects become more pronounced. The usual
rule of thumb is that they become apparent in a spectrum when the
difference in chemical shifts ($\Delta\nu$, measured in Hz) is less than five
times the coupling constant ($\Delta\nu < 5J$).

Consequences of second-order effects.   Large second-order effects produce the following:

- Signal intensities that are different from predicted values
- Additional signals beyond those predicted by simple splitting
  rules
- Coupling constants that cannot be directly measured from
  differences in signal positions

When the second-order effects are small, they can be useful, such as
when the differences in signal intensities produce "leaning" peaks,
which indicate the relative position of a coupling partner. Look back
at the quartet signal at 4.1 ppm in the 200-MHz NMR spectrum of
ethyl propanoate (Figure 19.7b) for an example of "leaning" peaks
(page 272). The pattern is not perfectly symmetrical. The right-hand
peaks are slightly higher than those on the left, an indication that
these protons are coupled with protons whose signals appear to the
right of that pattern. In this case, the coupling partner appears at 1.3
ppm. Notice that the signal at 1.3 ppm is "leaning" to the left
because its coupling partner appears downfield.

Complexities produced by second-order effects.   Examine the expanded
sections of the 60-MHz and 360-MHz spectra of cinnamyl alcohol,
which show the vinyl proton regions (Figure 19.35). On the
360-MHz NMR spectrum (Figure 19.35b), the individual vinyl
protons appear as well-defined signals at 6.3 ppm and 6.6 ppm,
separated by 100 Hz (0.28 ppm × 360 MHz); the coupling constant
is 15.9 Hz. On a 60-MHz instrument these signals are separated by
only 17 Hz (0.28 ppm × 60 MHz) and the coupling constant is
again 15.9 Hz. When the difference in chemical shifts is nearly the
same as the coupling constant between two protons, the NMR spectrum is almost useless for any analysis of NMR splitting.

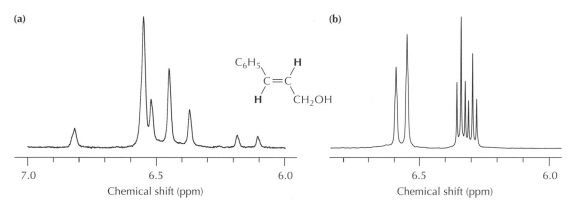

FIGURE 19.35
^{1}H NMR spectra of the vinyl protons of cinnamyl alcohol at (a) 60 MHz and (b) 360 MHz.

Even the 360-MHz spectrum exhibits small second-order effects. The patterns at 6.3 ppm and 6.6 ppm lean toward each other. In other words, the upfield portion of the 6.6-ppm signal and the downfield portion of the 6.3-ppm signal are larger than the other parts of the two patterns. On higher-field instruments, the frequency difference between signals is even larger, so that second-order effects, in many cases, become negligible.

Diastereotopic Protons

Subtle structural differences between protons in a molecule may not be obvious at first glance, which can be a source of confusion. For example, it is easy to assume that the two protons of a methylene group are always equivalent, and in most cases they are. However, if the methylene group is next to a stereocenter, such as an asymmetric carbon atom, the two protons of the methylene group become nonequivalent. They cannot be interchanged with one another by any bond rotation or symmetry operation, and they are said to be *diastereotopic.* They have different chemical shifts, and they also couple with each other. The appearance of diastereotopic protons is common in the NMR spectra of chiral molecules, those with stereocenters.

Consider the compound 2-methyl-1-butanol:

Diastereotopic protons

2-Methyl-1-butanol

If there is no coupling to the hydroxyl proton, you might expect the NMR signal for the adjacent methylene protons to appear as a doublet because of coupling with the vicinal methine proton. However, the protons of the methylene group are diastereotopic, which makes the NMR spectrum of 2-methyl-1-butanol much more complex.

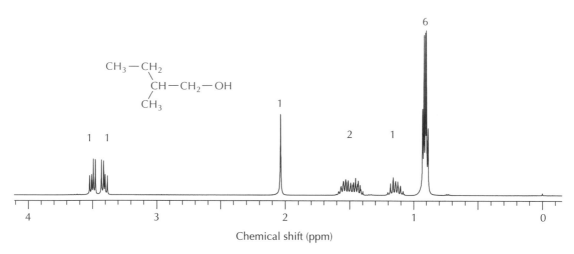

FIGURE 19.36
^{1}H NMR spectrum of 2-methyl-1-butanol at 360 MHz.

The spectrum shown in Figure 19.36 reveals an eight-line pattern for the methylene group of 2-methyl-1-butanol. The chemical shifts of the C-1 methylene protons are 3.4 ppm and 3.5 ppm. Because these two protons are not identical, they couple with each other, and each of the diastereotopic protons becomes a doublet of doublets. An expanded view of the C-1 methylene signals is shown in Figure 19.37. The coupling constants in one four-line set

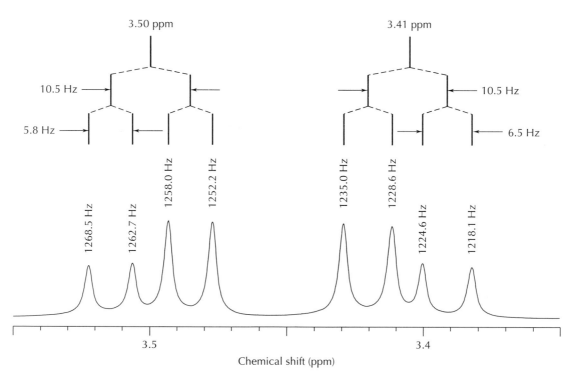

FIGURE 19.37
Splitting tree for the diastereotopic protons of 2-methyl-1-butanol.

are 6.5 Hz and 10.5 Hz, and the coupling constants for the other set are 5.8 Hz and 10.5 Hz. Notice that the two halves of the eight-line pattern in Figure 19.37 lean into each other. This leaning makes the central lines more intense than the outside lines, even though the splitting tree for this pattern shows equal intensities for all the lines. Use of a higher field NMR instrument for the spectrum would make all eight lines closer to equal intensity. The methylene protons of the ethyl group attached to the stereocenter are also diastereotopic, but the patterns are not distinct enough to analyze accurately.

Computer Simulation of NMR Spectra

Except for extremely simple compounds, typical organic compounds are likely to display NMR spectra containing second and higher-order effects. Because the exact chemical shifts and coupling constants cannot be directly measured from these spectra, computer programs are required to determine precise values. An initial estimate of the chemical shifts and coupling constants from an NMR spectrum are used to "simulate" a theoretical NMR spectrum. This calculated spectrum is then compared to the measured NMR spectrum. The chemical shifts and coupling constants are adjusted and the spectrum is recalculated. The iterative process of adjusting parameters and recalculating the spectrum is continued until the theoretical spectrum is identical to the measured NMR spectrum. The chemical shifts and coupling constants that produce the perfect match are the correct values. WINDNMR and RACCOON, available from the *Journal of Chemical Education—Software,* are simple simulation programs that simulate NMR spectra.

ChemDraw Ultra, HyperChem/HyperNMR, and Advanced Chemical Development/NMR Manager are more sophisticated programs that include chemical shift and coupling constant prediction along with calculation of theoretical spectra. The program gNMR automatically adjusts chemical shift and coupling constant values, carrying out the iteration process until the calculated spectrum matches the experimental spectrum.

^{13}C and Two-Dimensional NMR

19.13 ^{13}C NMR

Because ^{13}C NMR gives direct evidence about the carbon skeleton of an organic molecule, you might think that it would be the favored NMR technique. Why rely on the indirect evidence of ^{1}H NMR? In part, the answer is historical. ^{1}H NMR spectroscopy was developed first and had extensive early development. Most modern NMR spectrometers, however, are equipped with both ^{1}H and ^{13}C probes, and chemists routinely obtain both types of spectra.

Pulsed FTNMR Spectrometers

Carbon NMR became a useful and routine tool only when pulsed Fourier transform (FT) instrumentation was developed in the 1970s and 1980s. Because the magnetically active isotope of carbon, ^{13}C, is only 1.1% as abundant as ^{12}C in nature, the signal from carbon is extremely weak. Very few of the carbon atoms that are present in a compound provide a signal, so even concentrated samples have weak signals.

In addition, ^{13}C is much less sensitive because its gyromagnetic ratio is only one-fourth that of ^1H. Because the inherent sensitivity depends on the cube of the gyromagnetic ratio, the sensitivity of ^{13}C NMR relative to ^1H NMR is only $(0.25)^3$ or 0.016. This difference in the gyromagnetic ratio also explains why an instrument built to analyze protons at 300 MHz operates at 75 MHz, or one-fourth the frequency, for ^{13}C nuclei. The lower frequency usually presents no problems because carbon shifts occur over a 200-ppm range compared with the usual 10-ppm range for protons. Overlapping signals are not a significant problem with the broader range of ^{13}C NMR.

The use of pulsed FT NMR spectrometers allows NMR spectra to be acquired rapidly. By pulsing many times and adding together the NMR signals, the signals from the sample accumulate in a constructive fashion, whereas random noise signals more or less cancel each other out. Very good ^{13}C NMR spectra with a high signal-to-noise ratio can be obtained routinely with 25 to 50 mg of compound, using a modern high-field FT NMR instrument and a suitable number of pulses.

^{13}C Solvents

Samples are prepared for ^{13}C NMR much as they are for ^1H NMR. Table 19.1 (page 272) lists the suitable solvents. Deuterated chloroform ($CDCl_3$) is again the solvent of choice because it dissolves most organic compounds and is relatively inexpensive. Of course, each molecule of $CDCl_3$ has a carbon atom; its solvent signal always appears at 77.0 ppm in the ^{13}C NMR spectrum. The spin multiplicity of ^{13}C is the same as that for hydrogen (^1H), so the same splitting rules apply. For example, the ^{13}C signal of a methine group ($C—H$) will be split into a doublet by the attached proton. Unlike ^1H, a deuterium nucleus (^2H) has a spin of 1. Therefore, the ^{13}C signal of $CDCl_3$ is always a characteristic triplet in which the three lines have equal height.

Spin-Spin Splitting

The splitting of signals in a ^{13}C NMR spectrum can be used to identify the number of protons attached to a carbon atom. A methyl carbon signal appears as a quartet, a methylene signal appears as a triplet, a methine signal appears as a doublet, and a quaternary carbon signal appears as a singlet.

The ^{13}C spectrum of ethyl *trans*-2-butenoate shown in Figure 19.38 demonstrates these splitting patterns clearly. The carbonyl carbon at 166 ppm is a singlet because it has no attached protons. The vinyl carbons at 144 and 123 ppm are doublets because each one has only one attached proton. The signal at 60 ppm is produced

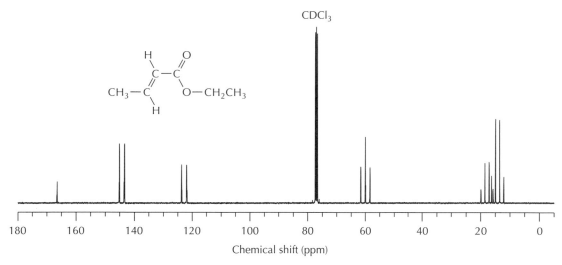

FIGURE 19.38
90-MHz ^{13}C NMR spectrum of ethyl *trans*-2-butenoate in CDCl$_3$.

by the methylene group attached to oxygen; it appears as a triplet because it has two attached protons. The complex pattern of signals between 12 and 20 ppm is two overlapping quartets because of the two methyl groups. Because the protons are directly attached to the ^{13}C atoms, their coupling is very large—on the order of 130 Hz. Coupling between adjacent ^{13}C nuclei is not observed because the probability that two attached carbons will both be ^{13}C is extremely small, about 1 in 10,000.

Broadband Decoupling

In molecules containing many carbon atoms, the coupled ^{13}C spectrum can become extremely complex because of multiple overlapping splitting patterns. The usual practice is to reduce this complexity by a technique called **broadband decoupling.** Irradiation of the sample with a broad band of energy during data acquisition decouples ^{13}C from ^{1}H nuclei and collapses the multiplets to singlets. The decoupled spectrum of ethyl *trans*-2-butenoate shown in Figure 19.39 consists of six sharp signals at 14, 18, 60, 123, 144, and 166 ppm, corresponding to the six different carbon nuclei in the molecule.

In addition to simplifying the spectrum, broadband decoupling enhances the signal-to-noise ratio and reduces the acquisition time dramatically. The coupled spectrum of ethyl *trans*-2-butenoate (see Figure 19.38) was acquired in 15,000 scans (overnight); the decoupled spectrum of the same sample (see Figure 19.39) was acquired in 450 scans (approximately 20 minutes).

You may have noticed that the signal at 166 ppm is much smaller than the other five signals for ethyl *trans*-2-butenoate. Peak areas for different carbon signals vary greatly, and quantitative ^{13}C integration is not easily accomplished. Integrals of routine ^{13}C spectra are generally unreliable. Excited ^{13}C nuclei relax back to

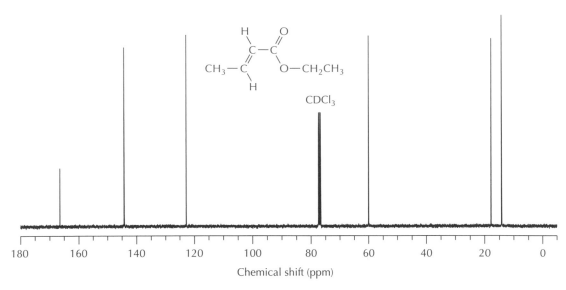

FIGURE 19.39
Decoupled 90-MHz ^{13}C NMR spectrum of ethyl *trans*-2-butenoate in $CDCl_3$. Protons in this sample have been irradiated by a broadband decoupling signal centered at 360 MHz to remove all $^{1}H-^{13}C$ coupling.

lower energy spin states at a slow rate. The size of the signal is influenced significantly by its proximity to protons, a phenomenon termed nuclear Overhauser enhancement (NOE). This effect explains why the signals of different types of carbon atoms can vary so much in amplitude and why the intensity of quaternary carbons is relatively low. In addition, because many scans are necessary to produce adequate signal-to-noise ratios in ^{13}C spectra, the time between scans is generally set at too short a time for complete relaxation to occur.

^{13}C Chemical Shifts

Table 19.10 reveals the same kind of chemical shift trends that we have seen before with ^{1}H NMR spectroscopy. The chemical shifts for carbon atoms are affected by electronegativity and anisotropy in similar ways. Signals of ^{13}C atoms that are in close proximity to electronegative atoms are moved downfield by diamagnetic deshielding. Signals resulting from C=C and C=O bonds are strongly deshielded because of anisotropic effects.

In the spectrum of ethyl *trans*-2-butenoate (see Figure 19.39), the chemical shifts of the carbons of the methyl groups attached to other sp^3 carbon atoms are upfield at 14 and 18 ppm. The methylene group attached to the electronegative oxygen atom, whose signal is at 60 ppm, is much more deshielded than the corresponding carbons attached only to sp^3 carbon atoms. The alkene carbons are also downfield; you may recall that we saw the same thing for protons on these sp^2 carbon atoms. The signals for

TABLE 19.10 Characteristic ^{13}C NMR chemical shifts	
Compound	Chemical shift (ppm)
TMS	0.0
$CDCl_3$ (t)	77
Alkane (C—**CH$_3$**)	7–30
Alkane (C—**CH$_2$**)	15–40
Alkane (C—**CH**) and (C—**C**)	15–40
Carboxylic acids, esters, and amides (**C**—C=O)	20–35
Allyl (**C**—C=C)	20–45
Alkane (**C**—Ar)	20–45
Ketones, aldehydes (**C**—C=O)	30–45
Amines (**C**—N)	30–65
Iodides (**C**—I)	–20–45
Bromides (**C**—Br)	25–65
Chlorides (**C**—Cl)	35–70
Fluorides (**C**—F)	80–95
Alcohols (**C**—OH), ethers (**C**—OR), esters (**C**—O[C=O]R)	50–80
Alkyne (**C**≡C)	65–90
Alkene (**C**=C)	80–160
Aromatic carbons	110–175
Nitriles (**C**≡N)	110–125
Carboxylic acids, esters, and amides (**C**=O)	160–180
Ketones, aldehydes (**C**=O)	185–220

the vinyl carbons appear at 123 and 144 ppm. Finally, carbonyl-carbon atoms show large downfield shifts. The chemical shift is dependent on the type of functional group containing the carbon atom (for example, ketone versus ester). The carbonyl-carbon atom of ethyl *trans*-2-butenoate appears at 166 ppm.

A few shielding effects are unique to ^{13}C NMR. Neighboring α and β carbon atoms deshield a carbon by about 9 ppm. That is, both the α and β carbons of the —C*—C$^\alpha$—C$^\beta$— structural unit increase the chemical shift of C* by 9 ppm compared with the shift of methane. Strongly electronegative halogens deshield carbon, but as the halogen atoms increase in atomic number, the deshielding of nearby carbon atoms is attenuated considerably. For example, iodomethane has a ^{13}C chemical shift of –21 ppm, which is over 20 ppm *upfield* of TMS. This shielding effect has been assigned to "steric compression." In such cases, steric factors apparently cause the electrons in the orbitals of the carbon atom to become compacted into a smaller volume closer to the nucleus, thus making the nucleus more highly shielded.

The chemical shifts of carbon atoms in substituted benzene rings can be estimated using the parameters listed in Table 19.11. This table can be used to calculate ^{13}C chemical shifts just as Table 19.4 is used for ^1H NMR chemical shifts of aromatic compounds.

TABLE 19.11	Additive parameters for predicting NMR chemical shifts of aromatic carbons in $CDCl_3$			
Base value for benzene		128.5 ppm		
Group	C-1	ortho	meta	para
—I	−34.1	8.9	1.6	−1.1
—Br	−5.8	3.2	1.6	−1.6
—Cl	6.3	0.4	1.4	−1.9
—F	34.8	−13.0	1.6	−1.1
—H	0.0	0.0	0.0	0.0
—(C=O)OCH$_3$	2.0	1.2	−0.1	4.3
—(C=O)OH	2.1	1.6	−0.1	5.2
—(C=O)H	8.2	1.2	0.5	5.8
—(C=O)CH$_3$	8.9	0.1	−0.1	4.4
—CH=CH$_2$	8.9	−2.3	−0.1	−0.8
—CH$_3$	9.2	0.7	−0.1	−3.0
—CH$_2$Cl	9.3	0.3	0.2	0.0
—C$_6$H$_5$	13.1	−1.1	0.5	−1.1
—CH$_2$CH$_3$	15.7	−0.6	−0.1	−2.8
—CH(CH$_3$)$_2$	20.2	−2.2	−0.3	−2.8
—C(CH$_3$)$_3$	22.4	−3.3	−0.4	−3.1
—NH(C=O)CH$_3$	9.7	−8.1	0.2	−4.4
—NH$_2$	18.2	−13.4	0.8	−10.0
—NO$_2$	19.9	−4.9	0.9	6.1
—O(C=O)CH$_3$	22.4	−7.1	0.4	−3.2
—OH	26.9	−12.8	1.4	−7.4
—OC$_6$H$_5$	27.6	−11.2	−0.3	−6.9
—OCH$_3$	31.4	−14.4	1.0	−7.7

Using methyl 3-nitrobenzoate as an example, the estimates of the chemical shifts calculated from Table 19.11 provide a useful guide for assigning the signals in the spectrum to the appropriate carbons.

	C-1	C-2	C-3	C-4	C-5	C-6
Base value	128.5	128.5	128.5	128.5	128.5	128.5
—(C=O) OCH$_3$	2.0	1.2	−0.1	4.3	−0.1	1.2
—NO$_2$	0.9	−4.9	19.9	−4.9	0.9	6.1
Estimated (ppm)	131.4	124.8	148.3	127.9	129.3	135.8
Observed (ppm)	131.9	124.5	148.3	127.3	129.6	135.2

EXERCISE

The ^{13}C NMR spectrum of 2-butanone shows signals at 8, 29, 37, and 209 ppm. The ^{13}C NMR spectrum of ethyl acetate shows signals at 14, 21, 60, and 171 ppm. Using Table 19.10, assign the NMR signals to the carbon atoms in each structure.

Answer:

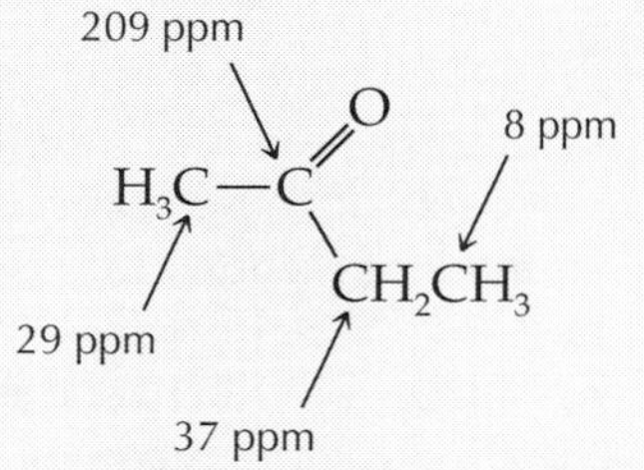

209 ppm

8 ppm

H_3C-C

O

CH_2CH_3

29 ppm

37 ppm

2-Butanone

171 ppm

14 ppm

H_3C-C

O

$O-CH_2CH_3$

21 ppm

60 ppm

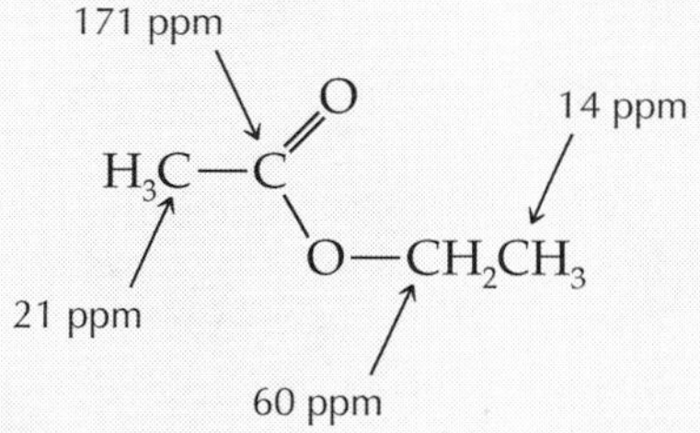

Ethyl acetate

Three useful observations can be made from the assignments:

- The methyl group directly attached to the carbonyl carbon of 2-butanone is shifted approximately 20 ppm downfield relative to the methyl group attached to the methylene group.
- The methylene group attached to the oxygen atom in ethyl acetate is approximately 25 ppm farther downfield relative to the methylene group attached to the carbonyl carbon in 2-butanone.
- The different shifts of the downfield signals of 2-butanone (209 ppm) and ethyl acetate (171 ppm) indicate different types of carbonyl-carbon atoms.

Symmetry and the Number of ^{13}C Signals

The number of signals in a ^{13}C NMR spectrum of a pure compound indicates the number of different types of carbon atoms in the molecule. If the number of signals in a compound's spectrum is less than the number of carbon atoms, there is probably some element of symmetry in the molecule, making some of the carbon atoms equivalent to one another. Symmetry elements can include mirror planes and axes of rotation.

Symmetry elements can be used advantageously to select a structure from several possibilities. For example, consider an aromatic compound with the molecular formula C_8H_{10}. There are four possible structures that are consistent with this formula:

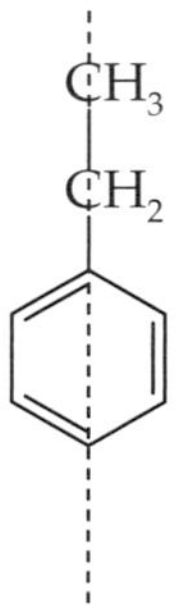

Ethylbenzene 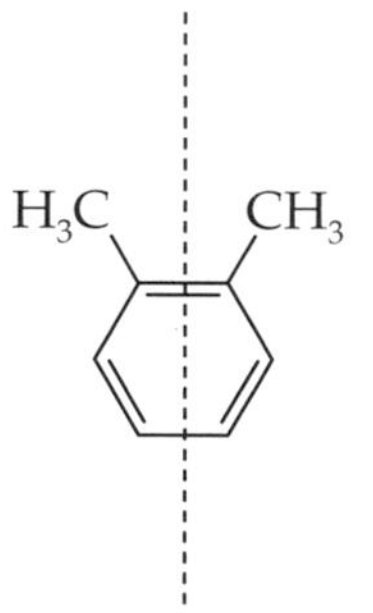   1,2-Dimethylbenzene 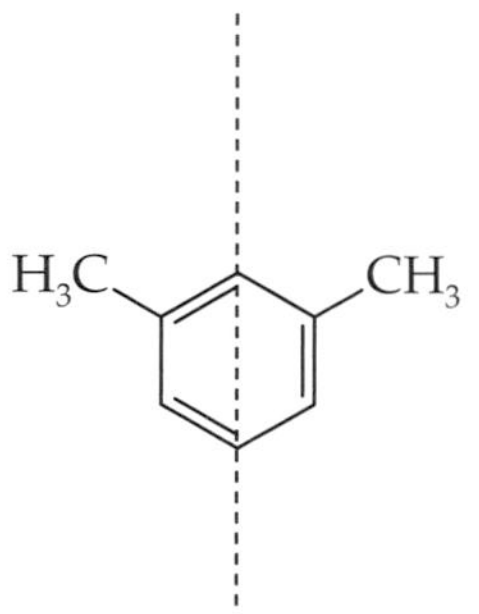   1,3-Dimethylbenzene 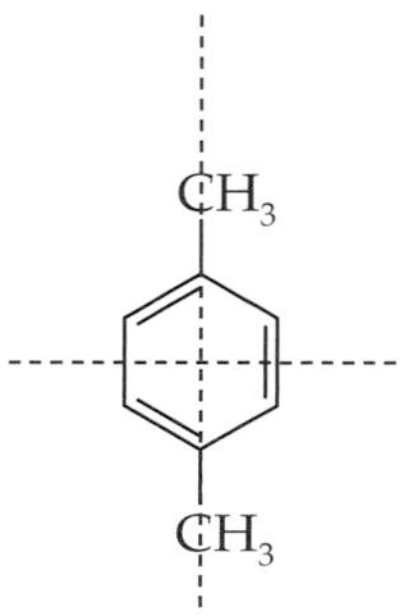   1,4-Dimethylbenzene

ethylbenzene, 1,2-dimethylbenzene, 1,3-dimethylbenzene, and 1,4-dimethylbenzene. Each structure possesses at least one plane of symmetry, and 1,4-dimethylbenzene has two planes of symmetry.

Each of these compounds contains a different number of non-equivalent carbon atoms: ethylbenzene has six, 1,2-dimethylbenzene has four, 1,3-dimethylbenzene has five, and 1,4-dimethylbenzene has three. Ethylbenzene is unique because it has two ^{13}C signals in the alkyl region, whereas each of the dimethylbenzenes has only one signal upfield. Obtaining the ^{13}C NMR spectrum and counting the number of signals can determine the structure of the compound.

EXERCISE

The ^{13}C NMR spectrum of an aromatic compound with the molecular formula C_8H_{10} is shown in Figure 19.40. (a) What is the structure of the compound? (b) Estimate the chemical shifts of the aromatic carbon atoms in your answer to (a) using Table 19.11 and assign the four aromatic ^{13}C signals in Figure 19.40 to the correct carbon atoms.

Answer: The ^{13}C NMR spectrum shows one upfield signal produced by the carbon atoms of two identical methyl groups and four signals produced by the carbon atoms in the aromatic ring. The compound that produced these five signals in the ^{13}C spectrum is 1,3-dimethylbenzene.

	C-1	C-2	C-3	C-4	C-5	C-6
Base value	128.5	128.5	128.5	128.5	128.5	128.5
—CH₃	9.2	0.7	−0.1	−3.0	−0.1	0.7
—CH₃	−0.1	0.7	9.2	0.7	−0.1	−3.0
Estimated (ppm)	137.6	129.9	137.6	126.2	128.3	126.2
Observed (ppm)	137.7	129.9	137.7	126.0	128.1	126.0

Although you would be hard-pressed to assign the ^{13}C signals with such precision directly from Figure 19.40, it is not difficult to order them by increasing chemical shift.

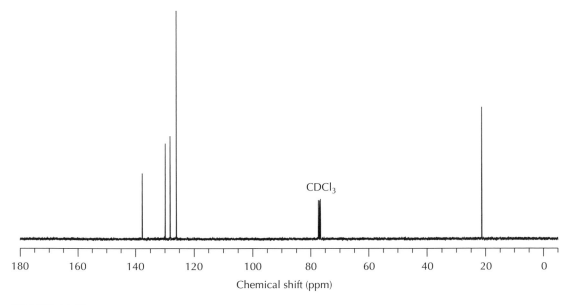

FIGURE 19.40
Decoupled 90-MHz ^{13}C NMR spectrum of an aromatic compound with molecular formula C_8H_{10} in CDCl₃.

19.14　Determining Numbers of Protons on Carbon Atoms

Typically, ^{13}C NMR spectra are obtained using broadband decoupling so that the carbon signals are collapsed into singlets. The cost of this simplification is the loss of information regarding the number of protons attached to carbon atoms. Numerous techniques have been developed to supply this important information. Two commonly used experiments provided with most modern FT NMR spectrometers are *APT* (Attached Proton Test) and *DEPT* (Distortionless Enhancement by Polarization Transfer). These experiments use complex pulse sequences at observation frequencies for both ^1H and ^{13}C nuclei.

APT

In a typical broadband-decoupled ^{13}C NMR spectrum, each different carbon atom in the sample appears as a single positive peak. In APT spectra, CH and CH_3 carbon nuclei give positive signals, whereas quaternary and CH_2 carbon nuclei give negative signals.

In the APT spectrum of ethyl *trans*-2-butenoate, shown in Figure 19.41, positive signals at 14, 18, 123, and 144 ppm are caused by the carbons of the methyl groups and the vinyl carbons. The negative signals at 60 and 166 ppm are caused by the carbon of the methylene group and the carbonyl carbon. With concentrated samples, an APT spectrum can be acquired in a short time. In some laboratories, APT spectra are preferred as the routine ^{13}C NMR spectra. However, this method is limited because one cannot normally distinguish between the signals of quaternary and CH_2 carbon nuclei or between signals of CH and CH_3 carbon nuclei.

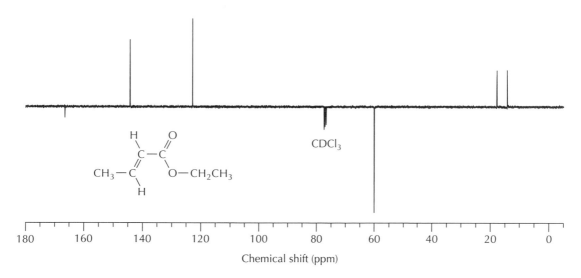

FIGURE 19.41
90-MHz ^{13}C NMR spectrum of ethyl *trans*-2-butenoate in $CDCl_3$ using an APT pulse sequence. Signals with positive amplitudes are assigned to CH or CH_3 groups, and those with negative amplitudes are assigned to quaternary carbons or CH_2 groups.

TABLE 19.12	Orientation of ^{13}C signals in DEPT NMR experiments			
Type of ^{13}C spectrum	CH_3	CH_2	CH	C
Broadband-decoupled ^{13}C	$+^a$	+	+	+
DEPT(45)	+	+	+	0^b
DEPT(90)	0	0	+	0
DEPT(135)	+	$-^c$	+	0

a. + = positive signal.　b. 0 = no signal.　c. − = negative signal.

DEPT

In a DEPT experiment, signals in the ^{13}C NMR spectrum may be suppressed or inverted depending on the number of protons attached to the carbon and the conditions set in the pulse program. The DEPT(45) version of the experiment provides a ^{13}C spectrum in which only carbon atoms that have protons attached to them appear. Signals caused by quaternary carbons are not observed. The spectrum produced by the DEPT(90) pulse program exhibits signals only from carbon atoms that have one hydrogen attached (methine carbons). Signals from all carbon atoms with attached protons are observed in the ^{13}C NMR spectrum from a DEPT(135) experiment; however, the signals from carbon atoms with two protons attached (methylene carbons) are inverted. Comparing the spectra from a set of DEPT experiments allows you to determine the number of protons attached to every carbon atom in a molecule. Table 19.12

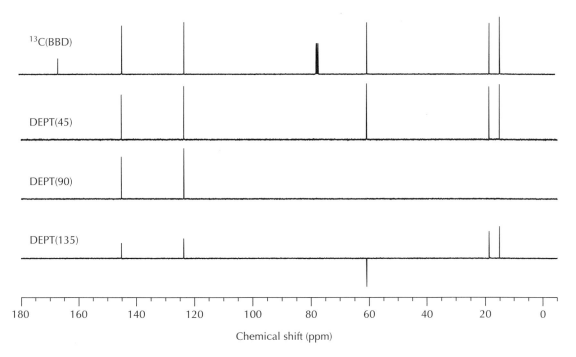

FIGURE 19.42
90-MHz DEPT spectra of ethyl *trans*-2-butenoate in CDCl$_3$. The broadband-decoupled 90-MHz ^{13}C NMR spectrum is shown at the top.

summarizes the information that can be obtained from a broad-band-decoupled ^{13}C spectrum and the three DEPT experimental spectra.

The broadband-decoupled ^{13}C, DEPT(45), DEPT(90), and DEPT(135) spectra of ethyl *trans*-2-butenoate are shown in Figure 19.42. The six signals in the ^{13}C spectrum were assigned earlier in the discussion following Figure 19.39. There is no $CDCl_3$ signal at 77 ppm in DEPT spectra because the carbon nucleus in $CDCl_3$ is attached to ^2H, not ^1H.

19.15   Two-Dimensional Correlated Spectroscopy (2D COSY)

In a typical ^1H or ^{13}C NMR spectrum, the positions of signals along the abscissa (*x*-axis) of the spectrum correspond to the frequencies of the signals, which are measured as chemical shifts. The intensities of the signals are measured along the ordinate (*y*-axis). This typical spectrum is referred to as a *one-dimensional (1D) spectrum* because only one axis is a frequency axis.

A *two-dimensional (2D) NMR spectrum* is one in which both the *x*-axis and the *y*-axis are frequency axes. The signal intensities in a 2D NMR spectrum are usually represented on the graph as a series of closely spaced contour lines, similar to a topographical map. The most important 2D correlated spectroscopy experiments are C,H-COSY (C,H-**CO**rrelated **S**pectroscop**Y**) spectra, in which one axis corresponds to ^{13}C chemical shifts and the other axis corresponds to ^1H chemical shifts, and H,H-COSY spectra, in which both axes correspond to ^1H chemical shifts.

2D COSY spectra indicate which nuclei are coupling with one another. They correlate the nuclei that are coupling partners. The correlations are shown by the presence of *cross peaks,* the contour line signals in 2D spectra that appear at the crossing of implicit vertical and horizontal lines connecting to the peaks on the *x*- and *y*-axes.

Two-Dimensional Heteronuclear (C,H)-Correlated NMR Spectroscopy (C,H-COSY)

An example of identifying a cross peak is shown in Figure 19.43, the 2D (C,H) COSY spectrum of ethyl *trans*-2-butenoate. The *x*-axis of the 2D spectrum displays the ^1H NMR spectrum, and the *y*-axis displays the ^{13}C spectrum. Each signal in this ^1H spectrum can be correlated to a signal in the ^{13}C spectrum. The cross peaks are the result of coupling between a ^1H nucleus and the ^{13}C nucleus to which it is attached, where the one-bond coupling constant is very large.

Thus, in the 2D (C,H) COSY spectrum the cross peak at 4.1 ppm along the chemical shift axis of the ^1H spectrum (*x*-axis) is located at 60 ppm along the chemical shift axis of the ^{13}C spectrum (*y*-axis). This cross peak shows that the protons producing the 4.1 ppm signal are attached to the carbon atom appearing at 60 ppm. The chemical shift of the ^1H spectrum thus correlates with the chemical shift of the ^{13}C spectrum. There are no cross peaks for quaternary

FIGURE 19.43
Two-dimensional
C,H-correlated
spectrum of ethyl
trans-2-butenoate in
CDCl₃. The one-
dimensional 360-MHz
¹H NMR spectrum is
shown at the top edge
and the 90-MHz
¹³C NMR spectrum is
shown at the left-hand
edge.

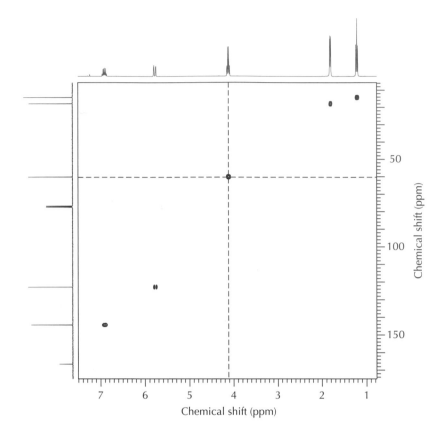

carbon nuclei because they have no attached protons. Therefore, in Figure 19.43 there is no cross peak on the 2D spectrum for the 166-ppm peak in the ¹³C spectrum, which is caused by the carbonyl carbon.

There are several 2D (C,H) COSY experiments that give basically the same information. They are identified by a variety of acronyms: HETCOR (**HET**eronuclear **COR**relation), HMQC (**H**eteronuclear **M**ultiple **Q**uantum **C**oherence), HSQC (**H**eteronuclear **S**ingle **Q**uantum **C**oherence), and others. The differences between the experiments are of little consequence when it comes to the interpretation of signals. The rationale for choosing particular experiments in many cases depends on the personal preferences of the people who obtain the spectra.

Two-Dimensional Homonuclear (H,H)-Correlated NMR Spectroscopy (H,H-COSY)

The 2D (H,H) COSY spectrum of ethyl *trans*-2-butenoate is shown in Figure 19.44. Here the ¹H NMR spectrum is displayed along the x-axis as well as the y-axis. For each signal in the ¹H NMR spectrum of ethyl *trans*-2-butenoate, there is a corresponding peak on the diagonal that runs from the lower left corner to the upper right corner. The presence of peaks on the diagonal is not useful in determining coupling patterns. Peaks on the diagonal appear because the magnetization is not completely transferred between coupling nuclei. The *off-diagonal cross peaks* are useful; they appear where ¹H nuclei are coupled.

FIGURE 19.44

Two-dimensional H,H-correlated spectrum of ethyl *trans*-2-butenoate in CDCl$_3$. The one-dimensional 360-MHz ^{1}H NMR spectra are shown at the top edge and at the left-hand edge.

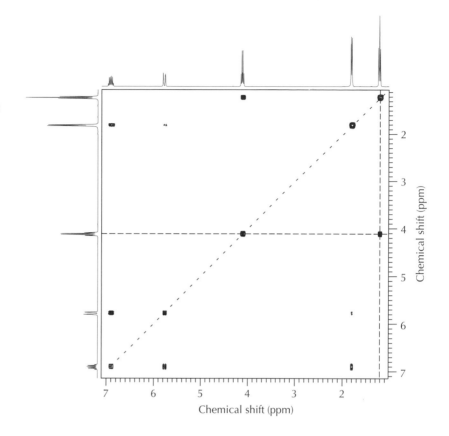

The off-diagonal cross peak at the intersection of the 1.2-ppm signal in the ^{1}H spectrum on the *x*-axis and the 4.1-ppm signal in the ^{1}H spectrum on the *y*-axis indicate that these two ^{1}H nuclei are coupled. Because of the symmetry of the 2D (H,H) COSY spectrum, another off-diagonal cross peak occurs at the intersection of the 1.2-ppm signal in the ^{1}H spectrum on the *y*-axis and the 4.1-ppm signal in the ^{1}H spectrum on the *x*-axis. No other off-diagonal cross peaks involve the 1.2-ppm and 4.1-ppm signals, so the protons they represent are coupled to each other but are not coupled to any other protons.

Figure 19.44 shows two off-diagonal cross peaks involving the 1.8-ppm ^{1}H signals on the two axes. The most intense one intersects with the 6.9-ppm signal. Thus, the ^{1}H nucleus at 1.8 ppm is coupled to the ^{1}H nucleus at 6.9 ppm. In addition, there is a less intense off-diagonal cross peak between the 1.8-ppm ^{1}H signal and the 5.8-ppm ^{1}H signal. Therefore, the proton at 1.8 ppm is also coupled to the proton at 5.8 ppm. The lesser intensity of the latter 2D peak suggests that the coupling constant is smaller due to long-range allylic coupling with a coupling constant of only 1.5 Hz. Last, the two off-diagonal peaks involving the 5.8-ppm signal and the 6.9-ppm signal show that these protons are coupled.

All these data confirm our original interpretation of the 1D ^{1}H NMR spectrum of ethyl *trans*-2-butenoate, which came in part from a detailed analysis of the spin-spin splitting patterns in the spectrum.

Several other variations of the 2D (H,H)-COSY experiment give basically the same information. By adjusting the parameters it is possible to enhance off-diagonal cross peaks resulting from long-range coupling and attenuate cross peaks resulting from the typical three-bond vicinal proton coupling.

References

1. Atta-ur-Rahman. *Nuclear Magnetic Resonance: Basic Principles*; Springer-Verlag, New York, 1986.
2. Duddeck, H.; Dietrich, W. *Structure Elucidation by Modern NMR*; 3rd ed.; Springer: Heidelberg, 1989.
3. Pouchert, C. J.; Behnke, J. (Eds.) *Aldrich Library of ^{13}C and ^1H FT-NMR Spectra*; Aldrich Chemical Co.: Milwaukee, WI, 1993; 3 volumes.
4. Pretsch, E.; Seibl, J.; Clerc, T.; Simon, W.; Biemann, K. (Trans.) *Tables of Spectral Data for Structure Determination of Organic Compounds*; 2nd English ed.; Springer-Verlag: New York, 1989.

5. Sanders, J. K.; Hunter, B. K. *Modern NMR Spectroscopy*; 2nd ed.; Oxford University Press: Oxford, 1993.
6. Silverstein, R. M.; Webster, F. X.; Kiemle, D. J. *Spectrometric Identification of Organic Compounds*; 7th ed.; Wiley: New York, 2005.
7. Crews, P.; Rodríguez, J.; Jaspars, M. *Organic Structure Analysis*; Oxford University Press: Oxford, 1998.
8. Friebolin, H. *Basic One- and Two-Dimensional NMR Spectroscopy*; 3rd ed.; Wiley-VCH: Weinheim, 1998.
9. Braun, S.; Kalinowski, H.-O.; Berger, S. *150 and More Basic NMR Experiments*; 2nd ed.; Wiley-VCH: Weinheim, 1998.

Questions

1. Given the ^1H NMR spectrum and molecular formula for each of the following compounds, deduce the structure of the compound, estimate the chemical shifts of all its protons using the parameters in Tables 19.3–19.5, and assign the NMR signals to their respective protons.

   a. $C_5H_{11}Cl$; ^1H NMR (CDCl$_3$): δ 3.33 (2H, s); 1.10 (9H, s).

   b. $C_5H_{10}O_2$; ^1H NMR (CDCl$_3$): δ 3.88 (1H, s); 2.25 (3H, s); 1.40 (6H, s).

   c. $C_6H_{12}O_2$; ^1H NMR (CDCl$_3$): δ 3.83 (1H, s); 2.63 (2H, s); 2.18 (3H, s); 1.26 (6H, s).

   d. $C_5H_{10}O$; ^1H NMR (CDCl$_3$): δ 9.77 (1H, t, $J = 2$ Hz); 2.31 (2H, dd, $J = 2$, and 7 Hz); 2.21 (1H, m); 0.98 (6H, d, $J = 7$ Hz).

   e. C_4H_8O; ^1H NMR (CDCl$_3$): δ 5.90 (1H, ddd, $J = 6$, 10, and 17 Hz); 5.19 (1H, d, $J = 17$ Hz); 5.06 (1H, d, $J = 10$ Hz); 4.30 (1 H, quintet); 2.50 (1H, bs); 1.27 (3H, d, $J = 6$ Hz).

2. The ^1H NMR spectrum of a compound of molecular formula $C_5H_8O_2$ is shown in Figure 19.45. Deduce the structure of the compound and assign its NMR signals.

3. A compound of molecular formula C_3H_8O produces the ^1H NMR spectrum shown in Figure 19.46. In addition, when this compound is treated with D$_2$O, the ^1H NMR signal at 2.0 disappears and another signal at 4.6 ppm appears. Moreover, when the C_3H_8O compound is highly purified and care is taken to remove all traces of acid in the NMR solvent, the singlet at 2.0 ppm is replaced by a doublet. Finally, the chemical shift of the 2.0-ppm signal is highly concentration dependent; an increase in the concentration of C_3H_8O in the NMR sample results in a downfield shift of this signal. Deduce the structure of C_3H_8O, assign its NMR signals, and explain the changes observed for the 2.0-ppm signal. Estimate the chemical

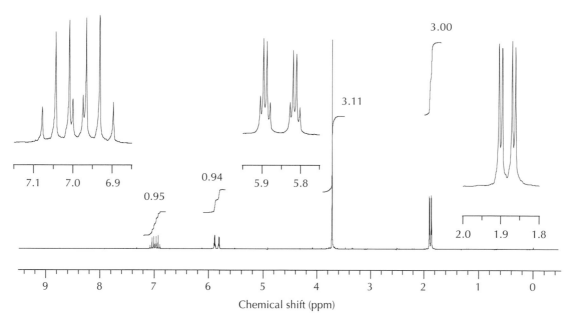

FIGURE 19.45
200-MHz ^{1}H NMR spectrum of a compound of molecular formula $C_5H_8O_2$.

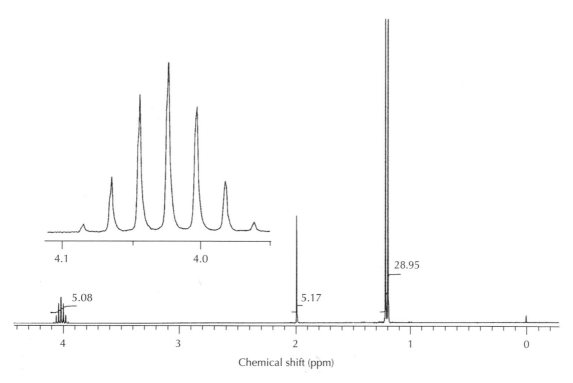

FIGURE 19.46
300-MHz ^{1}H NMR spectrum of a compound of molecular formula C_3H_8O.

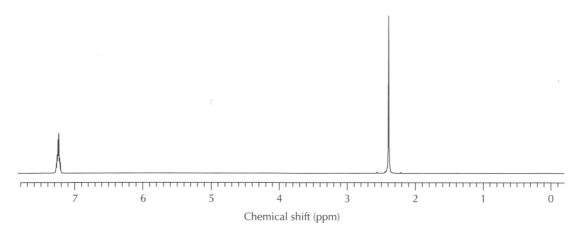

FIGURE 19.47
360-MHz ^{1}H NMR spectrum of a compound of molecular formula C_8H_{10}.

shifts of the different types of protons using the parameters in Tables 19.3–19.5 and compare them with those measured from the spectrum.

4. The ^{1}H NMR spectrum of an organic compound with the molecular formula C_8H_{10} is shown in Figure 19.47. The proton-decoupled ^{13}C NMR spectrum, DEPT(90) spectrum, and DEPT(135) spectrum are shown in Figure 19.48. Determine the structure of the compound and assign its signals in the ^{1}H and ^{13}C NMR spectra. Estimate the chemical shifts of all protons and carbon

atoms using the parameters in Tables 19.3–19.5 and 19.11 and compare them with those measured from the spectrum.

5. In solution, dimedone (5,5-dimethyl-cyclohexan-1,3-dione) is a mixture of keto and enol isomers. The ^{1}H NMR spectrum of a solution of dimedone in $CDCl_3$ is shown in Figure 19.49. In the sample, the two enol isomers are equilibrating very fast compared with the NMR time scale. Assign all the NMR signals and use NMR integrations to determine the composition of the keto/enol mixture.

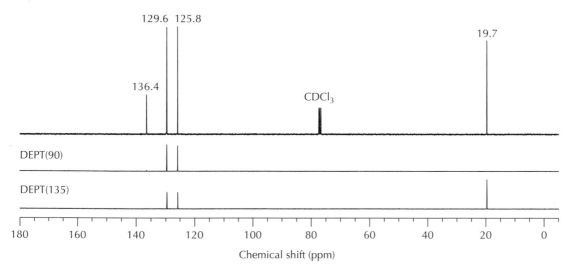

FIGURE 19.48
90-MHz ^{13}C NMR, DEPT(90), and DEPT(135) spectra of a compound of molecular formula C_8H_{10}.

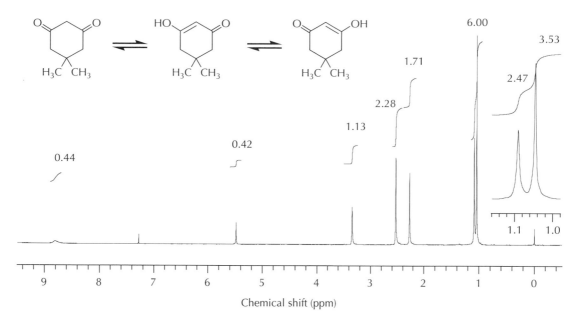

FIGURE 19.49
200-MHz ¹H NMR spectrum of 5,5-dimethylcyclohexan-1,3-dione.

6. A compound of molecular formula $C_{10}H_{14}$ produces the ¹H NMR spectrum shown in Figure 19.50. The proton-decoupled ¹³C NMR spectrum has signals at 145.8, 135.1, 129.0, 126.3, 33.7, 24.1, and 20.9 ppm. The ¹³C NMR spectrum, DEPT(90) spectrum, and DEPT(135) spectrum are shown in Figure 19.51. Deduce the structure of $C_{10}H_{14}$, estimate the chemical shifts of all protons and any aromatic carbon atoms using the parameters in Tables 19.3–19.5 and 19.11, and assign all the ¹H and ¹³C NMR signals.

7. A compound of molecular formula $C_8H_{14}O$ produces the ¹H NMR spectrum

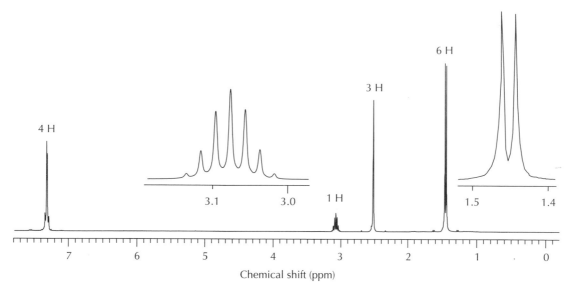

FIGURE 19.50
360-MHz ¹H NMR spectrum of a compound of molecular formula $C_{10}H_{14}$.

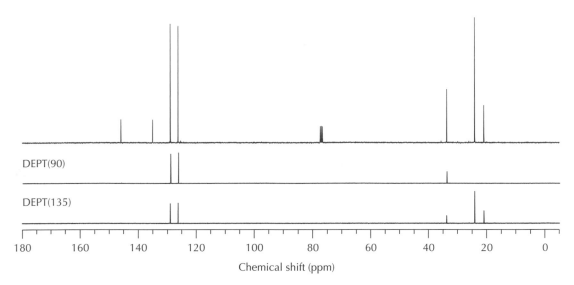

FIGURE 19.51
90-MHz ^{13}C NMR, DEPT(90), and DEPT(135) spectra of a compound of molecular formula $C_{10}H_{14}$.

shown in Figure 19.52. Its infrared spectrum shows a strong carbonyl stretching peak, which indicates that $C_8H_{14}O$ is either an aldehyde or a ketone. Deduce the structure of the compound, estimate the chemical shifts of the different types of protons using the parameters in Tables 19.3–19.5, and assign all the NMR signals.

8. Ibuprofen is the active ingredient in several nonsteroid anti-inflammatory drugs (NSAIDs). The ^{1}H NMR spectrum of the methyl ester of ibuprofen is shown in Figure 19.53a. Expansions of

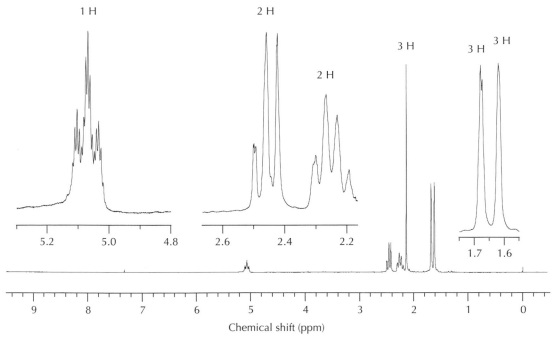

FIGURE 19.52
200-MHz ^{1}H NMR spectrum of a compound of molecular formula $C_8H_{14}O$.

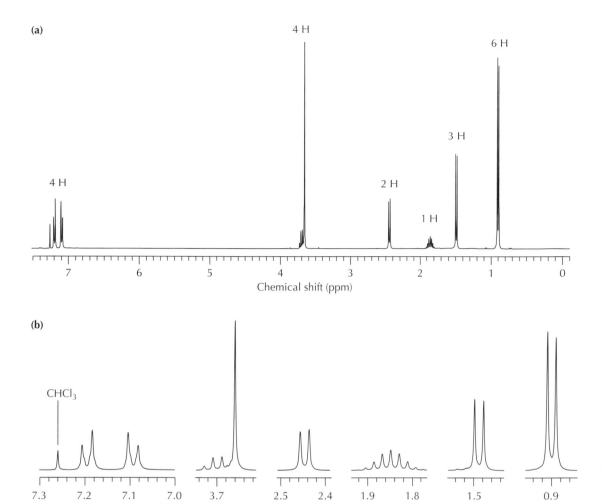

FIGURE 19.53
(a) 360-MHz ^{1}H NMR spectrum of the methyl ester of ibuprofen. (b) Expansions of the signals in the ^{1}H NMR spectrum of the methyl ester of ibuprofen.

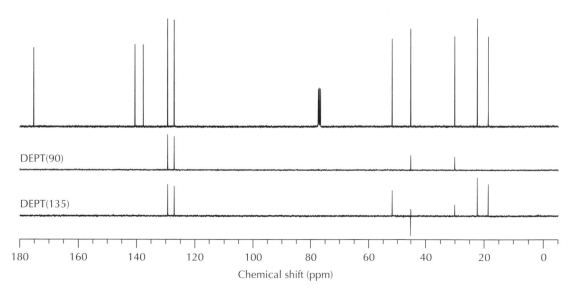

FIGURE 19.54
90-MHz ^{13}C NMR, DEPT(90), and DEPT(135) spectra of the methyl ester of ibuprofen.

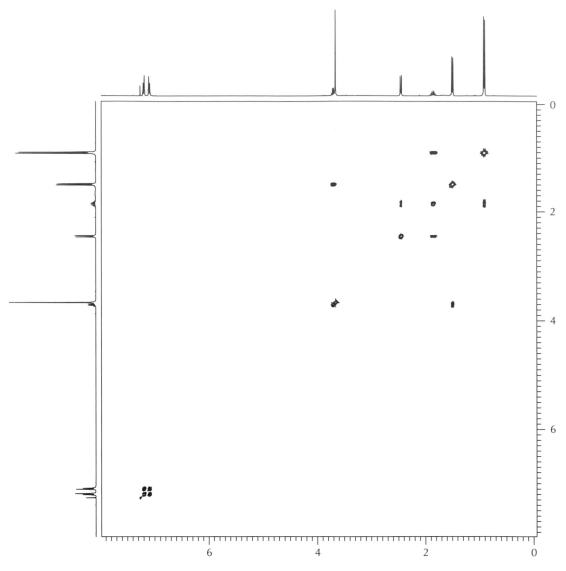

FIGURE 19.55
2D (H,H) COSY spectrum of the methyl ester of ibuprofen.

all its ^{1}H NMR signals are shown in Figure 19.53b. The molecular formula of the methyl ester of ibuprofen is $C_{14}H_{20}O_2$.

The broadband-decoupled ^{13}C NMR spectrum, DEPT(90) spectrum, and DEPT(135) spectrum for the methyl ester of ibuprofen are shown in Figure 19.54. **Hint:** The ^{13}C signal at 45 ppm is broader than the other signals in the ^{13}C spectrum and resolves into two separate signals at higher resolution. Pay careful attention to the

pattern of signals at 45 ppm in the DEPT(135) spectrum.

The 2D (H,H) COSY spectrum of the methyl ester of ibuprofen is shown in Figure 19.55, and its 2D (C,H) COSY spectrum is shown in Figure 19.56. Deduce the structure of the methyl ester of ibuprofen, estimate the chemical shifts of all protons and any aromatic carbon atoms using the parameters in Tables 19.3–19.5 and 19.11, and assign all the ^{1}H and ^{13}C NMR signals. Show your reasoning.

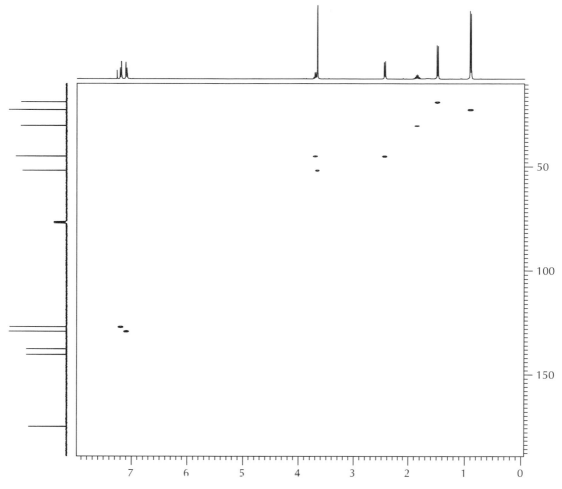

FIGURE 19.56
2D (C,H) COSY spectrum of the methyl ester of ibuprofen.

20 MASS SPECTROMETRY

Most spectrometric techniques used by organic chemists involve the ability of molecules to absorb light of various energies. Mass spectrometry (MS) is different: rather than the absorption of light, it normally involves energy transfer from energetic electrons. This energy produces ionization of the molecules, and mass spectrometry measures the masses of these ions. It is a very sensitive technique that can be carried out with microgram quantities of compounds. Mass spectrometry is used to determine the molecular weights and molecular formulas of compounds. Fragmentation of the initially formed ions in the mass spectrometer provides additional information that can be used to identify a compound or determine its structure.

20.1   Mass Spectrometers

In recent years, great strides have been made in instrumentation for mass spectrometry, and numerous types of mass spectrometers are now available. Even though they have functional differences, the basic components outlined in Figure 20.1 are common to all mass spectrometers. A sample is introduced into the mass spectrometer, where it is converted into a gas-phase ion through one of a variety of ionization techniques. These gas-phase ions are then sorted by their *mass-to-charge (m/z) ratios* in the mass analyzer. The sorted ions generate an electric current at the detector and a mass spectrum is created. Because the charge on the ions is typically +1, the *m/z* value for the molecular ion corresponds to the molecular weight of the compound.

Electron Impact (EI) Mass Spectrometry

The classic mass spectrometer ionizes the sample by electron impact and sorts the ions with a magnetic sector mass analyzer. To gain an appreciation of how all mass spectrometers work, it is worthwhile to examine this type of spectrometer in more detail. In the ionizer, shown in Figure 20.2, a stream of electrons with 70 electron volts (eV) of energy is created by heating a metal filament. The stream of electrons bombards the vaporized sample as it enters through a small hole in the vacuum chamber. A molecule struck by an external electron can become charged by either losing or gaining an electron; with electrons possessing 70 eV, the ionization produces many more positive than negative ions. The negative ions are attracted to the anode (electron trap), removing them from the ionization chamber. The positive ions are propelled toward the analyzer by the positively charged (+10,000 volts) repeller plate. Additional charged plates accelerate the ions to a constant velocity and focus the ion stream into the analyzer.

Molecules of the vast majority of organic compounds have only paired electrons, so when a single electron is lost from a molecule, a free radical is formed. Thus, the molecular ion formed by loss of an electron is a *radical cation;* it has an unpaired electron as well as a positive charge. The cation formed from an intact molecule is called a *molecular ion (M^+·).* Once formed, the highly energetic molecular ion often breaks apart, forming both charged and uncharged fragments. Uncharged

FIGURE 20.1
Basic components of a mass spectrometer.

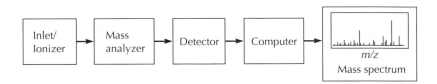

FIGURE 20.2
Electron impact (EI)
ionization chamber.

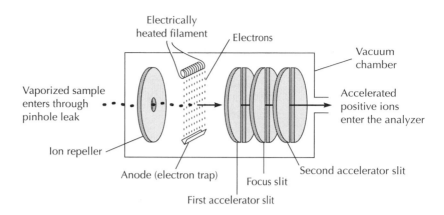

molecules and fragments are removed by the vacuum system.
Usually, only the positively charged ions are analyzed.

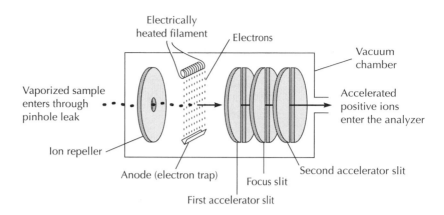

As shown in Figure 20.3, the application of a magnetic field
perpendicular to the flight path of the ions (perpendicular to the
page) allows us to separate ions by their masses. The magnetic field
causes the pathway of each ion to be curved. The amount of curva-
ture is a function of the mass of the ion and the strength of the
magnetic field. For an ion to strike the detector, it must follow a
particular path consistent with the radius of the mass analyzer
portion of the mass spectrometer. By adjusting the magnetic field

FIGURE 20.3
Magnetic sector mass
analyzer.

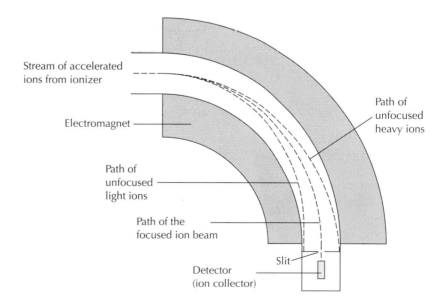

strength, a beam of ions with a specific mass-to-charge ratio can be focused on the detector. Ions with a larger m/z ratio do not bend enough to reach the detector, and ions with a smaller m/z ratio bend too much to hit the detector. The signal at the detector is recorded as a function of magnetic field strength. By varying the strength of the magnetic field, a scan over a range of m/z values can be collected. Because each ion usually bears a single positive charge, m/z simplifies to m, the mass of the ion.

GC-MS

Many research and teaching laboratories have acquired hybrid instruments combining a *gas chromatograph and a mass spectrometer (GC-MS).* In these instruments, small samples of the effluent stream from a gas chromatograph are directed into a mass spectrometer. The molecules in the sample are then ionized by electron impact, and the resulting ions are accelerated and passed into the mass analyzer. The result is a mass spectrum for every compound eluting from the gas chromatograph. This technique is very efficient for analyzing mixtures of compounds because it provides the number of components in the mixture, a rough measure of their relative amounts, and the possible identities of the components.

The mass analyzer in most GC-MS instruments is a *quadrupole mass filter,* diagrammed in Figure 20.4. The quadrupole filter consists of four parallel stainless steel rods. Each pair of rods has opposing direct current (DC) voltages. Superimposed on the DC potential is a high-frequency alternating current (AC) voltage. As the stream of ions passes through the central space parallel to the rods, the combined DC and AC fields affect the ion trajectories, causing them to oscillate. For given DC and AC voltages and frequencies, only ions of a specific mass-to-charge ratio achieve a stable oscillation. These ions pass through the filter and strike the detector. Ions with different mass-to-charge ratios acquire unstable oscillations, tracing paths that collide with the rods or otherwise miss the detector. Although this mass sorting method is very different from the mass analyzer in magnetic sector instruments, the resulting mass spectra are comparable. Quadrupole mass filters are compact and fast, making them ideally suited for interfacing with other instruments, such as gas chromatographs. These systems have excellent resolution in the mass range of typical organic compounds. The GC-MS has become a major workhorse in modern organic and analytical chemistry laboratories.

FIGURE 20.4
Quadrupole mass filter.

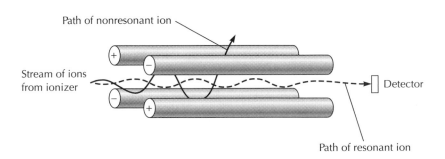

Path of nonresonant ion

Stream of ions from ionizer

Detector

Path of resonant ion

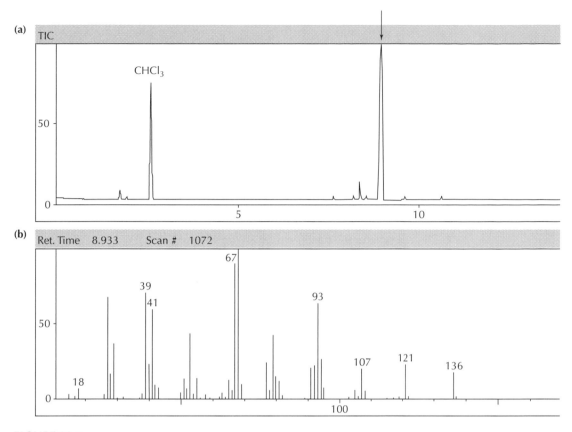

FIGURE 20.5
(a) Total ion current gas chromatogram of orange oil. (b) Mass spectrum of the major component.

An example of the data from a GC-MS is shown in Figure 20.5. The sample injected was orange oil purchased at a natural foods store. Figure 20.5a is a record of the total ion current (TIC) arriving at the detector of the mass spectrometer. This ion current corresponds to the gas chromatogram of the sample. The peak at approximately 2.5 min was caused by some residual chloroform that was used to clean the injection syringe. The peak at 8.9 min, designated with the arrow, is the major and virtually only compound in the sample. The mass spectrum of this major component of orange oil is shown in Figure 20.5b.

Advances in MS Instrumentation

Normally, samples are vaporized for simple mass spectrometric analysis. Thus, the electron impact technique is limited to compounds with significant vapor pressures. However, special ionization techniques can be used to ionize samples directly from the solid state or from solution. These techniques make it possible to study samples that have high molecular weights and very low vapor pressures, such as proteins and peptides. Special "softer" ionization techniques have also been developed that limit the amount of energy transferred to the molecules, thereby minimizing the amount of fragmentation the molecular ions undergo and providing a more precise molecular weight of the compound.

20.2 Mass Spectra and the Molecular Ion

Mass spectral data are usually presented in graphical form as a histogram of ion intensity (y-axis) versus m/z (x-axis). For example, in the mass spectrum of 2-butanone shown in Figure 20.6, the molecular ion peak appears at m/z 72. Because the highly energized radical cation breaks into fragments, peaks also appear at smaller m/z values. The intensities are represented as percentages of the most intense peak of the spectrum, called the **base peak.** In this spectrum, the base peak is at m/z 43.

When interpreting a mass spectrum, the first area of interest is the molecular ion region. If the molecular ion does not completely fragment before being detected, its m/z provides the molecular weight of the compound, a valuable piece of information about any unknown.

Rule of Thirteen

A method known as the **Rule of Thirteen** can be used to generate the chemical formula of a hydrocarbon, C_nH_m, using the m/z of the molecular ion. The integer obtained by dividing m/z by 13 (atomic weight of carbon + atomic weight of hydrogen) corresponds to the number of carbon atoms in the formula. The remainder from the division is added to the integer to give the number of hydrogen atoms. For example, if the molecular ion of a hydrocarbon appears at m/z 92 in its mass spectrum, we can find the number of carbon atoms in the molecule by dividing m/z by 13: n = 92/13 = 7 with a remainder of 1. The value of m is 7 + 1 = 8, and the molecular formula for the hydrocarbon is C_7H_8. If the compound contains oxygen or nitrogen as well, some carbon atoms must be subtracted and the number of hydrogen atoms must be adjusted to give the

FIGURE 20.6
Mass spectrum of 2-butanone.

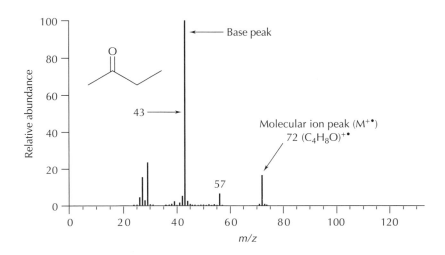

appropriate m/z value. One oxygen atom is the equivalent of CH_4; that is, each oxygen and CH_4 unit accounts for 16 atomic mass units. One nitrogen atom is the equivalent of CH_2, 14 atomic mass units.

To apply the Rule of Thirteen to 2-butanone, the m/z of 72 is divided by 13, which gives the value of n, and the remainder from the division is added to get the value of m; this calculation would provide the correct molecular formula if 2-butanone were a hydrocarbon. Because 2-butanone also contains an oxygen atom, an oxygen atom must be added to the formula and CH_4 subtracted. Dividing m/z by 13 yields $72/13 = 5$ (with a remainder of 7); C_nH_m would be C_5H_{12}. Including the presence of oxygen, the correct molecular formula of 2-butanone is $C_5H_{12}O - CH_4 = C_4H_8O$. If the number of oxygen or nitrogen atoms is unknown, a number of candidate molecular formulas would have to be considered. All heteroatoms in the periodic table can be set equal to a C_nH_m equivalent. If the molecular ion has an odd m/z value, the first heteroatom to consider is nitrogen. Tables that list formula masses of various combinations of C, H, O, and N, have been constructed (Ref. 3).

Fundamental
Nitrogen Rule

The fundamental nitrogen rule states that a compound whose molecular ion contains nothing other than C, H, N, or O atoms and that has an even m/z value must contain either no nitrogen atoms or an even number of nitrogen atoms. A compound whose molecular ion has an odd m/z value must contain an odd number of nitrogen atoms. The following compounds support the nitrogen rule:

$(CH_3CH_2)_3N$

CH_2NH_2

$H_2NCH_2CH_2NH_2$

Pyridine	Triethylamine	Benzylamine	Ethylenediamine
MW 79	MW 101	MW 107	MW 60

M+1 and M+2
Peaks

Most elements occur in nature as a mixture of isotopes. Table 20.1 provides a list of isotopes for elements commonly found in organic compounds. If the molecular ion (M) is reasonably intense, signals for M+1 and M+2 ions can also be observed. The ratios of the intensities of the M+1 and M+2 peaks to that of the M peak depend on the isotopic abundances of the atoms in a molecule and the number of each kind of atom. Isotopes of carbon, hydrogen, oxygen, and nitrogen, the elements that make up most organic compounds, make small contributions to the M+1 and M+2 peaks, and the resulting intensities can sometimes reveal the molecular formulas of organic compounds. For example, the intensity of the M+1 peak relative to the intensity of the M peak in 2-butanone (molecular formula C_4H_8O) should be $(4 \times 1.08\%) + (8 \times 0.012\%) + (1 \times 0.038\%) = 4.45\%$. Tables listing molecular formulas and the ratios expected for these formulas can be found in Refs. 1 and 2.

Unfortunately, practical experience has shown that for many C, H, N, and O compounds, the expected ratios are often in error,

TABLE 20.1   **Relative isotope abundances of elements common in organic compounds**

Elements	Isotope	Relative abundance	Isotope	Relative abundance	Isotope	Relative abundance
Hydrogen	^1H	100	^2H	0.012		
Carbon	^{12}C	100	^{13}C	1.08		
Nitrogen	^{14}N	100	^{15}N	0.37		
Oxygen	^{16}O	100	^{17}O	0.038	^{18}O	0.205
Fluorine	^{19}F	100				
Silicon	^{28}Si	100	^{29}Si	5.08	^{30}Si	3.35
Phosphorus	^{31}P	100				
Sulfur	^{32}S	100	^{33}S	0.79	^{34}S	4.47
Chlorine	^{35}Cl	100			^{37}Cl	32.0
Bromine	^{79}Br	100			^{81}Br	97.3
Iodine	^{127}I	100				

Adapted from J. R. de Laeter, J. K. Bohlke, P. de Bievre, H. Hidaka, H. S. Peiser, K. J. R. Rosman, and P. D. P. Taylor for the International Union of Pure and Applied Chemistry in "Atomic Weights of the Elements, Review 2000," *Pure and Applied Chemistry,* **2003,** *75,* 683–800.

which can occur for many reasons. For example, in the mass spectrometer the molecular ion may undergo ion/molecule collisions that provide additional intensity to the M+1 peak.

$$M^{+\bullet} + RH \longrightarrow MH^+ + R^\bullet$$

In addition, if the molecular ion has a low relative abundance, the precision of the M+1 data is insufficient to give reliable ratios.

Although it can be difficult to use M+1 and M+2 data to determine accurate molecular formulas, MS is highly valuable for qualitative elemental analysis. In particular, it is fairly easy to use the M+2 peak to identify the presence of bromine and chlorine in organic compounds. The appearance of a large M+2 peak in a mass spectrum is evidence for the presence of one of these elements. The relative intensities tell you which one. A good example is seen in the mass spectrum of 1-bromopropane (Figure 20.7). The two major

FIGURE 20.7

Mass spectrum of 1-bromopropane.

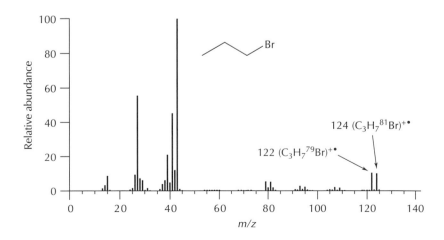

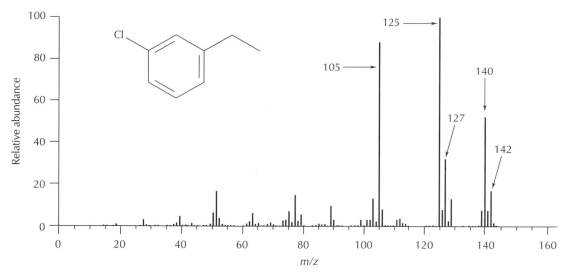

FIGURE 20.8
Mass spectrum of 3-chloroethylbenzene.

peaks in the molecular ion region are m/z 122 and 124 with an approximately 1:1 ratio. You can see from Table 20.1 that bromine exists in nature as a mixture of ^{79}Br and ^{81}Br in a ratio very close to 1:1. The peak at m/z 122, where the bromine atom has a mass of 79, is by convention defined as the molecular ion peak (M). Although the m/z 124 peak also corresponds to an intact molecule of $+1$ charge, it is referred to as the M+2 peak. Isotopic contributions of carbon, hydrogen, nitrogen, and oxygen to the M+1 and M+2 peaks are comparatively small. Thus a ratio of M/M+2 that is very close to 1:1 is a clear indication that the molecule contains a bromine atom.

A monochloro compound is expected to have an M+2 peak that is 32.0% as intense as the M peak. For example, the mass spectrum of 3-chloroethylbenzene, shown in Figure 20.8, has a peak at m/z 142 that is approximately one-third the intensity of the molecular ion peak at m/z 140. A small contribution of ^{13}C is shown in the M + 1 peak at m/z 141. The Rule of Thirteen can be used to calculate the molecular formula; the carbon equivalent of ^{35}Cl is C_2H_{11}. Dividing 140 by 13 yields $140/13 = 10$ (with a remainder of 10); C_nH_m would be $C_{10}H_{20}$. Including the presence of a chlorine atom, the correct molecular formula of 3-chloroethylbenzene is $C_{10}H_{20}Cl - C_2H_{11} = C_8H_9Cl$.

20.3 High-Resolution Mass Spectrometry

In modern research laboratories, molecular formulas are usually determined by *high-resolution mass spectrometry.* The expensive high-resolution instruments used for this purpose have both electric and magnetic fields for focusing the ion pathways. These double-resolution instruments measure masses to four figures beyond the decimal point. Table 20.2 provides the masses that should be used for this approach. The exact mass of an isotope is established using carbon-12 as the standard. The exact mass of a molecule is determined by summing the masses of all the isotopes in the molecule. For example, the exact mass of the molecular ion of 2-butanone is $(4 \times 12.00000) + (8 \times 1.00783) + (1 \times 15.9949) = 72.0575$. By looking at the exact masses of molecules whose nominal molecular weight is 72, it is obvious that the correct molecular formula can be determined from the masses measured to four decimal places.

Formula	Exact Mass
$C_2H_4N_2O$	72.0324
$C_3H_4O_2$	72.0211
$C_3H_8N_2$	72.0688
C_4H_8O	72.0575
C_5H_{12}	72.0940

TABLE 20.2	Atomic weights and exact isotope masses for elements common in organic compounds

Element	Atomic weight	Nuclide	Mass
Hydrogen	1.00794	^1H	1.00783
		$D(^2H)$	2.01410
Carbon	12.0107	^{12}C	12.00000 (std)
		^{13}C	13.00335
Nitrogen	14.0067	^{14}N	14.0031
		^{15}N	15.0001
Oxygen	15.9994	^{16}O	15.9949
		^{17}O	16.9991
		^{18}O	17.9992
Fluorine	18.9984	^{19}F	18.9984
Silicon	28.0855	^{28}Si	27.9769
		^{29}Si	28.9765
		^{30}Si	29.9738
Phosphorus	30.9738	^{31}P	30.9738
Sulfur	32.065	^{32}S	31.9721
		^{33}S	32.9715
		^{34}S	33.9679
Chlorine	35.453	^{35}Cl	34.9689
		^{37}Cl	36.9659
Bromine	79.904	^{79}Br	78.9183
		^{81}Br	80.9163
Iodine	126.9045	^{127}I	126.9045

Adapted from J. R. de Laeter, J. K. Bohlke, P. de Bievre, H. Hidaka, H. S. Peiser, K. J. R. Rosman, and P. D. P. Taylor for the International Union of Pure and Applied Chemistry in "Atomic Weights of the Elements, Review 2000," *Pure and Applied Chemistry,* **2003,** *75,* 683–800.

20.4   Mass Spectral Libraries

When a molecular ion breaks into fragments, the resulting mass spectrum can be complex because any one of a number of covalent bonds might be broken during fragmentation. Examination of Figures 20.5–20.8 shows that a large number of peaks arise even with relatively low-molecular-weight organic compounds. The array of fragmentation peaks constitutes a fingerprint that can be used for identification. Modern mass spectrometers are routinely equipped with computer libraries of mass spectra, some containing hundreds of thousands of spectra, for matching purposes. Typically, a computer program compares the experimental spectrum with spectra in the library and produces a ranked *"hit list"* of compounds with similar mass spectra. The ranking is based on how close the match is in terms of the presence of peaks and their intensities. At this point, the chemist intervenes. Mass spectra of highly ranked compounds on the hit list are compared with the acquired mass spectrum to determine the closest match.

The closest match does not necessarily prove the structure of a compound. Impurities that result from bleeding from the GC column can produce extra peaks in the mass spectrum and provide false hit-list candidates. In addition, the compound must be in the database, which is not always the case with research samples. Two comprehensive libraries of mass spectra are *NIST 05,* a collection of electron impact mass spectra of over 163,000 compounds from the National Institute of Standards and Technology (Ref. 5), and the *Wiley Registry of Mass Spectral Data,* a collection of over 338,000 EI mass spectra (Ref. 6). There are also a number of specialized mass spectral libraries available that are limited and targeted to specific types of compounds, such as drug metabolites or steroids.

The hit list for the major component of orange oil is shown in Figure 20.9. The second column, labeled *SI for "similarity index,"* corresponds to how well the mass spectra that are stored in the computer library match the acquired spectrum of the compound from the GC-MS. Notice that several of the compounds appear more than once in the list; there are several spectra for these compounds in the library because many laboratories contribute spectra to the collection. Slight differences in instrument conditions and/or configurations can lead to subtle differences in the acquired spectra—another reason to examine the hit list with a critical eye.

A computer screen printout for comparing spectra of the hit-list candidates is shown in Figure 20.10. The spectrum of hit 1 (Figure 20.10b) is virtually identical to the mass spectrum of the sample (Figure 20.10a). The spectrum of hit 2 (Figure 20.10c) is also similar, even though the compound's structure is different. However, on close examination some subtle differences can be

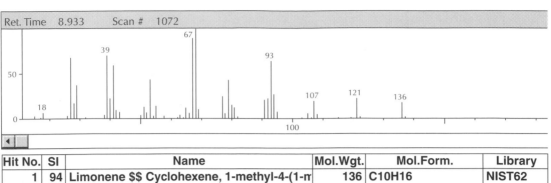

Hit No.	SI	Name	Mol.Wgt.	Mol.Form.	Library
1	94	Limonene $$ Cyclohexene, 1-methyl-4-(1-m	136	C10H16	NIST62
2	90	1,5-Cyclooctadiene, 1,5-dimethyl-	136	C10H16	NIST12
3	90	Cyclohexene, 1-methyl-4-(1-methylethenyl)	136	C10H16	NIST12
4	87	Camphene $$ Bicyclo 2.2.1 heptane, 2,2-di	136	C10H16	NIST62
5	86	Cyclohexanol, 1-methyl-4-(1-methylethenyl	196	C12H20O2	NIST62
6	86	Limomene	136	C10H16	NIST12
7	86	Cyclohexene, 1-methyl-4-(1-methylethenyl)	136	C10H16	NIST62
8	85	D-Limonene	136	C10H16	NIST12
9	85	Bicyclo 2.2.1 hept-2-ene, 1,7,7-trimethyl- $$	136	C10H16	NIST62
10	85	D-Limonene $$ Cyclohexene, 1-methyl-4-(1	136	C10H16	NIST62
11	84	Limonene	136	C10H16	NIST12
12	83	D-Limonene	136	C10H16	NIST12
13	83	Cyclohexanol, 1-methyl-4-(1-methylethenyl	196	C12H20O2	NIST12
14	83	1,5-Cyclooctadiene, 1,5-dimethyl- $$ 1,5-Di	136	C10H16	NIST62
15	83	Cyclohexene, 1-methyl-4-(1-methylethenyl)	136	C10H16	NIST62
16	83	Limonene	136	C10H16	NIST12
17	82	Cyclohexene, 4-ethenyl-1,4-dimethyl- $$ 1,4	136	C10H16	NIST62
18	82	Camphene	136	C10H16	NIST12
19	82	Cyclohexene, 1-methyl-5-(1-methylethenyl)	136	C10H16	NIST62
20	81	2,6-Octadien-1-ol, 3,7-dimethyl-, [Z]-	154	C10H18O	NIST12
21	80	4-Tridecen-6-yne, [Z]-	178	C13H22	NIST62
22	80	.alpha.-Myrcene	136	C10H16	NIST62
23	80	Bicyclo 2.2.1 heptane, 2,2-dimethyl-3-meth	136	C10H16	NIST62
24	80	2,6-Octadien-1-ol, 3,7-dimethyl-, [E]-	154	C10H18O	NIST12
25	80	D-Limonene	136	C10H16	NIST12

FIGURE 20.9

Hit list from a mass spectral library search for the major component of orange oil. The symbol $$ in the name denotes the start of a second name for the same compound.

discerned. A significant signal at *m/z* 108 is present in Figure 20.10c but not in Figure 20.10a. Also, the signal at *m/z* 92 in Figure 20.10a is missing in Figure 20.10c. By these observations, hit 2 can be ruled out as a match, and a tentative conclusion can be reached that the major component of orange oil is limonene. The hit list is more reliable in confirming a structural option if it is combined with other spectroscopic evidence. Infrared or NMR evidence and the history of the sample—for example, if it came from a chemical reaction—can help to ascertain the correct molecular structure.

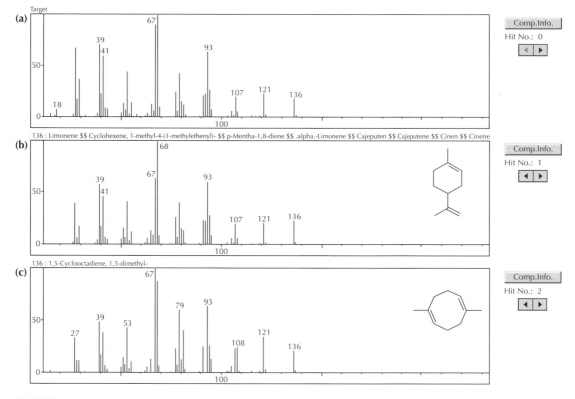

FIGURE 20.10
Computer comparison of two hit list compounds for orange oil. (a) MS of the compound from GC-MS run. (b) MS of hit 1. (c) MS of hit 2.

20.5 Fragmentation of the Molecule

In cases where you are working with a compound not included in a library or the hit list of which does not lead to a satisfactory structure candidate, fragmentation pathways can provide important clues to its molecular structure. Numerous fragmentation rules have been established, but the topic is too wide in scope to be adequately covered in this book. However, a few of the most useful fragmentation patterns for common functional groups are described in the following paragraphs. As a general rule, ions or free radicals that are more stable have a greater probability of forming from mass spectral fragmentation reactions.

Mechanisms of fragmentation processes are easier to understand if *"fishhooks"* (curved arrows with half-heads) are used to represent the migration of single electrons. This notation is similar to that used in free radical or photochemical processes:

$$\text{Molecule} \longrightarrow \underbrace{M^{+\bullet} = [f\!\!-\!\!\curvearrowleft\!\!f']^{+\bullet}}_{\substack{\text{Molecular}\\ \text{cation radical}}} \longrightarrow \underset{\text{Fragments}}{f^+ + f'^\bullet}$$

Forces that contribute to the ease with which fragmentation processes occur include the strength of bonds in the molecule (for example, the f—f′ bond) and the stability of the carbocations (f^+) and free radicals ($f'\cdot$) produced by fragmentation. Although these fragments are formed in the gas phase, we can still apply our "chemical intuition," based on reactions in solution.

The carbocation fragment is often an even-electron species; the odd electron in the molecular cation radical ends up on the free radical fragment. Of course, only ions are actually observed in the mass spectrum. When a molecular ion with an even m/z value gives a fragment ion that has an odd m/z value, a loss of a free radical by cleavage of just one covalent bond has occurred.

$$[H_3C—CH_3]^{+\cdot} \longrightarrow CH_3{}^+ \; + CH_3\cdot$$
m/z 30   m/z 15

Simple cleavage of a molecular ion that has an odd m/z value gives a fragment ion with an even m/z value:

$$(CH_3CH_2)_2\overset{\cdot\,+}{N}\!-\!CH_2\!-\!CH_3 \longrightarrow (CH_3CH_2)_2\overset{+}{N}\!=\!CH_2 + CH_3\cdot$$
m/z 101   m/z 86

Aromatic Hydrocarbons

Aromatic hydrocarbons are prone to fragmentation at the bond β to the aromatic ring, yielding a benzylic cation that rearranges to a stable C_7H_7 aromatic carbocation called a tropylium ion.

m/z 91

For mono-alkylbenzenes, the peak at m/z 91 is a very large signal, often the base peak. In the mass spectrum of ethylbenzene shown in Figure 20.11, the base peak of 91 (the tropylium ion) is the result of loss of a methyl group ($G\cdot = \cdot CH_3$).

Alkenes

Alkenes are similarly prone to fragmentation at the bond β to the double bond to give a stabilized allylic cation or allylic radical.

In the mass spectrum of 1-hexene shown in Figure 20.12, the allylic cation fragment (CH_2=CH—CH_2^+ is observed at m/z 41 and the propyl cation ($CH_3CH_2CH_2^+$) is observed at m/z 43.

FIGURE 20.11
Mass spectrum of ethylbenzene.

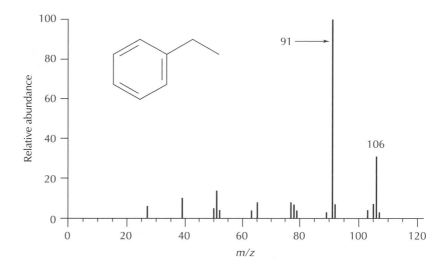

Alcohols

Alcohols fragment easily, and as a result, the molecular ion peak is often very small. In many cases, the molecular ion is not even apparent in the mass spectrum. One fragmentation pathway is the loss of hydroxyl radical (·OH) to produce a carbocation. However, the most important fragmentation pathway is the loss of an alkyl group from the molecular ion to form a resonance-stabilized oxonium ion. Primary alcohols show an intense m/z 31 peak, resulting from this type of fragmentation.

$$R\text{---}CH_2\text{---}\overset{+}{\underset{..}{O}}H \longrightarrow R\cdot + \left[CH_2=\overset{+}{\underset{..}{O}}H \longleftrightarrow \overset{+}{C}H_2\text{---}\overset{..}{\underset{..}{O}}H \right]$$
$$m/z\ 31$$

The mass spectrum of 2-methyl-2-butanol shown in Figure 20.13 provides examples of the various fragmentation pathways available to alcohols. Notice that the molecular ion (m/z 88) is not present in the spectrum.

FIGURE 20.12
Mass spectrum of 1-hexene.

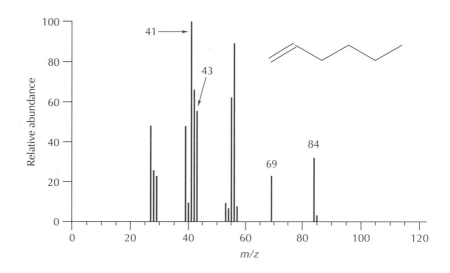

FIGURE 20.13
Mass spectrum of
2-methyl-2-butanol.

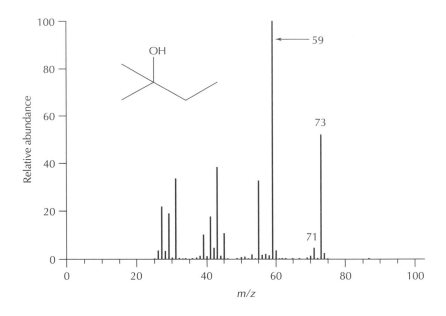

Other heteroatom-containing molecules undergo similar types of cleavage. Amines, ethers, and sulfur compounds can undergo fragmentations analogous to those exhibited by alcohols.

$$R\!\!-\!\!CH_2\!\!-\!\!\overset{+}{Y} \longrightarrow R\cdot + \left[CH_2\!\!=\!\!\overset{+}{Y} \longleftrightarrow \overset{+}{C}H_2\!\!-\!\!Y\!:\right]$$

Y = NH$_2$, NHR, NR$_2$,
OR, SH, or SR

*Carbonyl
Compounds*

Ketones and other carbonyl compounds, such as esters, fragment by cleavage of bonds α to the carbonyl group to form a resonance-stabilized acylium ion.

$$R' - C \equiv \overset{+}{O}: \quad \updownarrow \quad R' - \overset{+}{C} = \overset{..}{\overset{..}{O}}$$

$$R' - C \equiv \overset{+}{O}: \quad \updownarrow \quad R' - \overset{+}{C} = \overset{..}{\overset{..}{O}}$$

In the spectrum of 2-butanone shown earlier in Figure 20.6, we see fragmentations on both sides of the carbonyl group.

$$CH_3 - C \equiv \overset{+}{O}: \quad \updownarrow \quad CH_3 - \overset{+}{C} = \overset{..}{\overset{..}{O}} \qquad m/z\ 43$$

$$CH_3CH_2 - C \equiv \overset{+}{O}: \quad \updownarrow \quad CH_3CH_2 - \overset{+}{C} = \overset{..}{\overset{..}{O}} \qquad m/z\ 57$$

In the mass spectrum of the ester methyl nonanoate (MW 172), shown in Figure 20.14, there is a significant peak at m/z 141. This peak results from formation of an acylium ion by loss of a fragment with a mass of 31, corresponding to a methoxyl radical.

$$C_8H_{17} - C \equiv \overset{+}{O}: \quad \updownarrow \quad C_8H_{17} - \overset{+}{C} = \overset{..}{\overset{..}{O}} \qquad m/z\ 141$$

The base peak at *m/z* 74 in Figure 20.14 occurs through the loss of a fragment with a mass of 98—a mass that corresponds to the loss of a neutral molecule with a molecular formula C_7H_{14}. That a neutral molecule (not a free radical) is lost by fragmentation is apparent because the molecular ion has an even *m/z* value and gives a fragment ion that also has an even *m/z* value. Carbonyl compounds with alkyl groups containing a chain of three or more carbon atoms can cleave at the β bond. This pathway, called the

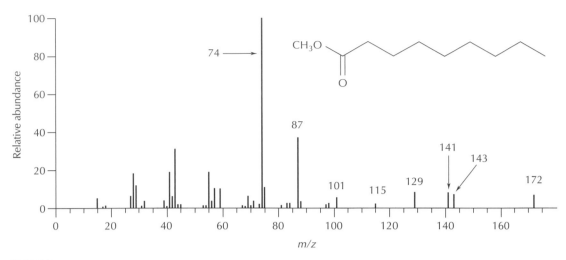

FIGURE 20.14
Mass spectrum of methyl nonanoate.

McLafferty rearrangement, requires the presence of a hydrogen atom on the γ (gamma) carbon atom.

The mass spectrum of methyl nonanoate demonstrates fragmentations also characteristic of other organic compounds with straight-chain alkyl groups. Carbon-carbon bonds can break at any point along the chain, leading to the loss of alkyl radicals.

m/z	Radical fragment lost from the molecular ion
143 (M−29)	$CH_3CH_2\cdot$
129 (M−43)	$CH_3CH_2CH_2\cdot$
115 (M−57)	$CH_3(CH_2)_2CH_2\cdot$
101 (M−71)	$CH_3(CH_2)_3CH_2\cdot$
87 (M−85)	$CH_3(CH_2)_4CH_2\cdot$

20.6   A Case Study

We have seen that if a molecular ion does not fragment completely before being detected, its *m/z* value provides the molecular weight of the compound, which is a significant clue to its structure. Moreover, the profile for the fragmentation of the molecular ion can establish its identity, particularly if the compound is listed in the instrument's mass spectral library. Determining a structure from

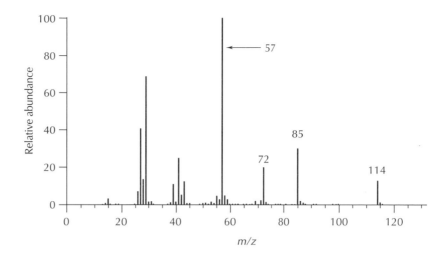

FIGURE 20.15
Mass spectrum of unknown for case study.

a mass spectrum alone, however, is challenging and in most cases requires supplementary spectroscopic information.

As an example of how to approach mass spectrometric analysis, consider the mass spectrum shown in Figure 20.15. The molecular ion peak appears at m/z 114. The even value of m/z suggests that the compound does not contain nitrogen, unless it contains more than one atom per molecule. The absence of a significant M+2 peak rules out the presence of chlorine or bromine. The IR spectrum of the compound has an intense peak at 1715 cm^{-1}, which indicates the presence of a C=O group. Knowing that we are working with a carbonyl compound suggests that the base peak at m/z 57 is a stabilized oxonium-ion fragment, formed by α cleavage and loss of a C$_1$H$_9$ radical (M−57).

$$C_2H_5 \diagdown \atop C_4H_9 \diagup \enspace C{=}\ddot{O}^+ \enspace \longrightarrow \enspace C_4H_9{\cdot} + C_2H_5{-}C{\equiv}\overset{+}{O}{:}$$

$$m/z\ 57$$

Application of the Rule of Thirteen can generate one or more candidate molecular formulas for the compound. Dividing 114 by 13, we have 114/13 = 8 (with a remainder of 10); if the compound were a hydrocarbon, its formula C_nH_m would be C_8H_{18}. Including the presence of an oxygen atom, we would have $C_8H_{18}O - CH_4 = C_7H_{14}O$. Our short list of possible molecular formulas is $C_7H_{14}O$ and perhaps $C_6H_{10}O_2$.

Next, an inventory of the significant MS peaks is put together, along with the masses lost on fragmentation.

m/z	Mass	Possible fragments
114	M	
85	M−29	$C_2H_5\cdot$
72	M−42	C_3H_6
57	M−57	$C_4H_9\cdot$

The mass spectral evidence gives no support for having more than one oxygen atom per molecule of the compound. It is likely that this carbonyl compound is a ketone, because the peak at m/z 85 is consistent with an α cleavage, with loss of an ethyl radical to give a stabilized oxonium ion.

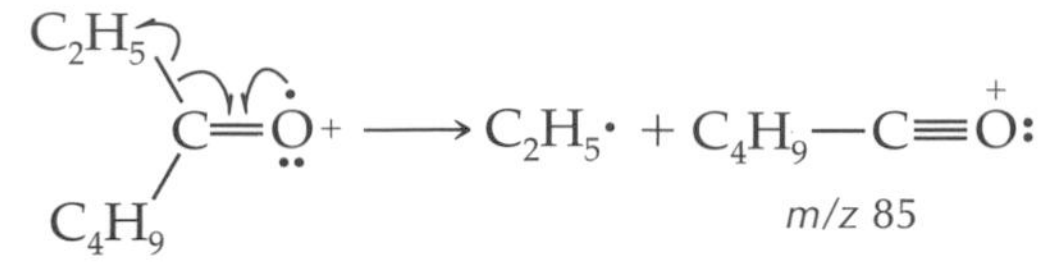

$$C_2H_5\cdot + C_4H_9{-}C{\equiv}\overset{+}{O}\!:$$
m/z 85

Another important clue in the mass spectrum shown in Figure 20.15 is the loss of a neutral molecule, C_3H_6, producing the peak at m/z 72. Loss of a neutral molecule results from a McLafferty rearrangement, which requires the presence of a carbonyl group and a γ hydrogen atom. This fact is also consistent with the presence of a C_4H_9 group.

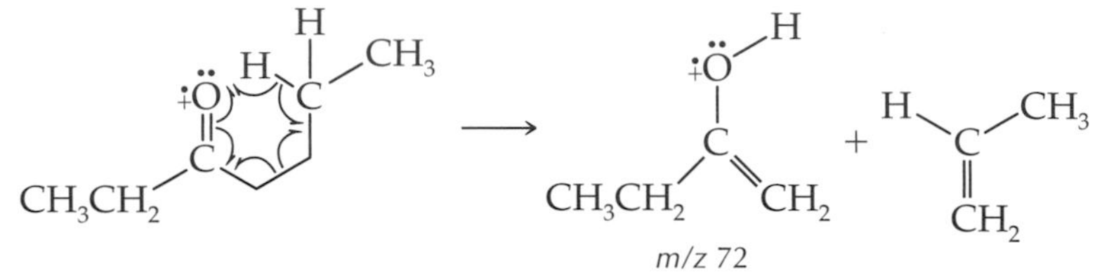

m/z 72

Propene could be lost if a methyl group were attached to the γ carbon atom or to the β carbon atom. There are two compounds consistent with all the evidence, 3-heptanone and 5-methyl-3-hexanone.

3-Heptanone   5-Methyl-3-hexanone

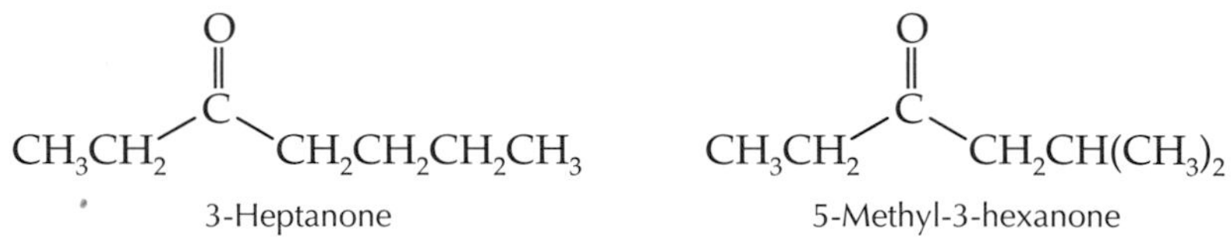

Referring to spectra in the *Wiley Registry of Mass Spectral Data*, the spectrum of the unknown is very similar to the spectra of both 3-heptanone and 5-methyl-3-hexanone. However, the spectrum of 5-methyl-3-hexanone has a signal at m/z 99 that is missing in the spectra of 3-heptanone and the unknown. The identity of the unknown is probably 3-heptanone. An NMR spectrum of the compound would establish the structure unambiguously.

| 20.7 | Sources of Confusion |

As with IR and NMR spectroscopy, it is important to be aware of factors that can lead to unexpected, confusing, or poorly defined peaks in a mass spectrum. Some of the problems can be avoided by proper sample preparation or modification of sampling conditions. Some "problems" are inherent features of the technique.

Presence of Impurities

Small amounts of impurities can produce MS peaks in regions of the mass spectrum that should be blank. This is particularly important when you are trying to determine the *m/z* of the molecular ion. In GC-MS the impurities may be residual material from a previous sample or from degradation of the GC column itself. Small peaks at *m/z* values higher than the molecular weight of any compound in your sample may appear. Thus it is necessary to be judicious in your assignment of a molecular ion peak. It is also important to allow enough time between GC injections to clear the previous sample from the instrument. A background scan can be used to identify peaks due to residual materials in the mass spectrometer.

Absence of a Molecular Ion

Many compounds fragment so easily that there is no discernible molecular ion in the mass spectrum. Examples of these types of compounds include tertiary alcohols, which dehydrate easily, and many alkyl bromides and chlorides, from which bromine or chlorine atoms are easily lost by fragmentation. Even without a usable molecular ion peak, use of the mass spectrometer's spectral library may provide a useful list of candidate molecular structures.

Complex Fragmentation Patterns

Sometimes a mass spectrum of a pure compound exhibits significant peaks that are difficult to rationalize. These fragments may be the result of multiple-step fragmentations or they may have been formed by complex rearrangements. Do not dwell on these peaks.

References

1. Beynon, J. H.; Saunders, R. A.; Williams, A. E. *The Mass Spectrometry of Organic Molecules;* Elsevier: New York, 1968.
2. Beynon, J. H.; Williams, A. E. *Mass and Abundance Tables for Use in Mass Spectrometry;* Elsevier: New York, 1963.
3. Silverstein, R. M.; Webster, F. X.; Kiemle, D. J. *Spectrometric Identification of Organic Compounds;* 7th ed.; Wiley: New York, 2005.
4. Crews, P.; Rodríguez, J.; Jaspars, M. *Organic Structure Analysis;* Oxford University Press: Oxford, 1998.
5. *NIST 05 Mass Spectral Library (NIST/EPA/NIH);* National Institute of Standards and Technology (NIST): Gaithersburg, MD, 2005.
6. McLafferty, F. W.; *Wiley Registry of Mass Spectral Data, Seventh Edition Database, with NIST Spectral Data CD-ROM;* Jossey-Bass: New York, 2000.

Useful Web Sites

1. NIST Standard Reference Database: http://webbook.nist.gov/chemistry
2. Spectral Database for Organic Compounds, SDBS, National Institute of Advanced Industrial Science and Technology (AIST), Japan: http://www.aist.go.jp/RIODB/SDBS

Questions

1. Match the compounds azobenzene, ethanol, and pyridine with their molecular weights: 46, 79, and 182. How does the fact that in one case the molecular weight is odd and in the other two cases the molecular weight is even help in the selection process?

2. The mass spectrum of 1-bromopropane is shown in Figure 20.7. Propose a structure for the base peak at *m/z* 43.

FIGURE 20.16
Mass spectrum of
unknown for
question 6.

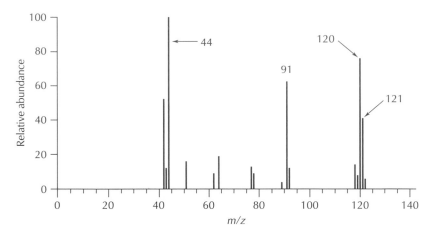

3. The base peak of 1-pentanol is at m/z 31, whereas that for 2-pentanol is at m/z 45. Explain briefly.

4. Similar types of cleavages give rise to two peaks for 4-chlorobenzophenone, one at m/z 105 (the base peak) and one at m/z 139 (70% of base). A clue to their identities is the fact that the m/z 139 peak is accompanied by a peak at m/z 141 that is about one-third of the m/z 139 peak

intensity. The m/z 105 peak has no such partner. What structures correspond to these peaks? Show your reasoning.

4-Chlorobenzophenone

Hit No.	SI	Name	Mol.Wgt.	Mol.Form.	Library
1	89	Phenol, 2-methoxy-4-(1-propenyl)- $$ Phen	164	C10H12O2	NIST62
2	88	Phenol, 2-methoxy-5-(1-propenyl)-, [E]-	164	C10H12O2	NIST12
3	87	Phenol, 2-methoxy-5-(1-propenyl)-, [E]- $$ I	164	C10H12O2	NIST62
4	86	Phenol, 2-methoxy-4-(1-propenyl)-, [E]- $$ I	164	C10H12O2	NIST62
5	86	Phenol, 2-methoxy-4-(1-propenyl)-, acetate	206	C12H14O3	NIST62
6	85	Eugenol	164	C10H12O2	NIST12
7	84	Eugenol	164	C10H12O2	NIST12
8	83	7-Benzofuranol, 2,3-dihydro-2,2-dimethyl- $	164	C10H12O2	NIST62
9	82	Phenol, 2-methoxy-4-(1-propenyl)-	164	C10H12O2	NIST12
10	82	Phenol, 2-methoxy-4-(1-propenyl)-	164	C10H12O2	NIST12
11	81	Eugenol $$ Phenol, 2-methoxy-4-(2-propen	164	C10H12O2	NIST62
12	81	Eugenol	164	C10H12O2	NIST12
13	80	Phenol, 2-methoxy-4-(2-propenyl)-, acetate	206	C12H14O3	NIST12
14	78	Phenol, 2-methoxy-6-(2-propenyl)- $$ o-Ally	164	C10H12O2	NIST62
15	77	3-Allyl-6-methoxyphenol $$ Phenol, 2-meth	164	C10H12O2	NIST62
16	76	Phenol, 2-methoxy-6-(1-propenyl)- $$ Phen	164	C10H12O2	NIST62
17	76	Eugenol	164	C10H12O2	NIST12
18	75	Carbofuran	221	C12H15NO3	NIST12
19	74	Phenol, 2-methoxy-4-(1-propenyl)-	164	C10H12O2	NIST12
20	73	Phenol, 2-methoxy-4-(1-propenyl)-	164	C10H12O2	NIST12
21	71	7-Benzofuranol, 2,3-dihydro-2,2-dimethyl-	164	C10H12O2	NIST12
22	71	3-Octen-5-yne, 2,2,7,7-tetramethyl-, [E]-	164	C10H12O2	NIST62
23	71	Benzene, 4-ethenyl-1,2-dimethoxy- $$ 3,4-E	164	C10H12O2	NIST62
24	71	5-Decen-3-yne, 2,2-dimethyl-, [Z]-	164	C10H12O2	NIST62
25	71	3-Octen-5-yne, 2,2,7,7-tetramethyl-	164	C10H12O2	NIST62

FIGURE 20.17
Hit list from a mass spectral library search of a major clove oil component.

5. What fragmentations lead to the peaks at m/z 127, 125, and 105 in Figure 20.8?

6. An unknown compound has the mass spectrum shown in Figure 20.16. The molecular ion peak is at m/z 121. The infrared spectrum of the unknown shows a broad band of medium intensity at 3300 cm^{-1}. Determine the structure of the unknown. What fragmentations lead to the peaks at m/z 120, 91, and 44?

7. The exact m/z of a sample of aspirin was determined to be 180.0422. What is the molecular formula that corresponds to that exact mass?

8. A sample of clove oil was analyzed using a GC-MS. A search of the mass spectral library for a match to the mass spectrum of the major component of clove oil produced the hit list shown in Figure 20.17. A computer screen printout comparing the mass spectra of hit 1, hit 2, and hit 6 with the mass spectrum of the major component of clove oil is shown in Figure 20.18. Which of the three hit-list candidates is the best match with the MS of the major component of clove oil? Show your reasoning.

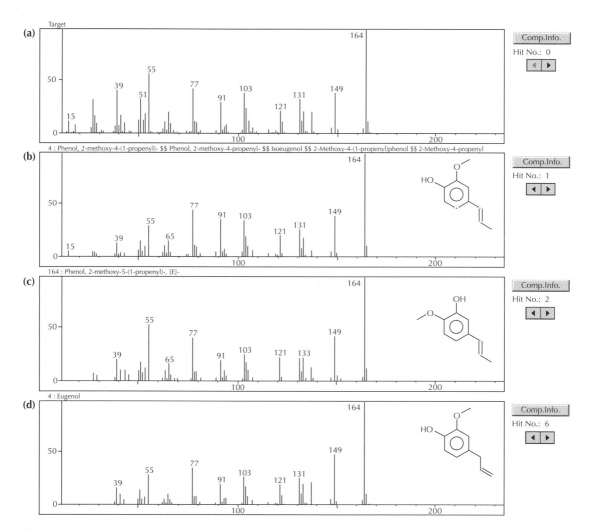

FIGURE 20.18

Computer comparison of three hit list compounds for clove oil. (a) MS of the compound from the GC-MS run. (b) MS of hit 1. (c) MS of hit. (d) MS of hit 6.

APPENDIX

A

Computational Chemistry

Computational chemistry is the calculation of physical and chemical properties of compounds using mathematical relationships derived from theory and observation. It is often referred to as *molecular modeling.* However, we use the term *computational chemistry* to avoid confusion with molecular model sets, which you may have already used to create three-dimensional structures of chemical compounds. Once the exclusive domain of mainframe and supercomputers, computational chemistry has migrated to desktop and laptop computers. Advances in computer hardware provide massive amounts of memory, high computational speed, and high-resolution graphics displays.

Picturing Molecules on the Computer

Computational chemistry can be used to create three-dimensional images and two-dimensional projections of chemical structures. The computer images that result are completely interactive. In this way they are similar to a molecular model set, but computational chemistry is also much more. In molecular model sets, the bond lengths and bond angles are fixed at certain "standard values," such as 109.5° for the bond angle of a tetrahedral (sp^3) carbon atom. Anyone who has built a molecule containing a cyclopropane ring is well aware of the limitations of using a 109.5° bond angle for its "tetrahedral" carbon atoms. The structure of a molecule created on the computer can be optimized so that it represents the lowest energy conformation of the molecule. Optimization means that the bond lengths and bond angles of the structure may deviate from the "standard values" of bond angles and bond lengths. Thus, the molecule created on the computer is a more accurate picture of the actual molecule than can be obtained from using a molecular model set.

Most computational chemistry programs consist of interacting modules that carry out specialized tasks such as building a molecule, displaying a graphical image of the molecule, optimizing the molecular structure, and extracting physical properties from the calculation. The computer image of a molecule can be shown in a variety of ways—wire frame, ball and stick, and space filling, to mention a few. Wire frame images are best to represent bond angles, lengths, and direction. A molecule's size and shape are probably best represented by a space-filling model. The rendering methods can be mixed to emphasize steric interactions in a specific portion of a molecule. The electron density surface can be displayed, providing a view of its overall shape. The electrostatic potential can be mapped onto the molecular surface, highlighting regions of potential reactivity within the molecule.

Molecular orbitals can also be superimposed onto a molecular structure. In many reactions, important insights can be gained by examining the *highest occupied molecular orbital (HOMO)* and the *lowest unoccupied molecular orbital (LUMO).* The HOMO and LUMO can be thought of as the valence shell of a molecule where most of the chemical reactions occur. Many physical and chemical properties can be extracted from an optimized molecular structure. These properties include bond lengths, bond angles, dihedral angles, interatomic distances, dipole moments, electron densities, and heats of formation. The computed properties are often very good approximations of the actual values determined by experiments.

Computational Chemistry Programs

The first and more rigorous of the two major types of computational methods is based on quantum mechanics, which can be used to describe the physical behavior of matter on a very small scale. The other type is derived from a classical mechanical model, which treats atoms as balls and bonds as springs connecting the balls.

Following are some of the packages available for modern microcomputers:

- MacSpartan and PC Spartan from Wavefunction
- CAChe for Macintosh and CAChe for PC from Fujitsu
- HyperChem from HyperCube

We will describe in brief and general terms the types of calculations that are possible and their limitations. Because the operation of the program to build a molecule and invoke the calculation modules differs from one package to another, details are left to the instructions included with the computational chemistry package you use. Materials included with the packages also provide more comprehensive descriptions of the methods the programs use.

Ab Initio Quantum Mechanical Molecular Orbital (MO) Methods

Quantum mechanical molecular orbital (MO) methods are based on solving the *Schrödinger wave equation,* $\hat{H}\Psi = E\Psi$, in which $\hat{H}$ is the Hamiltonian operator describing the kinetic energies and electrostatic interactions of the nuclei and electrons that make up a molecule, E is the energy of the system, and Ψ is the wavefunction of the system. Although simple in expression, the solution is exceedingly complex and requires extensive computational time. Even an organic molecule as simple as methane defies exact solution. The key to obtaining useful information from the Schrödinger relationship in a reasonable length of time lies in choosing approximations that simplify the solution. There are tradeoffs, however. When more approximations are used, the calculation is faster but the accuracy of the result may be degraded.

Quantum mechanical MO models with the least degree of approximation are called *ab initio methods.* *Ab initio* is a Latin phrase that means "from the beginning" or "from first principles." Following are some common approximations that are used even in *ab initio* MO theory:

1. Nuclei are stationary relative to electrons, which are fully equilibrated to the molecular geometry (Born-Oppenheimer approximation).
2. Electrons move independently of each other, and the motion of any single electron is affected by the average electric field created by all the other electrons and nuclei in the molecule (Hartree-Fock approximation).
3. A molecular orbital is constructed as a linear combination of atomic orbitals (LCAO approximation).

Ab initio calculations use a collection of atomic orbitals called a *basis set* to describe the molecular orbitals of a molecule. There are numerous basis sets of varying complexity in use. The choice affects the accuracy of the calculation and the amount of time required for a solution. Normally, you should use the lowest degree of complexity that will answer your question or solve the problem. The smallest basis set in common use is STO-3G. This basis set uses three Gaussian functions to describe the orbitals and works reasonably well with first- and second-row elements that incorporate *s*- and *p*-orbitals. An *ab initio* calculation using an STO-3G basis set can often provide good equilibrium geometries. Much of the time, the medium-sized 3-21G basis set is a good starting point. The 6-31G(d) basis set, using more Gaussian functions and a polarization function on heavy atoms, provides better answers and is more flexible for elements that have various kinds of bonding. However, it requires more calculation time, typically 10–20 times more than the same calculation using an STO-3G basis set.

Semiempirical Molecular Orbital (MO) Approach

The geometries and energies of small organic molecules can be optimized by the *ab initio* MO method using a 3-21G basis set with a desktop computer. For example, the optimization of methylcyclohexane with the methyl group in an axial position takes approximately 50 min using MacSpartan on a Macintosh G4-400. For most practical purposes, a faster method of calculation is needed. The *semiempirical molecular orbital approach* introduces several more approximations that dramatically speed up the calculations. A geometry optimization using a semiempirical molecular orbital method is typically 300 or more times faster than one using an *ab initio* MO method with a 3-21G basis set.

The approximations generally used with semiempirical molecular orbital methods are as follows:

1. Only valence electrons are considered. Inner shell electrons are not included in the calculation (this is also an option with *ab initio* MO calculations).
2. Only selected interactions involving at most two atoms are considered. This is called the *neglect of diatomic differential overlap*, or NDDO.
3. Parameter sets are used to calculate interactions between orbitals. The parameter sets are developed by fitting calculated results with experimental data.

Several popular versions of semiempirical methods follow:

- MNDO or Minimum Neglect of Differential Overlap
- AM1 or Austin Method 1
- PM3 or Parameterized Model 3

In many cases, AM1 is the method of choice for organic chemists; it should be used whenever possible before resorting to an *ab initio* calculation. The PM3 method is often used for inorganic molecules because it has been parameterized for more chemical elements. The MOPAC or **M**olecular **O**rbital **PAC**kage combines these three semiempirical methods in a single program. As you become more familiar with computational chemistry, you will be able to experiment with the various methods to find the one that works best for the molecules you are working with.

Density Functional Theory (DFT)

In contrast to molecular orbital theory, the quantum mechanical *density functional theory (DFT)* optimizes an electron density rather than a wave function. Because the electron correlation energy as a function of the electron density can be included in the functional, DFT is more robust than MO theory with respect to calculating the electron-electron interaction term. DFT has become increasingly popular in the computational chemistry community within the last decade and is now a part of the standard packages that are available. The use of wave functions has slightly broader utility, but DFT is often the method of choice to achieve a particular level of accuracy in the least amount of time for an average problem.

To determine a particular molecular property using DFT, such as the energy of a molecule, one needs to know how the property depends on the electron density.

$$E[\rho(r)] = T_{ni}[\rho(r)] + V_{ne}[\rho(r)] + V_{ee}[\rho(r)] + E_{xc}[\rho(r)]$$

In this equation, $\rho(r)$ is the electron density at a specific position in space, and $E[\rho(r)]$ is called the *energy functional*. The electron density integrated over all space gives the total number of electrons. The equation allows the electrons to interact with one another and with an external potential, the attraction of the electrons to the nuclei.

$T_{ni}[\rho(r)]$ = the kinetic energy of the noninteracting electrons.
$V_{ne}[\rho(r)]$ = the interaction of the nucleus and the electron.
$V_{ee}[\rho(r)]$ = the classical electron-electron repulsion.
$E_{xc}[\rho(r)]$ = the exchange-correlation energy, a combination of the correction to the kinetic energy deriving from the interacting nature of the electrons and all nonclassical corrections to the electron-electron repulsion energy.

As with MO calculations, for DFT a basis set or sets with which to construct the density is chosen and a molecular geometry is also chosen. Then one guesses an initial electron density matrix and iteratively solves the basic DFT equation. After repeated iterations to minimize the ground state electronic energy and optimization of the molecular geometry, the desired molecular property is calculated.

Molecular Mechanics

The molecular mechanics method, which is not based on quantum mechanics, treats molecules as an assemblage of classical balls (atoms) and springs (bonds, bond angles, and so on) connecting the balls. The total energy of a molecule, called the *steric energy,* is the sum of contributions from bond stretching, angle strain, strain resulting from improper torsion, steric or van der Waals interactions, and electronic charge interactions.

$$E_{steric} = E_{bonds} + E_{angles} + E_{torsion} + E_{vdW} + E_{charge}$$

The contributions are described by empirically derived equations. For example, the energy of a bond is approximated by the energy of a spring described by Hooke's law,

$$E_{bond} = 1/2k\,(x - x_0)^2$$

in which k is a force constant related to bond strength and $(x - x_0)$ is the displacement of an atom from its equilibrium bond length (x_0). The force constants for various types of bonds can be derived from experimental data and are incorporated into the parameter set. The energy of the bonds in the molecule is the sum of the contributions from all the bonds.

$$E_{bonds} = \sum_{i=1}^{i=n\,bonds} \tfrac{1}{2}k_i\,(x - x_0)_i^2$$

Other energy contributions are developed in a similar fashion. The collections of equations describing the various energies and their associated parameter sets are called *force fields.* Following are some frequently used force fields:

- MM2, MM3, MM4
- MMX
- MMFF
- SYBYL

The absolute value of the steric energy of a structure has no meaning by itself. Its calculated value varies greatly from one force field to another. Steric energies are useful only for comparison purposes. The comparisons can be meaningful for conformers, such as chair and twist-boat cyclohexane, and diastereoisomers, such as *cis-* and *trans-*1, 3-dimethylcyclohexane. Most molecular mechanics programs can also be used to predict relative heats of formation for isomeric molecules, such as 2-methyl-1-butene and 2-methyl-2-butene. They do so by associating with each atom type an unstrained heat of formation. The molecular heat of formation is then the sum of the heats for the unstrained atom types plus the strain energy.

2-Methyl-1-butene   2-Methyl-2-butene

Differences in steric energies can also be used to estimate equilibrium constants between interconverting conformers. At room temperature methylcyclohexane is a mixture of *axial*-methylcyclohexane and *equatorial*-methylcyclohexane that are rapidly interconverting by way of a ring flip.

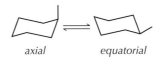

axial   *equatorial*

The relative amount of each conformer at equilibrium can be determined by the difference in energy between the two conformers, which is related to the equilibrium constant, K_{eq}, by the following relationships:

$$K_{eq} = \frac{\text{number of } eq\text{-methylcyclohexane molecules}}{\text{number of } ax\text{-methylcyclohexane molecules}}$$

$$\Delta G° = -RT \ln K_{eq} = -2.303 \, RT \log K_{eq}$$

where $\Delta G°$ is the change in Gibbs standard free energy in going from *ax*-methylcyclohexane to *eq*-methylcyclohexane, R is the gas constant (1.986 cal deg$^{-1} \cdot$ mol^{-1}) and T is the absolute temperature in degrees kelvin (K).

Using the MM2 force field with CAChe, the steric energy of *axial*-methylcyclohexane is calculated to be 8.69 kcal $\cdot$ mol^{-1}, and the steric energy of *equatorial*-methylcyclohexane is calculated to be 6.91 kcal $\cdot$ mol^{-1}. If the difference in steric energy approximates the difference in free energy between the conformers, the free energy difference is -1.78 kcal·mol^{-1}. The negative value for $\Delta G°$ signifies a release of energy in going from ax-methylcyclohexane to eq-methylcyclohexane. At room temperature (25°C, 298 K), the preceding equation becomes

$$-1.78 = -1.36 \log K_{eq}$$
$$\log K_{eq} = 1.31$$
$$K_{eq} = 20.4$$

At equilibrium, there would be approximately 20 molecules of *equatorial*-methylcyclohexane present for each molecule of *axial*-methylcyclohexane—close to the experimental value.

Which Computational Method Is Best?

The best computational method depends on the question you are asking and on the resources at your disposal. Determination of molecular geometry is one of the easier aspects of computational chemistry. If you are simply trying to find the optimum (lowest energy) structures of organic molecules, molecular mechanics provides reasonable structures, and it is very fast. Good values for bond angles, bond lengths, dihedral angles, and interatomic distances can be determined from an optimized structure. In general, you are limited to typical organic compounds; for instance, there are few good parameter sets for carbon-metal bonds.

The energy differences between conformers determined by molecular mechanics are often very close to experimentally determined values, and they can be used to determine equilibrium ratios of the conformers. Since the calculations are fast, the energies of many conformers can be determined in a short time. This is especially useful when examining *rotamers*, conformations related by rotation about a single bond.

As a classical mechanical model, molecular mechanics says nothing about electron densities and dipole moments. It also says nothing about molecular orbitals. However, the optimized structure from molecular mechanics can provide input data for other programs. Using a molecular mechanics calculation is often an efficient way to get an approximation that can be further refined with a quantum mechanical method, often saving computational time.

Semiempirical methods, which are significantly faster than *ab initio* calculations, provide reliable descriptions of structures, stabilities, and other properties of organic molecules. They often do a good job in calculating thermodynamic properties, such as heats of formation. The heats of formation can be used to compare energies of isomers such as 2-methyl-1-butene and 2-methyl-2-butene with greater accuracy than molecular mechanics may provide. The calculated heats of formation can also be used to approximate the energy changes in balanced chemical equations.

Sources of Confusion

Computational chemistry is inherently complex, but most of the commercially available packages have been "human engineered," making it relatively easy to get started. When you get to a point in the process where you have a choice, a default option is usually provided. It is beneficial to acquaint yourself with the information provided with the package so that you can make the best choices.

Two things can cause a good deal of confusion and should be avoided. The first occurs if you start with the wrong structure, and the second deals with the problem of local rather than global energy minima.

Starting the Computation with the Correct Structure

Starting with the correct structure is closely related to the method you use in building a molecule. In many packages, the user draws a two-dimensional projection, similar to the line formulas printed in a book, and the program translates it into a rough three-dimensional structure. However, if the projection is ambiguous, the program may create an unsuitable structure. For example, suppose you wanted to create *axial*-methylcyclohexane. The projection entered on the computer might look like this:

Viewing the structure created by this projection on the computer screen and then rotating it, you would probably observe a flat molecule, clearly unsuitable for optimizing the molecule's structure. To turn this projection into a three-dimensional structure usually requires invoking some sort of "clean-up" or "beautifying" routine. The routine creates a three-dimensional structure using "normal" bond lengths and bond angles. In the case of methylcyclohexane, the structure typically becomes a cyclohexane in the chair conformation with a methyl group in an equatorial position.

Building a cyclohexane with a methyl group in the axial position usually requires the creation of the structure in stages. In this case, you need to create a chair cyclohexane and then replace one of the axial hydrogens with a methyl group. As you can see, the process involves building the framework first and then adding the necessary attachments at specific locations. Most computational chemistry packages contain templates or molecular fragments to assist in creating complicated structures.

Local and Global Minima

Another potential source of confusion encountered in attempting optimization of a structure is the *global minimum* problem. During the optimization of the geometry, the program tries to find the

FIGURE A.1
Local and global minima resulting from energy minimization.

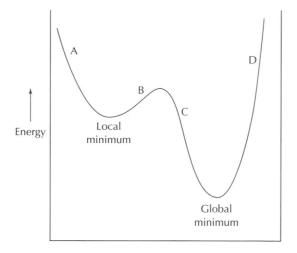

structural conformation with the lowest energy. At each point, it calculates the gradient or first derivative of the energy with respect to the motion of each atom in each Cartesian direction, and the geometry is perturbed in the direction of the resulting gradient vector. Each individual perturbation depends on the history of the energies and gradients from prior steps. This process is repeated until the gradient is computed to be zero, at which point a *local minimum* is likely to have been found.

Global minimum. The global minimum problem arises because the energy profile is not a smooth curve with one minimum. The energy surface is uneven, with lumps, bumps, ridges, and several low spots. The low spot that a minimization falls into depends on where you start on the energy surface. In Figure A.1, a start from point A or B will end up at the local minimum. A start at point C or D will end up at the desired global minimum. The calculation of *axial-* and *equatorial*-methylcyclohexane illustrates this point. The two structures are conformers that can be interconverted by way of a ring flip. *ax*-Methylcyclohexane is a local minimum and *eq*-methylcyclohexane is the global minimum. The barrier represents the strain energy required to flip the ring.

Local minima. If you are looking for a global energy minimum for a molecule, the possibility of encountering local minima adds uncertainty and doubt to any result. How does one know if the structure represents a local minimum or a global minimum? That question has led to many research projects. For our purposes, the answer is to create several different starting structures, carry out minimizations on each of them, and use the lowest energy as the global minimum. One of the several methods for systematically creating possible starting structures is conformational searching. Several conformations of a structure are created by rotating portions of the molecule connected by single bonds. Some modeling packages have routines

FIGURE A.2
Output of a molecular dynamics simulation plotted as a graph of energy versus the conformation of the structure, changing with time.

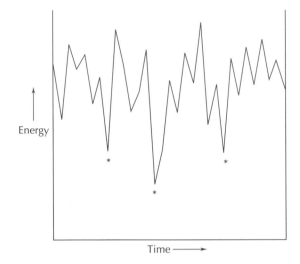

called *sequential searching* which automate this process. Other packages have methods such as Monte Carlo routines for generating random structures.

Yet another method of generating candidate structures for minimization is to use a molecular dynamics simulation program. This program simulates the motions of atoms within a structure. The molecule is given increased kinetic energy, the amount depending on the designated temperature. As the atoms move around, energy "snapshots" are taken at regular intervals. The structures with the lowest energies are used as starting structures for minimization. This method often propels molecules over energy barriers that are caused by steric interactions, bond strain, or torsional strain. The results of a molecular dynamics simulation can be plotted as the internal energy of a molecule versus time. In Figure A.2, structures corresponding to low-energy conformers are designated with asterisks. These conformers can be used as initial structures for energy minimizations by molecular mechanics or quantum mechanical calculations. Even using these methods, there is no guarantee that the global minimum will always be found with systems of fairly modest size. The situation is completely hopeless with a large molecule, such as a protein.

Relationship of Computations and Experiments

Computational chemistry is based on theoretical models using approximations and parameter sets derived from theory and experiment. Thus, it is important to keep a firm grip on reality at all times. You need to evaluate the result, especially a surprising result, and determine whether it makes sense chemically and physically. It is all too easy to accept the results of the calculations as physical truth. In spite of this caveat, computational chemistry has become a highly valuable tool for gaining insights into organic chemistry.

References

1. Goodman, J. M. *Chemical Applications of Molecular Modeling;* Royal Society of Chemistry: Cambridge, 1998.
2. Hehre, W. J.; Shusterman, A. J.; Huang, W. W. *A Laboratory Book of Computational Organic Chemistry;* Wavefunction, Inc.: Irvine, CA, 1998.
3. Hehre, W. J. *A Guide to Molecular Mechanics and Quantum Chemical Calculations;* Wavefunction, Inc.: Irvine, CA, 2003.
4. Cramer, C J. *Essentials of Computational Chemistry: Theories and Models;* 2nd ed.; Wiley: New York, 2004.

APPENDIX

B

The Literature of Organic Chemistry

Before organic chemists carry out reactions in the laboratory, they must be aware of what is known about the chemicals and reactions they will be using, information that can be found in the chemistry library. Carrying out any laboratory project requires the use of chemical information sources in both printed and electronic forms.

The great change in chemistry libraries within the last ten years is the transition from printed to electronic materials. Electronic access has revolutionized the way many libraries do business and the way scientists access information. Journal articles and reference works can now be delivered directly to a scientist's desktop computer. Electronic searches of the chemistry literature can be completed far more rapidly and comprehensively than manual searches.

Four types of information sources are found in all chemistry libraries: textbooks, reference works, chemistry journals, and chemical databases. Textbooks and reference works are frequently called secondary sources of information because they compile information from the primary sources, the journals in which the original research was published. Chemical databases are invaluable for locating journal articles on a topic or compound and looking up specific information about chemical compounds.

Textbooks

Textbooks contain general information. For example, an undergraduate organic chemistry text will tell you that aldehydes can be oxidized to carboxylic acids with chromic acid or that aromatic amines are weak bases. Advanced textbooks provide a more comprehensive treatment of organic reactions and their applications, and they usually include numerous references to the original literature and to review articles. Some particularly useful advanced texts are the following:

1. Smith, M. B.; March, J. *March's Advanced Organic Chemistry: Reactions, Mechanisms and Structures*; 5th ed.; Wiley: New York, 2001.

2. Carey, F. A.; Sundberg, R. M. *Advanced Organic Chemistry; Part A: Structure and Mechanisms; Part B: Reactions and Synthesis;* 4th ed.; Kluwer: New York, 2000/2001.

3. Lowry, T. H.; Richardson, K. S. *Mechanism and Theory in Organic Chemistry;* 3rd ed.; Benjamin-Cummings: Menlo Park, CA, 1997.

Reference Works

Specific information about a compound, such as its melting point, its ^{1}H NMR spectrum, or how it can be synthesized, is not commonly found in textbooks. Reference works that contain this kind of information include the following.

Handbooks

Handbooks provide compilations of physical data, such as the boiling point of 2-butanol or the density of benzaldehyde. A quick, straightforward electronic handbook resource is www.chemfinder.com. Some handbooks, including *The Merck Index* and the *Aldrich Catalog Handbook,* list references. Many of these handbooks are also available in electronic form.

1. Lide, D. R. (Ed.) *CRC Handbook of Chemistry and Physics;* 85th ed.; CRC Press: Boca Raton, FL, 2004.

2. Lide, D. R.; Milne, G. W. (Eds.) *Handbook of Data on Common Organic Compounds;* CRC Press: Boca Raton, FL, 1995.

3. Speight, J. (Ed.) *Lange's Handbook of Chemistry;* 16th ed.; McGraw-Hill: New York, 2004.

4. *Aldrich Catalog Handbook of Fine Chemicals and Laboratory Equipment;* Aldrich Chemical Co.: Milwaukee, WI, published annually.

5. O'Neill, M. J.; Smith, A.; Heckelman, P. E.; Oberchain, J. R. Jr., (Eds.) *The Merck Index: An Encyclopedia of Chemicals, Drugs and Biologicals;* 13th ed.; Merck & Co., Inc.: Whitehouse, NJ, 2001.

6. Buckingham, J. B. (Ed.) *Dictionary of Organic Compounds;* 6th ed.; 9 vols.; CRC Press: Boca Raton, FL, 1997.

7. Gordon, A. J.; Ford, R. A. *The Chemist's Companion: A Handbook of Practical Data, Techniques and References;* Wiley: New York, 1972.

Spectral Information

The following reference books contain spectra of thousands of organic compounds.

1. Pouchert, C. J.; Behnke, J. (Eds.) *Aldrich Library of ^{13}C and ^{1}H FT-NMR Spectra;* 3 vols.; Aldrich Chemical Co.: Milwaukee, WI, 1992. Print or CD-ROM.

2. *Aldrich Library of FT-IR Spectra;* 2nd ed.; 3 vols.; Aldrich Chemical Co.: Milwaukee, WI, 1997.

3. *Sadtler Collection of High-Resolution (NMR) Spectra;* Sadtler Research Laboratories: Philadelphia, 1992.

4. *Sadtler Reference (IR) Spectra;* Sadtler Research Laboratories: Philadelphia, 1992.

Reactions, Synthetic Procedures, and Techniques

Following are examples of reference books that contain descriptions of techniques, methods, procedures, and reactions used in organic synthesis. They all include extensive references to the primary literature.

1. Furniss, B. S. (Ed.) *Vogel's Textbook of Practical Organic Chemistry;* 5th ed.; Prentice Hall: Upper Saddle River, NJ, 1996.
2. Loewenthal, H. J. E. *A Guide for the Perplexed Organic Experimentalist;* 2nd ed.; Wiley: New York, 1992.
3. Larock, R. C. *Comprehensive Organic Transformations: A Guide to Functional Group Preparations;* 2nd ed.; Wiley: New York, 1999.
4. Fieser, L. F.; Fieser, M. *Reagents for Organic Synthesis;* 19 vols.; Wiley: New York, 1967–2000. Also available as Pearson, A. J.; Roush, W. R. *The Wiley Fieser's Reagents for Organic Synthesis Database.*
5. *Organic Syntheses;* Wiley: New York, 1932–present. Collective Volumes 1–10 (2004) combine and index five or ten volumes each through Volume 79, 2002. The preparations have been carefully checked in two separate research laboratories.
6. *Organic Reactions;* Wiley: New York, 1932–present.
7. *Handbook of Reagents for Organic Synthesis;* 4 vols.; Wiley: New York, 1999.
8. Coffey, S.; Rood, E. H.; Ansell, M. F.; Sainsbury, M. *Rodd's Chemistry of Carbon Compounds;* 2nd ed.; Elsevier: New York, 1964–present.
9. Katritzky, A. R.; Meth-Cohn, M.; Rees, C. W. (Eds.) *Comprehensive Organic Functional Group Transformations;* 7 vols.; Elsevier: New York, 1995.
10. Katritzky, A. R.; Taylor, R. J. K. (Eds.) *Comprehensive Organic Functional Group Transformations II, A Comprehensive Review of the Synthetic Literature 1995–2003;* 2nd ed.; 7 vols.; Elsevier: San Diego, CA, 2004.
11. Harrison, I. T.; Wade, Jr., L. G.; Smith, M. B. (Eds.) *Compendium of Organic Synthetic Methods;* 8 vols.; Wiley: New York 1971–1995.
12. Mackie, R. D. *Guidebook to Organic Synthesis;* 3rd ed.; Prentice Hall: Upper Saddle River, NJ, 2000.
13. Paquette, L. A. (Ed.) *Encyclopedia of Reagents for Organic Synthesis;* 11th ed.; Wiley: New York, 2003.
14. Sandler, S. R.; Karo, W. *Sourcebook of Advanced Laboratory Preparations;* Academic Press: San Diego, CA, 1992.
15. Trost, B. M. (Ed.) *Comprehensive Organic Synthesis;* 9 vols.; Elsevier: New York, 1992.

Chemistry Journals

Primary journals publish original research results, and as such they are the fundamental source of most information about organic chemistry. References to primary chemistry journals are found in the chapters of the general reference works listed in the previous section.

Important current journals that publish original papers in organic chemistry include the following:

Angewandte Chemie (International Edition in English)
Bulletin of the Chemical Society of Japan
Canadian Journal of Chemistry
Chemical Communications
European Journal of Organic Chemistry
Journal of the American Chemical Society
Journal of Heterocyclic Chemistry
Journal of Medicinal Chemistry
Journal of Organic Chemistry
New Journal of Chemistry
Organic & Biomolecular Chemistry
Organic Letters
Organometallics
Synthesis
Synthetic Communications
Tetrahedron Letters

All these journals are available online, and a few are available only in electronic form.

Electronic Abstracts and Indexes

Because the literature of chemistry is so vast, finding specific information, such as the preparation of a particular compound or reactions where a specific type of catalyst has been used, is difficult and time consuming without a survey of the entire literature of chemistry. *Chemical Abstracts (CA)*, published by the American Chemical Society, is such a survey and is the most complete source of information on chemistry in the world.

Chemical Abstracts condenses the content of journal articles into abstracts and indexes the abstracts by subject, author's name, chemical substance or reaction, molecular formula, and patent numbers. Each chemical compound is also assigned a number, called a registry number, which can facilitate finding references to the compound. In evaluating an abstract you need to keep in mind that it gives only a brief summary of an article; you should always consult the original journal article as the final source.

Chemical Abstract Services (CAS), the publishers of *Chemical Abstracts,* provides a number of databases. The newest of these databases, called *SciFinder Scholar,* is an excellent search engine (Figure B.1). If it is available on your campus, you will find it invaluable. In addition to *SciFinder Scholar,* CAS provides *STN,* a more limited but nonetheless helpful database for *Chemical Abstracts.* Today, most college and university libraries are equipped to search *Chemical Abstracts* using these computerized databases. It is important to obtain assistance and training before undertaking an online search. Consult the library at your college or university.

Science Citation Index contains all articles published in the more prominent journals and also lists all the articles that were cited or

FIGURE B.1
Initial search menu of *Scifinder Scholar.* (*SciFinder Scholar* and the SciFinder Scholar logo are trademarks and/or registered trademarks of the American Chemical Society. All graphics relating to *SciFinder Scholar* software have been reproduced with permission of the American Chemical Society. All rights are reserved.)

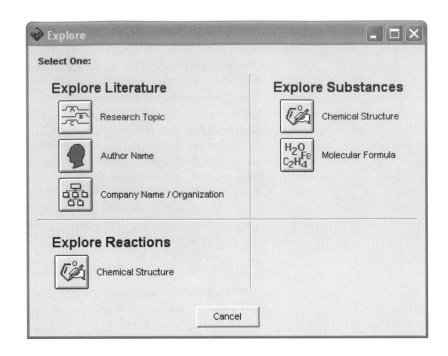

referred to in current articles. It is available in the online *ISI Web of Knowledge* through its Web of Science, which can be searched by subject, author, journal, and cited references.

The *Beilstein CrossFire* databases are drawn from *Beilstein's Handbook of Organic Chemistry* and over 170 leading journals. *Beilstein* is an excellent though expensive database for locating information about organic compounds. It contains records on almost 10 million organic substances. For each compound, the database contains the name (or names), formula, physical properties, methods of synthesis, chemical reactions, and biological properties. Every piece of information has a reference to the primary literature and thus data may be checked. The database continues to add information on many compounds that were reported in *Beilstein's Handbook.* Thus, corrections and updating continue. The entry for an organic compound in the *CRC Handbook* and the *Aldrich Catalog* also gives the location of the compound in *Beilstein.* If *Beilstein* is available at your university, it is well worth learning how to use it effectively.

It is difficult to provide complete current database information in this book because many databases undergo changes on a regular basis. However, the *Journal of Chemical Education* in its *JCE Online* site (www.jce.divched.org) maintains a list of reviewed Web sites.

More Information About the Chemistry Library

We urge you to consult the library at your college or university for assistance in conducting a search for information in the books and journals and online. The following books and journal articles contain more information about chemistry information sources, how to use them, and how to plan and carry out an online search.

1. Maizell, R. E. *How to Find Chemical Information: A Guide for Practicing Chemists, Educators, and Students;* 3rd ed.; Wiley: New York, 1998.
2. Poss, A. J. *Library Handbook for Organic Chemists;* Chemical Publishing Company: New York, 2000.
3. Bottle, R. T.; Rowland, J. F. B. *Information Sources in Chemistry;* 4th ed.; Bowker: New Providence, NJ, 1993.
4. Smith, M. B.; March, J. *March's Advanced Organic Chemistry;* 5th ed.; Wiley: New York, 2001, Appendix A.
5. Wienbroer, D. R. *Guide to Electronic Research and Documentation;* McGraw-Hill: New York, 1997.
6. *Using CAS Databases on STN: Student Manual;* American Chemical Society: Washington, DC, 1995.

APPENDIX

C

Integrated Spectroscopy Problems

The three major spectroscopic methods presented in Techniques 18–20 have revolutionized structure determinations of organic compounds. Although for the most part these methods were considered separately in Techniques 18–20, the connections were made apparent from time to time. In practice, organic chemists generally solve structure problems by using an integrated spectroscopic approach. The mass spectrum is usually a good starting point because it can provide the molecular weight of the compound. Next, consider the IR spectrum, which provides data for the identification of the functional groups present. Interpretation of the ^1H NMR spectrum and perhaps the ^{13}C NMR spectrum usually allows the structural analysis to be completed.

Many chemists believe that NMR is the most versatile source of structure data, and we have emphasized it more than infrared spectroscopy and mass spectrometry. However, to be efficient in tackling structure determinations, organic chemists need to be proficient in all three spectroscopic methods. One method may reveal features about a compound that are not clear from another. Experienced researchers are alert to when extra emphasis should be placed on a few pieces of data chosen from a large data set, a skill that comes from experience. The following problems highlight the use of an integrated approach to using spectroscopy for organic structure determination.

1. The base peak of a compound with a molecular ion at m/z 72 in its mass spectrum is at m/z 43. An infrared spectrum of this compound shows, among other absorptions, four bands in the 2990–2850 cm^{-1} range and a strong band at 1715 cm^{-1}. There are no IR peaks at greater than 3000 cm^{-1}. The ^1H NMR spectrum contains a triplet at 1.08 ppm (3H), a singlet at 2.15 ppm (3H), and a quartet at 2.45 ppm (2H). The magnitudes of the splitting of both the quartet and triplet are identical. Deduce the structure of this compound and assign all the MS, IR, and NMR peaks.

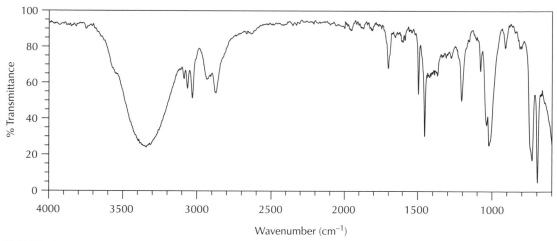

FIGURE C.1
Infrared spectrum (thin film) of unknown compound for problem 2.

2. The infrared spectrum of a compound is shown in Figure C.1. Its ^{1}H NMR spectrum contains a somewhat broadened singlet at 7.3 ppm (5H), a singlet at 4.65 ppm (2H), and a broadened singlet at 2.5 ppm (1H). Deduce the structure of this compound and assign the NMR and important IR peaks.

3. A compound shows a molecular ion peak in its mass spectrum at m/z 92 and a satellite peak at m/z 94 that is 32% the intensity of the m/z 92 peak. The ^{1}H NMR spectrum contains only one signal, a singlet at 1.65 ppm. The proton-decoupled ^{13}C NMR spectrum reveals a strong peak at 35 ppm and a weaker peak at 67 ppm. Deduce the structure of this compound and assign all MS and NMR peaks.

4. The mass spectrum, infrared spectrum, and ^{1}H NMR spectrum of a compound are shown in Figures C.2–C.4. Deduce the structure of this compound from the spectral data and show your reasoning.

FIGURE C.2
Mass spectrum of unknown compound for problem 4.

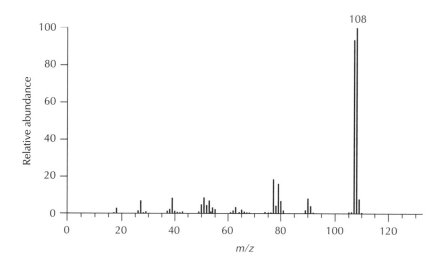

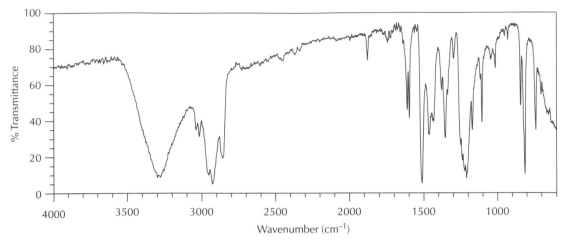

FIGURE C.3
Infrared spectrum (KBr pellet) of unknown compound for problem 4.

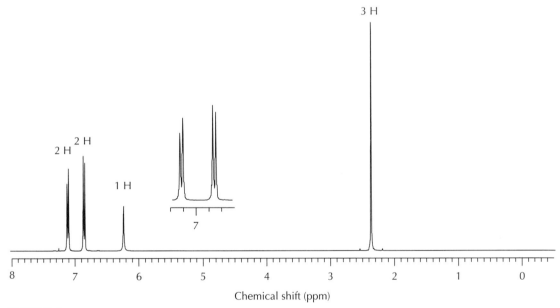

FIGURE C.4
360-MHz ^{1}H NMR spectrum of unknown compound for problem 4.

5. Figure C.5 shows the ^{1}H NMR spectrum of a compound of molecular formula $C_3H_6Cl_2$. The proton-decoupled ^{13}C NMR spectrum, as well as the DEPT(90) and DEPT(135) spectra, are shown in Figure C.6. Deduce the structure of this compound and assign all peaks. Estimate the chemical shifts of all protons in the compound using Table 19.3 (see the foldout inside the back cover) and compare them with the chemical shifts measured from the ^{1}H NMR spectrum.

6. The infrared spectrum of a compound of molecular formula $C_7H_{16}O$ is shown in Figure C.7. Its 360-MHz ^{1}H NMR spectrum is shown in Figure C.8, and its proton-decoupled

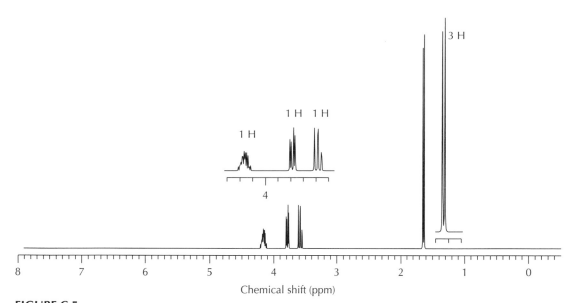

FIGURE C.5
360-MHz ^{1}H NMR spectrum of unknown compound for problem 5.

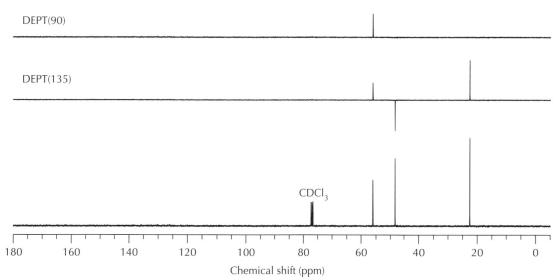

FIGURE C.6
90-MHz ^{13}C NMR, DEPT(90), and DEPT(135) spectra of unknown compound for problem 5.

^{13}C NMR and DEPT(135) spectra are shown in Figure C.9. Deduce the structure of this compound, assign all of the NMR and important IR peaks, and explain your reasoning. Estimate the chemical shifts of all of the compound's protons using Table 19.3 (see the foldout inside the back cover) and compare them with the chemical shifts measured from the ^{1}H NMR spectrum.

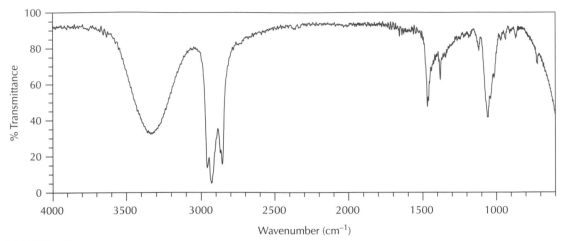

FIGURE C.7
Infrared spectrum (thin film) of unknown compound for problem 6.

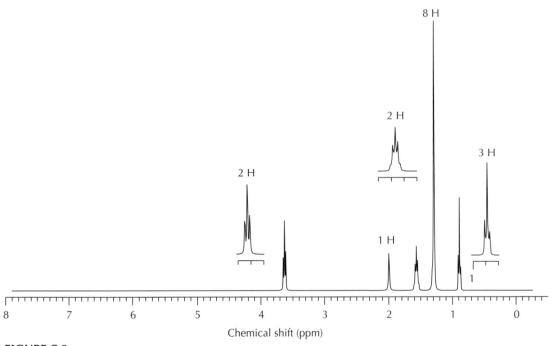

FIGURE C.8
360-MHz ^{1}H NMR spectrum of $C_7H_{16}O$ for problem 6.

7. The mass spectrum, infrared spectrum, 360-MHz ^{1}H NMR spectrum, and proton-decoupled ^{13}C NMR and DEPT(90) and DEPT(135) spectra of a compound are shown in Figures C.10–C.13. Deduce the structure of this compound and show your reasoning. Assign all the NMR peaks and all important MS and IR peaks.

DEPT(135)

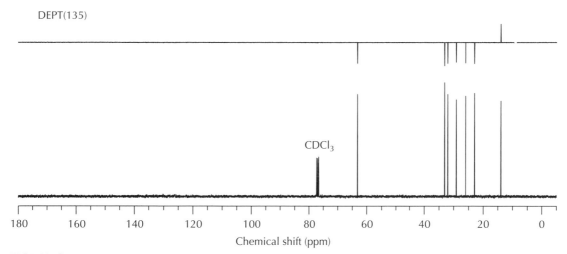

FIGURE C.9
90-MHz ^{13}C NMR and DEPT(135) spectra of $C_7H_{16}O$ for problem 6.

FIGURE C.10
Mass spectrum of unknown compound for problem 7.

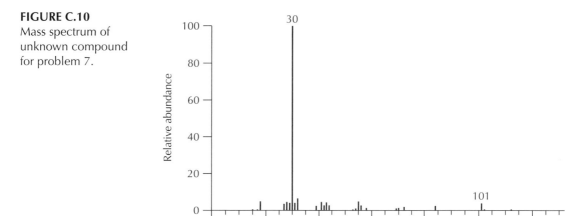

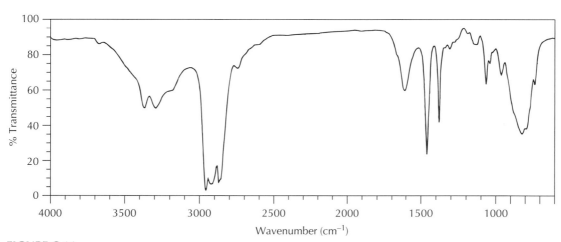

FIGURE C.11
Infrared spectrum (thin film) of unknown compound for problem 7.

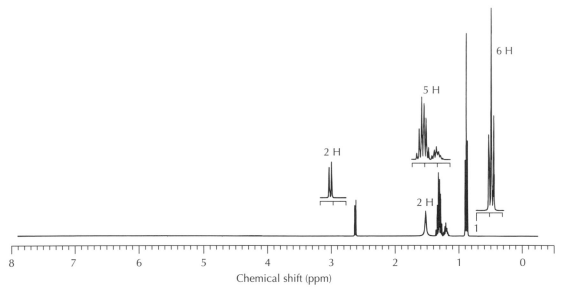

FIGURE C.12
360-MHz ^{1}H NMR spectrum of unknown compound for problem 7.

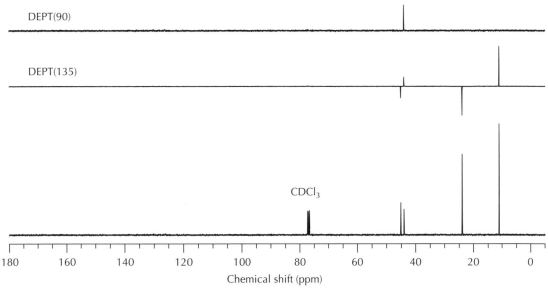

FIGURE C.13
90-MHz ^{13}C NMR, DEPT(90), and DEPT(135) spectra of unknown compound for problem 7.

8. The mass spectrum, infrared spectrum, 360-MHz ^{1}H NMR spectrum, and proton-decoupled ^{13}C NMR and DEPT(90) and DEPT(135) spectra of a compound are shown in Figures C.14–C.17. The molecular ion is not discernible in the mass spectrum. Deduce the structure of this compound and show your reasoning. Assign all the NMR peaks and all important MS and IR peaks.

FIGURE C.14
Mass spectrum of unknown compound for problem 8.

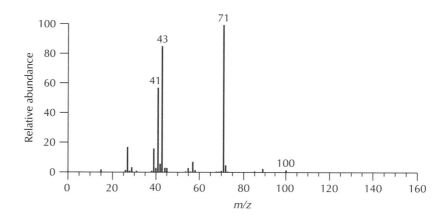

FIGURE C.15
Infrared spectrum (thin film) of unknown compound for problem 8.

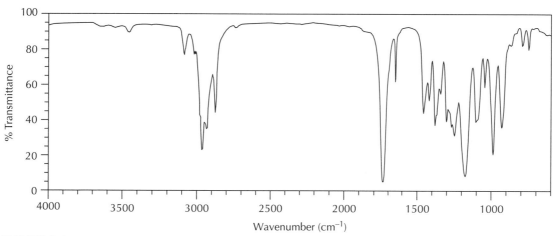

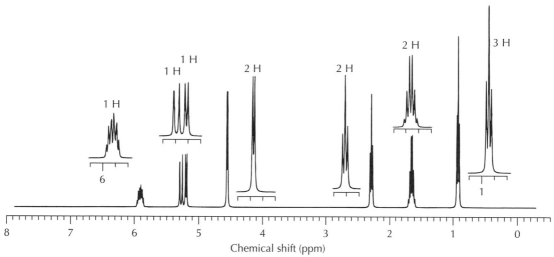

FIGURE C.16
360-MHz ^{1}H NMR spectrum of unknown compound for problem 8.

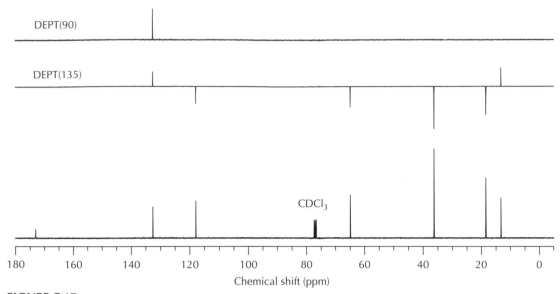

FIGURE C.17
90-MHz ^{13}C NMR, DEPT(90), and DEPT(135) spectra of unknown compound for problem 8.

INDEX

Note: Page numbers followed by f indicate figures; those followed by t indicate tables. Page numbers preceded by AP indicate appendices.

Quick reference for other important tables

	Page
^{13}C DEPT signals (19.12)	330
^{13}C chemical shifts (19.10)	325
Chemical resistance of gloves (1.1)	11
Drying agents (8.2)	94
^1H chemical shifts (19.2)	281
^1H coupling constants (19.6)	303
NMR Solvents, deuterated (19.1)	272
Recrystallization solvents (9.1)	101
TLC solvent polarities (15.1)	186

Quick reference for other important figures

	Page
Distillation	
simple (11.6)	134
short-path (11.7)	137
standard taper miscroscale (11.9)	139
Williamson microscale (11.12)	141
Extraction	
microscale (8.10, 8.11)	89, 91
miniscale (8.3)	80
Glassware	
standard taper (2.3)	25
standard taper miscroscale (2.5)	27
Williamson miscroscale (2.7)	28
Pasteur pipet calibration (5.8)	45

Quick reference for sections on sources of confusion

	Page
Distillation	156
Drying organic liquids	99
Extraction	92
Gas-liquid chromatography (GC)	200
IR Spectroscopy	259
Liquid chromatography (LC)	221
Melting points	125
Mass spectrometry (MS)	360
NMR Spectroscopy	304
Recrystallization	114
Thin-layer chromatography (TLC)	188

Approximate regions of chemical shifts for different types of protons in organic compounds

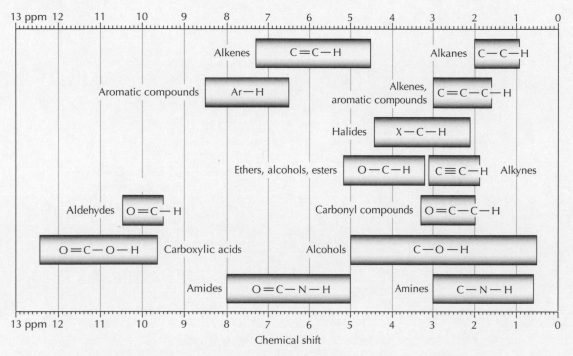

Chemical shift

NMR signals of common solvents

Solvent	NMR signals[a]
Acetone	2.1 (s)
Benzene	7.4 (s)
Chloroform	7.3 (s)
Cyclohexane	1.4 (s)
Dichloromethane	5.3 (s)
Diethyl ether	1.2 (t), 3.5 (q)
Dimethyl sulfoxide	2.5 (s)
Ethanol (anhydrous)	1.2 (t), 3.0 (t) or (s), 3.7 (m) or (q)
Ethyl acetate	1.3 (t), 2.0 (s), 4.1 (q)
Hexane	0.9 (t), 1.3 (m)
2-Propanol	1.2 (d), 2.6 (d) or (s), 4.0 (m)
Methanol (anhydrous)	2.3 (q) or (s), 3.4 (d) or (s)
Tetrahydrofuran	1.9 (m), 3.8 (m)
Toluene	2.3 (s), 7.2 (m) or (s)
Water (dissolved)	1.6 (s)
Water (bulk)	4.5 (br s)

a. In $CDCl_3$. Multiplicity of signal is shown in parentheses.

Additive parameters for predicting NMR chemical shifts of alkyl protons in CDCl$_3$[a]

	Base values	
	Methyl	0.9 ppm
	Methylene	1.2 ppm
	Methine	1.5 ppm

Group (Y)	Alpha (α) substituent	Beta (β) substituent	Gamma (γ) substituent
	H—C—Y	H—C—C—Y	H—C—C—C—Y
—R	0.0	0.0	0.0
—C=C	0.8	0.2	0.1
—C=C(C=O)OR	1.0	0.3	0.1
—C≡C—R	0.9	0.3	0.1
—C≡C—Ar[b]	1.2	0.4	0.2
—Ar	1.4	0.4	0.1
—(C=O)OH	1.1	0.3	0.1
—(C=O)OR	1.1	0.3	0.1
—(C=O)H	1.1	0.4	0.1
—(C=O)R	1.2	0.3	0.0
—(C=O)Ar	1.7	0.3	0.1
—(C=O)NH$_2$	1.0	0.3	0.1
—(C=O)Cl	1.8	0.4	0.1
—C≡N	1.1	0.4	0.2
—Br	2.1	0.7	0.2
—Cl	2.2	0.5	0.2
—OH	2.3	0.3	0.1
—OR	2.1	0.3	0.1
—OAr	2.8	0.5	0.3
—O(C=O)R	2.8	0.5	0.1
—O(C=O)Ar	3.1	0.5	0.2
—NH$_2$–	1.5	0.2	0.1
—NH(C=O)R	2.1	0.3	0.1
—NH(C=O)Ar	2.3	0.4	0.1

a. There may be differences of 0.1–0.5 ppm in the chemical shift values calculated from this table and those measured from individual spectra.

b. Ar = aromatic group.